The Prairies and the Pampas

In Memoriam
Carl E. Solberg, 1940–1985

This book was published with financial assistance from friends, colleagues, and the family of the late author.

Carl E. Solberg

THE PRAIRIES
AND THE PAMPAS

Agrarian Policy in
Canada and Argentina, 1880–1930

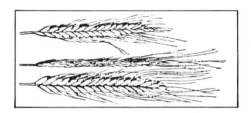

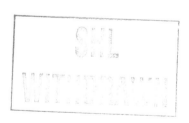

STANFORD UNIVERSITY PRESS
Stanford, California   1987

Comparative Studies in History, Institutions, and Public Policy
John D. Wirth and Thomas C. Heller, Editors

Stanford University Press, Stanford, California
©1987 by the Board of Trustees of the
Leland Stanford Junior University
Printed in the United States of America

CIP data appear at the end of the book

To Robert E. Burke

Contents

Tables

Foreword

◂•▸

Carl Solberg died on April 7, 1985, just weeks after he wrote the last words of this book. It stands as he wanted it: a well-developed comparative study that links his reputation as a leading historian of Argentina to a newer field for him, Canadian history. Aside from the copy-editing, this work is wholly Carl's—his third and best book.

Studies comparing Canada's historical development with Latin America are still rare, so this is both a path-breaking study with respect to the juxtaposition of themes and areas covered and a major contribution to the political economy of wheat, an important New World export crop. Drawing on his extensive research in Argentine and Canadian archives, Solberg sheds fresh light on several issues that are familiar to Canadianists and Argentinists, but have been treated in isolation. This book also explains why the Argentine wheat economy stagnated while the Canadian wheat economy rose to world leadership. Comparison brings the two cases into sharp focus, enabling Solberg to show how different state policies allowed two "new countries" with similar access to world wheat markets to produce such varying outcomes.

Canada outpaced Argentina, a country that had better natural conditions and a much shorter haul to port, to become the world's second-largest producer. Absent external market constraints on either nation, it was better public policy and a more responsive political system that enabled Canada to pull ahead. With his interest in political institutions and economic history, this interpretation suited Carl. He reached conclusions that are quite different from the dependency approach, which has been so influential, especially in Latin American studies. What gives force and power to his arguments is the judicious use of comparison.

The book is rich in leads for future research on topics of this sort. For example, it is abundantly clear that the neighbors of Canada and Argentina posed quite different challenges. To keep their prairies from being absorbed into the United States, the Canadians—who are known as the

Japanese of North America—adopted many U.S. ideas, including a homestead act and agrarian cooperativism, in order to build their own nation. In the period under review, neither Chile nor impoverished Brazil posed this kind of threat to Argentina, the innovating and dynamic nation of South America. Argentines discussed reforms in land tenure and in production and marketing techniques. There was some legislation but little follow-through, in part because there was no comparable challenger with a stock of innovations nearby. This may help to explain why Canada's political elite devised a national development plan, but Argentina's did not.

Solberg avoided cultural explanations for such divergent outcomes in the wheat economies of Canada and Argentina. Instead, he probed into the differences in their land-tenure systems; in their techniques of production and marketing; in their transportation networks; in the play of politics in the two countries. His analysis of the political economy is richly textured and convincing. Yet he did not dismiss the importance of one essentially cultural fact: that Canadians expected far more from their government than the Argentines, who were wedded to laissez-faire.

What aspects of their colonial pasts might explain these different expectations? Was it the legacy of a mercantile, town-centered society in British North America in contrast to the less-governed but more hierarchical ranching society of the Argentine interior? Carl always preferred more straightforward explanations. Both nations achieved national unity at roughly midcentury and then welcomed immigrants. But Canada with its two founding cultures needed government to create a nation in ways that Argentina did not.

For its excellent research and conceptualization, for its careful and exacting use of comparison, and for its contribution to both Latin American and Canadian studies, this book does credit to Carl Solberg. Those of us who discussed the project with him and commented on earlier drafts are grateful that he was able to complete his work and see it accepted for publication.

John D. Wirth

Acknowledgments

I would like to thank the American Council of Learned Societies, the American Philosophical Society, and the University of Washington Graduate School Research Fund for fellowships and grants that made the research for this book possible. I would also like to express my gratitude to the staff at Glenbow-Alberta Archives in Calgary, the Biblioteca Tornquist in Buenos Aires, and the library of the Federación Agraria Argentina in Rosario. Special thanks go to Ruth Kirk and her staff at the University of Washington Inter-Library Loan office. And, finally, I wish to express my appreciation to John Wirth, Dauril Alden, David Tamarin, and my colleagues in the History Research Group at the University of Washington. Their comments and criticism proved most helpful.

C.E.S.

Abbreviations

The Prairies and the Pampas

1

Introduction

Late in the nineteenth century a spectacular economic transformation began in the vast, empty grassland countries of the New World and Australia. Within two decades, roughly between 1880 and 1900, the once desolate prairies of North America, pampas of Argentina, and plains of Australia became breadbaskets of the world. Millions of immigrants swept into these regions to break the sod, begin cultivation, and produce mountains of cereals to provision a rapidly industrializing Europe. Simultaneously, capital worth billions of dollars flowed from European financial centers to build the railways, develop the ports, and provide the commercial and marketing systems that these agricultural regions required to transport and handle their vast production.

Nature had favored these new countries—Australia, Canada, Argentina, and the United States. They were "land surplus" countries rather than the "labor surplus" type so common on the peripheries of the expanding Western European economy. As Melville Watkins puts it, these new countries had "an enormous advantage over the typical underdeveloped country" because they did not "start their development with population pressing against scarce resources."[1] With their abundant land and sparse populations, the new countries could easily feed their own people and at the same time produce massive food exports. And ready markets awaited these exports in the rapidly urbanizing countries of Northern and Western Europe.

Although these new countries had much in common, their economic development and particularly the development of their agrarian sectors moved along quite distinct paths. This book analyzes why such deep contrasts emerged in two countries, Argentina and Canada, that became big agricultural exporters at about the same time. Because wheat was the largest single export of both countries, the analysis will revolve primarily around the wheat economies of the Canadian prairies and the Argentine pampas.

Prior to 1910 Argentina produced and exported more wheat than Canada, but after that year, Canada surged ahead and became the world's leading wheat exporter. What is more, the Canadian prairies were not only outpacing the Argentine pampas in quantity, but also surpassing them in the quality of wheat production. While Canadian wheat was becoming what the Australian economic historian John Fogarty calls a "super-staple," an export product whose quality and reliability earned worldwide respect, Argentine wheat did not enjoy a particularly distinguished reputation.[2] Thus, two New World wheat economies that emerged at about the same time, and that once competed vigorously with each other, diverged in very different directions.[3] These trends have continued. Canada has maintained its position as a world leader in the wheat trade, but Argentine wheat—and pampa agriculture in general—has languished, and Argentina has lost its historic role as a major wheat exporter. By the early 1970's the pampa agrarian situation had become so bleak as to lead H. S. Ferns to comment that "the rural economy of Argentina has tended to resemble that of the USSR in terms of stagnant productivity."[4]

The central problem, then, that this book addresses is why Canada surged ahead while Argentina stagnated and declined. Why did wheat agriculture in the prairies and the pampas, regions whose economic history began along parallel lines, diverge in such distinct directions? To analyze these questions, we will utilize the concept of political economy. The basic assumption is that government policy and the role of the state decisively affected the paths the Canadian and Argentine wheat economies took. As John Kirby has demonstrated in his comparative study of agrarian development in New Zealand and Uruguay, politics is the principal variable that explains the different performances of new countries producing the same staple exports for the same markets.[5] Following this line of reasoning, we will examine the agrarian policies that Argentina and Canada adopted and their impact on pampa and prairie agricultural development.

The role the state played in Argentine and Canadian agricultural development is closely related to social class structure in both countries. As Donald Denoon has pointed out in his recent comparative history of several "new countries," the role of the state is a crucial factor in explaining why these "settler societies" differed among themselves, but the role of the state in turn reflects social class interest and class interaction in the political arena.[6] For this reason, we will investigate who held power in Argentina and Canada and examine how this power was used in the agrarian sector. Although our focus will be on the half century beginning in 1880, when the prairies and pampas opened to development and when

Argentina and Canada put their basic agricultural policies into effect, the impact of these policies in many cases endures to the present day.

The theories of regional development formulated by the American economic historian Douglass C. North have helped to shape the conceptual basis for this comparative study. North notes that "as regions grew up around the export base, external economies developed which improved the competitive cost position of the exportable commodities," and that "the concerted effort to improve the technology of production has been equally important."[7] These comments suggest a number of points of investigation. In what ways did Canadian and Argentine wheat growers and policy makers attempt to improve their "competitive cost positions?" Did they introduce mechanization, efficient marketing systems, and modern transport? Did Argentina and Canada promote agricultural education, research, and technology? In sum, did the state consciously intervene to promote more efficient crop production? If so, why; if no, why not?

It would be premature to discuss these questions in detail here, but it may be useful to the reader to present an overview of the book's analysis. On the one hand, Canadians overcame the numerous natural disadvantages of the prairies and made that region into a premier wheat producer. On the other hand, Argentines took for granted the great natural advantages of the pampas and failed to build a modern agricultural economy on the basis of these advantages. The Argentine pampas had a warmer and more benign climate than the prairies. The soil of the pampas was fabulously fertile—at least as rich as that of the prairies. Argentine pampa farms were close to the ocean, generally within 200 miles of a port, and thus pampa farmers did not face the long and expensive rail haul that imposed a major economic burden on prairie farmers. For decades, the huge flow of Southern European emigrants to Argentina gave pampa agriculture access to a cheap harvest labor force, while labor shortages at harvest time constantly plagued prairie agriculture. And although Argentina was not a low tariff country, the tariff did not place the heavy burden on pampa agriculture that it did on prairie farmers. Climate, soil, location, an abundant labor supply, and national tariff policy long combined to keep Argentine costs of wheat production competitive with Canada's. Throughout the 1900–1930 period Argentina's political leaders assumed that these advantages would perpetually guarantee the pampas a major share of the world wheat market. During these same years, however, Canadians were busy devising ways to overcome the disadvantages a cold and isolated environment posed for massive wheat-export agriculture. Large-scale agricultural research and education programs, efficient rail transport, sophisticated wheat grading and standardization methods, and excellent systems of storage and marketing gradually emerged in Canada as the result of the foresight of the political elite and the political and co-operative organizations of the farming population.

Some interesting new interpretations of long-discussed problems arise from this comparative study. In the case of Canada the analysis suggests that the federal government at Ottawa was not as negligent of the prairies or as exploitative of its interests as many Western Canadian agrarian leaders and scholars have claimed.[8] From a comparative perspective, it becomes clear that Canadian policy prior to 1930 developed the economic and technical base for a wheat-export economy strong and enduring enough to withstand the double catastrophe of the Depression and the Dust Bowl and to play a major role in the world wheat trade after the Second World War. The comparative perspective also suggests that traditional interpretations of Argentina's decline as a major wheat exporter after the Second World War are short-sighted. Numerous writers have argued that the economic policies of Juan Perón undermined the competitive position of Argentine export agriculture.[9] There is a good deal of evidence to support the position that Perón was hardly a friend of agriculture, but serious structural weaknesses plagued Argentina's agrarian economy long before he came to power. In contrast to Canada, the Argentine elite since the late nineteenth century had failed to develop agricultural research and education, to provide an efficient transport system, and to build an effective marketing structure. Argentine agriculture was built on the systematic exploitation of the nation's tenant farmers rather than on modern production and marketing systems that could keep pace with the republic's international competition.

Comparative analysis of the political economy of wheat also provides critical insight into dependency theory. Much discussed in recent years as an explanation of the causes of economic and social underdevelopment, dependency theory is based on a number of central assumptions. First, the level of prices and demand in the world market will have a major economic and social impact on the raw-material exporting countries. When commodity prices drop sharply, according to this reasoning, social and political upheaval is likely to result in the countries that specialize in producing and exporting these commodities. A second and related assumption is that economic relationships between raw-material exporters and their industrialized trading partners are inherently unequal and exploitative. The result, according to the theory, is that the commodity exporters are relatively deprived economically, and that while the "metropolis" (the developed, industrialized countries) develop their economies, the "satellites" (raw-material exporters) become more underdeveloped. And third, many dependency theorists argue that the raw-material exporting countries lack real political independence; their governments are not in fact autonomous because they are subject to external economic forces. According to this line of reasoning, foreign markets and foreign investors, often with the help of the governments of the metropolis, decisively influence policy formulation in the "satellites."[10]

Dependency theory has made some important contributions to Latin American economic historiography, but comparative study of the Argentine and Canadian wheat economies suggests that the theory needs refinement and must be used with great care. On the one hand, there is no question that the prairie and pampa economies were very sensitive to fluctuations in world wheat prices, and that periods of agrarian unrest accompanied major slumps in the world market. On the other hand, one cannot seriously argue that Canada and Argentina were economically deprived countries or became more "underdeveloped" during the great wheat boom. In fact, in per capita terms, they were among the richest countries in the world. Although income was seriously maldistributed in both Argentine and Canadian society, the same was true in the industrialized countries at that time.

But perhaps the most important objection that this study raises against dependency theory concerns politics and the role of the state in agricultural development. Both Argentina and Canada exported their wheat to the same markets. During the half-century this study covers, the same world wheat price (except during the First World War) prevailed for both countries. Even during the Imperial Protection period of the 1930's, Great Britain—the major wheat market—never excluded Argentine wheat or gave Canadian wheat any real economic advantage over pampa wheat. Because both countries faced the same market circumstances (except during the war), one would expect, if one followed the logic of dependency analysis, that agricultural development in Canada and Argentina would proceed along roughly similar lines. But this is not what happened. In fact, one country pulled ahead to become a world leader in the wheat trade, and the other country fell increasingly farther behind. The reason was that the Argentine and Canadian governments had enough autonomy to make certain key policy choices, and that they usually made these choices not in response to external pressures, but in response to domestic political alignments. Of course, neither the Canadian nor the Argentine governments was an entirely independent, autonomous actor, free from all foreign economic and political influence. This could be said of no country in modern history. But analysis of Argentine and Canadian wheat policy suggests that, at least in some cases, individual exporting countries are able to act effectively to improve their competitive position in the world market.

The book is organized into three parts. The first, consisting of two chapters, provides the essential background and context for understanding the political economy of agriculture in the prairies and the pampas. These chapters provide the reader who is unfamiliar with one or the other (or both) of these countries with a basic political and economic overview of Argentine and Canadian history between 1880 and 1930. Here we also will examine and compare basic relationships between social-class structure and economic policy formulation in Argentina and Canada.

The second part (Chapters 4–6) analyzes and compares the basic agricultural development policies that the two countries adopted during our period. These policies on land distribution, immigrant recruitment and settlement, railway construction and rates, and agricultural technology were critical to the patterns of development that prairie and pampa agriculture would follow. In most cases the two countries adopted these policies early (in some cases before 1880). Because they established the structure of social-class relationships in the countryside as well as the infrastructure for agrarian development, these policies deserve careful analysis.

In the third part (Chapters 7–9) the focus moves away from the topical emphasis of the preceding chapters and shifts to an analysis of major agricultural policy issues in the two countries. These chapters focus primarily on relationships among the state, the political elites, and the farming populations between 1900 and 1930. It was during these years that sustained agrarian unrest emerged and farmers' organizations became important forces in Argentine and Canadian politics. Analysis of how the state dealt with this unrest and this political pressure will enable us to understand more fully why one country (Canada) introduced major reforms that made its wheat economy a world leader while the other country (Argentina) continued to rely on outdated and inefficient techniques in many phases of the wheat economy. Finally, Chapter 10 presents some thoughts on the different paths of agrarian development in the two countries.

Part I

Political and Economic Background

2

An Overview of Argentine and Canadian History, 1860-1930

At a casual glance Argentina and Canada, two countries separated by many thousands of miles and by very different ethnic and cultural heritages, might not seem to have much in common. But, in fact, between 1880 and 1930, Argentina and Canada shared numerous similarities, particularly in the patterns of their economic development. These similarities were sufficiently pronounced that the Argentine economist Alejandro Bunge could refer to the "Argentine-Canadian economic parallel," in the title of an influential 1929 article.[1]

The basis for this economic comparison was that both countries contained huge grasslands, the prairies and the pampas, that became enormously productive agricultural regions. Both regions attracted massive immigration. Both were tied closely to Great Britain, their leading export market and the leading source of their massive foreign investments. The agricultural bounty of the two regions enabled Argentina and Canada to rise to preeminence among world agricultural exporters.

The demographic and economic trends that gave Argentina and Canada certain important commonalities also demarcated them from their continental neighbors. Indeed (except for Uruguay) Argentina was an anomaly in South America. In a continent that was predominantly mestizo or mulatto, Argentina was ethnically primarily European. While most South American countries were densely populated (at least in relation to their arable land), Argentina was not. While the other South American countries relied on mining or labor-intensive plantation agriculture for their exports, Argentine exports were the product of extensive agriculture on farms and ranches. The volume of this trade made Argentina far more prosperous than any other Latin American country. Rich, white, and apparently successful, the Argentines tended to ignore or disdain their Latin American neighbors and instead to identify with Western Europe, the United States, or Britain's white "settler dominions."

Canadian economic and demographic history also was clearly demarcated from that of its huge North American neighbor. For one thing, about 30 percent of the Canadian population spoke French rather than English, a proportion that gave Canadian history a distinctive flavor in North America. And second, the Canadian economy was much more closely tied to international trade and foreign markets than was the U.S. economy. While it is true that both Canada and the United States contained huge grasslands and became large wheat exporters, the United States had over ten times the population of Canada, and this enormous domestic market consumed the large bulk of U.S. agricultural production. Unlike Canada (and Argentina), the United States did not have to rely on the export trade to sell the majority of its rural output. And while Canada was a more predominantly export economy than the United States, Canada's big southern neighbor consumed few of the prairie provinces' agricultural exports. High protective tariffs on both sides of the border divided the two North American economies, kept Canadian produce out of the United States, and ensured that the prairies' huge agrarian output would move primarily across the Atlantic to Europe.

This chapter lays groundwork for our comparative study of the political economy of Argentine and Canadian wheat. Because the author assumes that many of his readers may be unacquainted with Argentine or Canadian history (or both), we will begin with a brief survey of the political history of the two countries. Our purpose is to analyze the dominant political groups in both societies and to examine the economic policies they adopted to develop their vast grassland regions. We will survey politics in both countries over a long time period, from the 1860's, when both countries emerged as distinct political entities, to the onset of the Great Depression in 1930. Because of this broad scope, the historical analysis necessarily relies on generalizations, but they will serve to introduce the reader to subsequent chapters, which will examine many of these introductory summaries in greater detail.

Political Elites and Economic Policy

Argentina and Canada began their existence as organized states during the 1860's. A long series of civil and foreign wars followed Argentina's declaration of independence from Spain in 1816; it was not until 1862 that the national capital effectively established its dominance over the interior provinces, previously largely autonomous. Canada became a self-governing Dominion in 1867 and acquired Rupert's Land, the vast territory of the Hudson's Bay Company that included the prairies, in 1869. By the mid-1880's both the Argentine and the Canadian government had subdued the nomadic Indian peoples who roamed the prairies and the pampas—although the Canadian Indians were placed on reserves while the

Argentine pampa Indians were largely exterminated. Both governments also met opposition from mestizo horsemen (called métis in Canada) who long had inhabited the grassland regions. The Canadian government crushed métis resistance during the Northwest Rebellion of 1885; in Argentina, the mestizo horsemen, better known as the gauchos, had ceased organized opposition to the advance of settlement by the early 1870's. In this way, both governments established unquestioned control over their open grasslands by the mid 1880's and could then devote their attention to large-scale agricultural development.

Argentina and Canada pursued similar policies of rapid frontier expansion during the 1870's and 1880's, but once the two governments subdued the Indians and conquered the grasslands, they would pursue very different policies of economic development. These differences, which the bulk of this book will discuss, reflected the distinct political systems and structures of political power in the two countries. Before reviewing the general patterns of Argentine and Canadian history between the 1860's and 1930, it will be useful to sketch out some of the principal political differences between Argentina and Canada.

Very different groups of people composed the political elites in the two countries. In Canada merchants, primarily from Montreal and Toronto, formed the economic and political elite that founded the modern nation. Largely of Scottish descent, the Canadian merchants who formed the core of the elite were able to accumulate enough capital from trade to stake out an economic sphere that was relatively autonomous from Britain and from the United States. They controlled the large Canadian banks, which gave them a great deal of influence over the course and direction of economic development. By the mid-nineteenth century these merchants and bankers were beginning to invest in industry, and to safeguard this venture, they became ardent protectionists. Although they strongly supported the tariff and Canadian industrial development, the financial elite clearly recognized that this protected system rested on Canada's earnings from exports of its raw material staples, particularly wheat. The economic strategy of the Canadian elite thus involved balancing prairie agrarian development on the one hand and eastern industrial development on the other.[2]

In Argentina, by contrast, large landowners of the pampa provinces formed the dominant elite. This landed class dated back to the colonial period, when the king of Spain had made enormous grants of the pampa to various favorites. In the late eighteenth and early nineteenth centuries, merchants from Buenos Aires, realizing the enormous potential of the pampas and valuing the social prestige that landowning conferred, also bought land. Still others became landowners in the mid- and late nineteenth century, when the government rolled back the Indian frontier.

Whatever the source of their land, the elite families, some of whom owned tens of thousands of hectares, became wealthy through hides and dried beef exports in the early nineteenth century, through wool at mid-century, and then from the breeding and fattening of beef cattle after the 1870's. Eventually they turned to agriculture as well, and by the 1920's wheat had become Argentina's leading export. But the landed elite remained fundamentally tied to cattle raising.[3]

Because the landownership yielded huge speculative as well as productive profits, land became and remained the key focus of Argentine capital investment. Although numerous Argentines did enter business, most aspiring Argentine capitalists were content to leave trade and banking to foreign investors (who were primarily British) and to concentrate their investments in land and cattle. Very few Argentines before the end of the nineteenth century invested in industry or supported a general protective tariff. The Argentine gentleman, wrote a perceptive early-twentieth-century British traveler, "knows only of one way in which to invest the surplùs of his income—in land or the things intimately connected with land and its immediate productivity."[4]

In terms of comparative historical development, the Argentine elite was in an "exceptional situation," as H. S. Ferns has expressed it. The contrast with Canada is particularly instructive, for in Canada, a British Dominion, British bankers and merchants held much less economic power than they did in Argentina. The Argentine elite was "unlike that in the United States or Canada or Australia or any European country." In these agrarian nations, the creditor classes—bankers and merchants—held political power. But in Argentina the landed class was, as a distinguished traveler observed, "all powerful. Nothing important happens in that country without their participation."[5]

This sharp contrast between the economic bases of elite power in Argentina and Canada meant that the two countries would follow very different economic policies. The Canadian elite would give grain growing high priority, while the Argentine elite would subordinate it to cattle raising. The Canadian elite would use tariffs to industrialize, while the Argentines would remain resolutely wedded to classical free-trade theory. And perhaps most significantly, the Canadian economic elite would formulate a coherent national economic development plan that placed the state at the center of the economic policy-making arena, while the Argentines had no coherent development plan other than following the dictates of the marketplace. In matters of economic policy the Argentine state remained much weaker than the Canadian.[6]

The political voice of the nineteenth-century Canadian elite was the Conservative Party of Sir John A. Macdonald, Canada's first prime minister. The Conservatives strongly promoted Confederation in 1867 to keep the West out of the clutches of the United States and to create a viable Canadian transcontinental economy. Or, as the Western Canadian historian Vernon Fowke has put it, the merchants of central Canada aimed to use the prairies as "a basis for commercial and financial empire."[7] To achieve this goal, the Conservative Party, which held power from 1867 to 1873 and again from 1878 to 1896, carried out Macdonald's famous "National Policy."

The National Policy consisted of three related developmental strategies. One was to promote Canadian industrialization through a protective tariff. This strategy had a long history in Canada. The protectionist theories of the German economist Friedrich List had attracted a strong Canadian following by the 1850's, and the Canadian colonial parliament passed a protective tariff as early as 1859.[8] Second, the National Policy aimed to build a transcontinental, all-Canadian railway to tie Canada's vast new Western possessions to the East with bonds of steel. And third, the National Policy envisaged a massive immigration program to settle the prairies. The immigrants would produce staples for export and consume products of the protected Eastern industries. The new transcontinental railway would link the prairie settlers with Eastern markets and ports. As a blueprint for building a second transcontinental North American nation, the National Policy was easy to understand.

On the basis of this policy, Macdonald's Conservative Party built a powerful political base, which consisted not only of the merchant-banker elite but also of the industrialists and their workers. The powerful Quebec Roman Catholic Church long supported the Conservative Party and thus gave the party and the National Policy a mass voting base in French Canada. The opposition Liberal Party, which was based largely on farmers in central Canada (and later in the West), in principle stood for free trade rather than protectionism. Initially it contained strong anticlerical elements, which of course alienated the party from the Catholic Church. But when in power between 1896 and 1911 and again in the 1920's, the Liberals shelved their anticlericalism and continued to carry out the National Policy, with some minor modifications. The Canadian industrial base had become too well entrenched politically, and the federal government too dependent on the tariff revenues, to permit the Liberals to challenge the basics of the National Policy. By acting early and decisively, the Macdonald government had defined the main directions that Canadian economic life would follow until the 1930's and in some respects until the present.[9]

One aspect of the National Policy needs to be highlighted in this discussion of comparative Canadian-Argentine history. It concerns the role of the state in the process of national economic development. The National Policy, it is safe to say, virtually institutionalized the Canadian federal government as an active developmental promoter. The protective tariff made the state and the private industrial sector close partners.[10] The federal government was an equally close ally of the Canadian Pacific Railway, the transcontinental line that was completed in 1885 only after receiving vast government grants and loans. And the Ottawa government maintained tight control over political and economic affairs in the prairie West after it was acquired from the Hudson's Bay Company in 1869, keeping most of this vast region in the political status of territories (they were known as the Northwest Territories) ruled directly from the federal capital. Except for Manitoba, which became a province in 1870 (but with only a small fraction of its present area), this situation endured until Alberta and Saskatchewan became provinces in 1905. Even after that, Ottawa retained power over the public lands in the prairie provinces until 1930. Between 1869 and 1930, a period that spanned the great immigration and settlement booms in the prairie West, the central government controlled the critical power of land distribution and used this power to build a society of individual farm owners.[11] Even after the new provinces of Alberta and Saskatchewan were organized, the prairies remained politically a subordinate region, for the size of each province's delegation in the all-important federal House of Commons depends on its population. The two largest provinces, Quebec and Ontario, held two-thirds or more of the total Canadian population and thus received about two-thirds of the seats.

The National Policy, then, not only made the prairies a "hinterland region to St. Lawrence capital," but also made it a kind of political colony of central Canada.[12] It is for this reason that many, if not most, Western Canadians long have viewed the National Policy as an instrument of domination, designed to keep the West strictly subordinate to the industrialized East. We will examine this attitude many times in this book, but here we should emphasize that the National Policy tariff raised the cost of wheat production in Western Canada substantially because Canadian-produced farm implements, clothes, and tools were higher priced than imported products. But prairie farmers had to sell their wheat on the world market, in direct competition with Argentine and U.S. farmers, who had access to cheaper manufactured goods.[13] This situation led to bitter resentment in the prairie West.

There was nothing in Argentine history remotely similar to the National Policy. The Canadian state had far more power over the economy and assumed a much more active role in the developmental process than the Argentine state for two principal reasons. For one thing, the Canadian state was constitutionally more powerful, a fact that reflected the different early histories of Argentina and Canada. Bitter civil wars between the interior provinces and the rich coastal province of Buenos Aires continually wracked nineteenth-century Argentina until 1862, when Buenos Aires finally agreed to enter a federal government whose constitution had been written earlier, in 1853. This was an event of major importance for Argentine economic history because the 1853 constitution granted some key economic powers to the provinces and thus weakened the national state's authority as a developmental agency. In contrast to the situation in the Canadian prairies, the Argentine pampa region was already organized into provincial governments in 1862. In Argentina the national state could not maintain the close supervision over land distribution and settlement that characterized the Canadian prairies because the 1853 constitution specifically gave the provinces jurisdiction over the public lands.

The other reason why the Argentine state was a weak partner in the process of economic development was that the landed elite wanted it that way; there was little political support for protected industrialization in the pampa provinces, which contained the bulk of the nation's population by the 1890's. As we have already seen, the Argentine economic elite was composed of landowners rather than bankers or industrialists, and this landed elite's economic prosperity was directly linked to the smooth functioning of the free-trade export-import economy. Those few lone wolves who wished the Argentine state to follow a consistent policy of industrial protection or to pursue an activist course of economic development found little political support, nor did they find much backing among Argentine intellectuals.[14]

Classical liberalism prevailed in Argentine economic thought in this era. The "Generation of '80," the group of intellectuals and political leaders who dominated Argentine public affairs in the late nineteenth and early twentieth centuries, was convinced that the free and unregulated operation of market economics would bring the fastest possible development to Argentina, and that government intervention in the economy would weaken economic growth rather than strengthen it. The Generation of '80 embraced the doctrine of unlimited progress and believed that the best proof that liberalism was the key to progress was the dynamism the Argentine economy displayed by the late nineteenth century.[15]

Two Political Traditions: An Introduction

We will be referring frequently to political parties and political leaders throughout the chapters to come in both countries. Readers unfamiliar with one national history or the other (or both) may find these names bewildering, and for that reason, this section, which surveys Argentine and Canadian history between the 1860's and 1930, attempts to give readers a basic introduction to the history of the two countries. The section makes no pretense at providing a comprehensive overview but merely aims to set the political scene for the analysis in the subsequent chapters.

We are of course discussing two very distinct political systems that reflect the heritage of Britain, on the one hand, and of Spain on the other. Canada's political heritage was parliamentarian; Argentina's was authoritarian. Although after independence from Spain Argentine constitutions adopted the procedures of political democracy, the tradition of absolutism remained very powerful there as in the rest of Latin America. The 1853 Argentine constitution, which was largely modeled on the U.S. constitution, established three separate branches of government, but in fact the Argentine president dominated the legislative and judicial branches. Although the constitution established the principles of popular sovereignty and democratic elections, in fact the executive branch regularly and openly interfered with elections for the two houses of Congress (the Senate and the Chamber of Deputies) and often rigged them until 1912, when Argentina enacted major electoral reforms. Like every republic in the Americas, Argentina was born in a political revolution, but the revolution against Spain did not fundamentally alter the country's political tradition. Although all the Latin American revolutions proclaimed that they were aimed at severing ties with the political past in order to create a new and democratic polity, in fact the political traditions as well as the social structures of colonial Latin America endured and kept power in the hands of the few. The tradition of political democracy did not take root in Argentina until the 1890's, and after that it grew slowly.[16]

Canada alone among the American nations never experienced a revolution. It was instead the product of centuries of slow political evolution in Britain and of gradual, incremental change in North America. The direction of this change was clear between 1763, when Britain acquired New France, and 1867, when Canada became a self-governing Dominion within the British empire. That direction was toward responsible government—the primacy of the House of Commons, which the voting public elected, over the Crown's appointed officials. Under the system of responsible government, in effect since 1848, the leader of the party that wins the majority of seats in the House of Commons becomes the prime minister and forms a Cabinet, which becomes the executive branch of the government. The British North American Act, which established the new Canadian Dominion in 1867, incorporated these principles (although not always explicitly) and added an upper house of Parliament called the Senate. But this Senate, whose members the prime minister appoints for life, has few real powers and is a far weaker body than the Senate in the American republics.

Electoral corruption was no stranger to Canadian politics before or after the British North American Act, but it was less extensive than in Argentina. The tradition of honest elections was far better entrenched in Canada, at least before the 1912 Argentine electoral reforms. This is not to say that economic elites did not exercise a great deal of political influence in Canada, only that popular sovereignty was a more active principle than in Argentina.[17]

Two Political Traditions: An Overview

Argentina

Argentina became a modern organized state in 1862, over half a century after the first rebellions against Spanish colonialism began in the old Viceroyalty of La Plata in 1810.[18] The long, bitter wars of independence dragged on until 1826, but in Argentina another cycle of violence began before those wars ended. A series of civil wars between centralists, who demanded a strong national government, and "federalists," who insisted on more provincial autonomy, began in the late 1820's and dragged on for years.* One of the results of the federalist-centralist wars was the tyranny of the dictator Juan Manuel de Rosas, who dominated the country from 1829 until 1852. This cattlemen's government did not solve the federalist issue but merely repressed it. After Rosas fell, the federalists and unitarians compromised many of their differences in the 1853 national constitution, which on the one hand gave the provinces jurisdiction over the public lands, but on the other established a powerful central government that, under the "intervention clause," had the authority to intervene in the provinces under rather vaguely defined circumstances.

Despite this political compromise, the new nation fell into more turmoil because the province of Buenos Aires insisted that its capital, the city of Buenos Aires, should also be the national capital. The interior, which was convinced that this arrangement gave Buenos Aires, already the largest and wealthiest province, too much power, insisted that the city would have to be separated from the province if it was to be made the national capital. This dispute produced two Argentine governments, one for Buenos Aires, and the other (at Paraná) for the rest of the country, between 1853 and 1862. That year Buenos Aires settled the issue on the battlefield, and Argentina began the period of national organization.

Enormous obstacles faced the new nation. Its population was small (1,700,000 in 1869), its resources were still mostly untapped, and its people had no practical democratic experience. Under these circumstances, it is hardly surprising that the Spanish authoritarian heritage prevailed in

---

*In Latin America, unlike the United States, a federalist has historically stood for the decentralization of power and greater provincial autonomy.

politics. Between 1862 and 1880 a series of presidents, all allied closely with the landed elite, gave Argentina leadership that was capable and energetic but that ignored the democratic provisions of the 1853 constitution. This oligarchic system, which controlled the presidency until 1916, manipulated elections to dominate politics at both the provincial and the national level.

Tension still existed over the federalist question, particularly over the location of the national capital. During a political crisis and brief civil war in 1880, the interior prevailed over Buenos Aires province and converted the city of Buenos Aires into a federal district, which it has remained until the present. Buenos Aires province then built a new provincial capital at nearby La Plata.

The year 1880 is a watershed in Argentine history not only because it marked a formal end to the federalist political question and inaugurated a long period of civil peace, but also because it marked the beginning of a half-century of rapid economic development. The political elite, which organized itself as the Partido Autonomista Nacional (PAN), was essentially an alliance of pampa landowning oligarchs and conservative machines from the interior provinces. Although the PAN maintained a staunchly autocratic regime that paid only lip service to constitutional rule and popular sovereignty, its leaders were strongly committed to the economic liberalism that the intellectual Generation of '80 supported. The PAN strongman Julio Roca, Argentina's late-nineteenth-century military hero and twice president (1880–86; 1898–1904), summed up this political approach with his dictum that the government's duty was to provide "peace and administration."

Roca aimed to open the pampas to rapid development. He had already taken a major step toward this goal when, as minister of war in the late 1870's, he had led a major military campaign against the remaining pampa Indians. This "Conquest of the Desert," which exterminated most of the Indians, opened up the southern pampas to settlement.* But the Conquest of the Desert did not presage the rise of a small farmer class. Some of the newly acquired lands the government distributed to soldiers and officers in Roca's army; the rest went on the auction block, where speculators quickly bought it up. Most of it ended in the hands of the landed elite.

When Roca became president in 1880, he resolutely followed a laissez-faire economic policy, a policy that, with few exceptions, was to characterize Argentina's economic development for the next 35 years. In contrast to the Canadian situation, where the central government played a major role in the economic and social development of the prairies, in Ar-

---

*Later military campaigns were able to mop up the last resisters in Patagonia by 1885.

TABLE 2.1
An Overview of Argentine and Canadian Economic Growth,
1870–1900

Country	Exports[a]	Railways[b]	Population
Argentina			
1870	22,367,312	732	1,737,076
1900	154,600,412	16,563	4,607,000
Canada			
1870	59,043,000	4,187	3,625,000
1900	168,972,000	28,251	5,301,000

SOURCES: Ernesto Tornquist & Cía., pp. 1, 110–11, 133–34; Argentina [4], p. 159; Urquhart & Buckley, pp. 14, 173, 528.
[a]In current Argentine gold pesos and Canadian dollars.
[b]In kilometers.

gentina the government had no overall development plan for the pampas. Both the provincial and the central governments relied primarily on market forces to determine the course and the nature of pampa development. On the crucial question of land policy, the federal government was unable (even if it had been willing) to pursue a homestead-type land-distribution policy, because the provinces administered the public lands, and none of the provincial governments was much interested in democratic land tenure. Roca's immigration policy consisted primarily of opening the nation's doors to virtually unrestricted immigration, although the PAN governments did abandon their laissez-faire policies to the extent of advertising for immigrants in Europe. Despite the difficulty of obtaining land, millions of immigrants, primarily poor southern Europeans, responded to the chance to find a new future in Argentina and poured into the pampas seeking work. Argentina's population grew very rapidly (see Table 2.1). Until the era of easy international migration ended in 1914, Argentina's open-door policy gave it access to a vast labor supply that made pampa development possible.

The PAN governments also extended virtually an open-door policy to railway investors, although here again the state exceeded the limits of laissez-faire by guaranteeing the profits of new railway concerns for a fixed number of years. Railway investors rushed in, and the network grew rapidly (Table 2.1). But because the government had no railway development plan, the system expanded in a haphazard fashion. The foreign companies that built the lines did not even use the same gauge tracks, with the result that the pampas contained lines of three different gauges.

The difference between prairie and pampa development was not that the Argentine government stayed entirely out of the development process—for it did promote both immigration and railway building—but rather that, unlike Canada, Argentina had no coherent development program for the pampas. There was no plan for land distribution to immigrants or for the systematic development of a railway network. Nor was the Argentine government interested in agricultural technology. While the Canadian government had an active Ministry of Agriculture, which began to carry out important plant genetics research in the prairies as early as 1886, Argentina did not create a federal Ministry of Agriculture until 1898, and even after that the ministry did little scientific research.

In Argentina, leaving the pampa's development to the private sector really meant leaving it in the hands of the cattle oligarchy and the British merchants, and so it is hardly surprising that when agriculture did begin to develop during this period, its role was strictly subordinate to cattle raising. Nonetheless, the growth of pampa meat and cereal production was spectacular. Exports multiplied seven times between 1870 and 1900 (Table 2.1). But the absence of any comprehensive government policy meant that property ownership remained highly concentrated, that the educational system was primitive, that the roads were abominable, and that the agricultural marketing system was left in the hands of grasping grain merchants.

The oligarchic regime brought rapid material progress to Argentina, but the political system was beginning to decay by Roca's last term in office. The export boom, along with massive immigration and urbanization, was transforming Argentine society rapidly. By the turn of the century, an urban proletariat and middle class were emerging in the rapidly growing Buenos Aires metropolitan area. In this fluid social situation new political parties formed to challenge the PAN's dominance. Since the mid-1890's a small but vigorous Socialist Party had been attempting to establish an electoral base in the capital city. The party scored its first victory when it elected a candidate to the national Chamber of Deputies in 1904. But the Socialists remained a relatively small group, for their emphasis on gradualist and democratic procedures failed to appeal to the anarchist and syndicalist unions that controlled the bulk of Buenos Aires' rapidly growing labor movement. Dedicated to the use of the strike as a tactic for change, the anarchist and syndicalist unions grew increasingly militant after 1900. The several general strikes they mounted deeply alarmed the upper class and stimulated elite politicians to search for some sort of political reform to dispel the specter of mass rebellion.

Reformists in the political elite began to seek an accommodation with another emerging force in Argentine politics, the Unión Cívica Radical, or Radical Party, a coalition of members of the urban middle classes and newly rich landowners that had been organizing nationwide since the early 1890's. The Radicals concentrated their appeal on the middle classes, which, at least when defined in functional terms, constituted nearly 30 percent of the Argentine population by 1914. This was, as David Rock has termed it, a "dependent" middle class, reliant on the export-import trade and the big foreign firms that owned so much of the Argentine economy. Although the economic boom and the rapidly growing educational system stimulated their aspirations for upward mobility, the middle classes found that they could not rise to the higher management levels in the big foreign firms. They were left with lower level clerical positions, careers in the government bureaucracy, or the limited entrepreneurial opportunities of petty commerce. This middle class was fertile political ground for the Radicals, who promised to vindicate the common man, although the party was vague about how it would do this.

Hipólito Yrigoyen, the tireless and domineering leader of the Radicals, avoided potentially divisive economic and social issues, and concentrated instead on attacking oligarchic rule and demanding effective suffrage and constitutional democracy. A brilliant organizational strategist, Yrigoyen built up the Radical Party throughout the nation and emphasized that his party would abstain entirely from the political process until the government guaranteed honest elections. If all else failed, Yrigoyen threatened violent revolution.

By 1910 the old PAN machine was falling apart, and the new president elected that year, Roque Sáenz Peña, made it clear that he intended to support electoral reform and to integrate the Radicals into Argentine political life. This experienced diplomat and son of a former president negotiated an agreement with Yrigoyen and successfully prodded Congress into passing legislation in 1912 known ever since as the "Sáenz Peña Laws." These reforms created a system of fair elections based on the secret ballot and obligatory voting for all male citizens over the age of eighteen. This marked the beginning of Argentina's first period of political democracy, which was to endure until 1930.

The Radicals captured the presidency with the election of Yrigoyen in 1916, and from then until 1930, the Radical chief would be the central figure in Argentine politics. He blended timely concessions to the working classes with populist and nationalist appeals, and in this way he developed a near-mystical reputation as the people's champion. But although many lionized Yrigoyen, others reviled him. The oligarchs of the Conservative Party (the successor to the PAN) scoffed at Yrigoyen's humble social origins and strongly distrusted his mass-based political appeals; and on the left the Socialists bitterly opposed Yrigoyen, whose Radical Party was capturing much of the working-class vote that they wanted. In Congress Socialists and Conservatives teamed up in an unlikely alliance against the Radicals until Yrigoyen's party won control of the Chamber of Deputies in 1920. But the Conservatives, who still controlled powerful political machines in several interior provinces, remained in control of the Senate throughout the 1916–30 period and

maintained an effective veto over Argentine politics. Scorned by the opposition parties, Yrigoyen also faced opposition within the Radical Party,
particularly from its patrician segments, who distrusted his populism.
This opposition produced a party schism in 1924, when the Radicals split
into two separate and hostile groups, the "Personalists" (the pro-Yrigoy-
enists) and the "Antipersonalists." Constitutionally unable to succeed
himself, Yrigoyen had hoped that he would be able to dominate the government of his Radical successor, Marcelo T. de Alvear, a wealthy aristocrat (1922–28). But Yrigoyen's plans failed, and the party split, much
to the detriment of Argentine democracy.

Despite this schism both wings of the Radical Party shared substantially the same economic policy—a policy that had not changed significantly from the period of Conservative dominance. Yrigoyen and Alvear, like the Conservatives, wanted to maintain Argentina's traditional
role as massive agrarian exporter. They too opposed protected industrialization (except in special cases) and favored retaining close economic
ties with the British. Although (as we shall see) Alvear did come out
rather timidly for land reform, which was more than either Yrigoyen or
the Conservatives had done, none of these political groups had any consistent policy of agrarian development. Closely tied to the powerful cattlemen and the landed elite, the Conservatives, as well as both wings of
the Radicals, continued to give agriculture less priority than cattle raising
even though agriculture earned far more foreign exchange than the products of the cattle industry.

Argentina's laissez-faire economic policy continued until the onset of
the Depression. Although with one major exception (the nationalization
of the oil industry) Yrigoyen's economic policies were orthodox, the
elites never had liked or respected the Radical leader. They particularly
resented his populism and his free-spending patronage, and they feared
that Yrigoyen might unleash social forces that he would be unable to control. When the venerable leader returned to the presidency in 1928 at the
age of seventy-six, the opposition found more to criticize, for the president was becoming senile, and corruption as well as inefficiency stamped
his government. The economic crisis of 1930 greatly weakened the Yrigoyen government and made it an easy target for a military coup, which
toppled it in September 1930.

Canada

The contours of Canadian history differed substantially from the Argentine pattern during the 1880–1930 period.[19] We have already discussed
some of the roots of these differences, including the constitutional provisions over the public lands in the two countries. We have seen that the
pampas region was already organized into provincial governments when

Argentina became a nation in 1862 while most of the Canadian prairies held only territorial status as late as 1905. Another important difference in the history of the two countries was that Canada entered the First World War, sent half a million troops to Europe, and industrialized greatly as a result of wartime munitions contracts, while Argentina stayed neutral and suffered economically because of Yrigoyen's determination to remain at peace. Certainly one of the main differences between the political-economic histories of the two countries has already been mentioned: the fact that the Canadian political elite had a national development plan and Argentina did not. Unlike the unplanned growth of the pampas, the National Policy made the central government an active participant in the agrarian development of the prairie region.

The National Policy was an enormous project for a country with limited finances and a small population (Canada had only 3,500,000 people in 1867). Nonetheless Macdonald was determined to shape Canada's destiny by means of protectionism, a transcontinental railway, and prairie settlement—the three components of the National Policy. Although the prime minister did not formally announce the National Policy until 1878, he actively implemented parts of it during the Conservative Party's initial term in office (1867–73). First and foremost, Canada acquired the prairies in 1869 and made them into the Northwest Territories. Then, in 1872, the Macdonald government moved to implement its settlement policy by passing the Homestead Act, which provided free land grants to aspiring settlers in the Territories. That same year, Parliament created a special national police force, the Royal Northwest Mounted Police, to keep order in the vast prairie region.

The construction of an all-Canadian transcontinental railway was vital to the National Policy, and Macdonald during his first term in office sought to award a contract and to begin building the new Canadian Pacific Railway. The transcontinental was now politically feasible—and indeed politically essential—for British Columbia had entered the Confederation in 1871 on Macdonald's promise that the railway would be completed within ten years. But the corruption and scandals that surrounded the Macdonald government's initial railway contract became so blatant that the Conservative government fell in 1873, and the Liberals came to power for the next five years.

Tied to a laissez-faire economic ideology that distrusted Macdonald's activist policies, the Liberal government of Alexander Mackenzie did little to promote the Canadian Pacific. The Liberals were unpopular, the public forgot or forgave the colorful Macdonald's transgressions, and the Conservatives returned to power in 1878 on the platform of the National Policy, which Macdonald now systematically set out for the first time.

The Conservatives' long period of political dominance, which lasted from 1878 to 1896, began with Macdonald moving quickly to implement the tariff and railway policies that were central to the National Policy. Legislation of 1879 raised tariffs to protectionist levels and, as we shall see, stimulated the early emergence of key Canadian industries. In 1880 Macdonald's government signed a new Canadian Pacific contract with a group of Montreal investors. It committed the government to grant the new company a subsidy of 25,000,000 dollars, as well as 25,000,000 acres of prairie land.* Before the transcontinental was completed in 1885, the Macdonald government would have to loan the company millions more to save it from bankruptcy. Thus the federal government was intimately involved with financing the company that became the most powerful economic institution in the prairies, and one hated by generations of prairie farmers.

Although the tariff and railway policies were in operation by 1885, the National Policy was still incomplete, because few settlers moved to the prairies—despite the liberal 1872 Homestead Act. The slow settlement of the prairies resulted partly from the 1870's depression and partly from the attraction of other, less remote frontier regions, but it also reflected the strong political discontent that pervaded the prairies in the early and mid-1880's. Particularly volatile were the 34,000 Indians and 15,000 métis who inhabited the Northwest Territories. The extermination of the buffalo left the Indians near starvation, and Macdonald's blundering Indian policies, which violated important provisions of the treaties made when the Indians had gone on reservations, added fuel to their discontent. The Saskatchewan métis, who were predominantly French-speaking, were also bitter critics of the government, which had done nothing to resolve their land claims and had repeatedly ignored their petitions for redress. In 1885 this resentment boiled over in the short but violent Northwest Rebellion, which featured a Cree Indian uprising and a métis revolt under the French-speaking leader Louis Riel. At the same time that the Roca government was putting down the last Indian resistance in the Argentine pampas, the Macdonald government rushed thousands of troops west, quickly crushed the Northwest Rebellion, and effectively ended all Indian and métis resistance to the National Policy.

Macdonald refused to pardon Riel, who died on the scaffold in 1885. This decision produced a major political crisis, for Quebec public opinion sympathized with the plight of the francophone métis, and numerous prominent French-Canadian leaders pleaded for amnesty for Riel. Macdonald's refusal to grant it gave the opposition Liberal party a major issue in Quebec and eventually cost Macdonald's Conservatives heavily in Canada's second-largest province. Macdonald grimly held on to power

*All dollar amounts in the text are in Canadian dollars unless otherwise noted.

until he died in office in 1891, but the revitalized Liberal Party, now led by Wilfrid Laurier of Quebec, made rapid gains and rose to power in the 1896 elections. A fifteen-year period of Liberal rule had begun.

Although the Liberals long had criticized the statist aspects of the National Policy, once in power, Laurier's party did not attempt to reverse Macdonald's economic policies. The agrarian population—particularly in the prairies—strongly opposed the protective tariff, but the Laurier government found that the tariff was a critical revenue source, and that the industrialists had influence in the Liberal Party. Under these circumstances, Laurier made only a few politically strategic reductions in the tariff. High tariffs continued to protect most Canadian industries under the Liberals, just as they had under the Conservatives.

The Liberal government did move to limit the enormous economic power the Canadian Pacific Railway held in the West by following two new railway policies. First, the Laurier cabinet encouraged private firms to build new rail systems across the prairies to compete with the Canadian Pacific, and heavily subsidized two new transcontinentals. Second, in a move aimed primarily at the Canadian Pacific but ultimately extended to all prairie railroads, the Laurier government sharply cut freight rates on grain between prairie points and the port facilities on the Great Lakes. The Canadian Pacific received a land grant in return for agreeing to this "Crow's Nest Rate," but the Laurier government obtained an important concession for prairie farmers.

Laurier's prairie development policy placed its strongest emphasis on the one aspect of the National Policy that Macdonald had not lived to see fulfilled—the massive settlement of the prairies. The Liberal tenure in power from 1896 to 1911 coincided with a buoyant world economy, with generally rising wheat prices, and with the virtual closing of the agrarian frontier in the United States. (See Table 2.1 for an overview of late-nineteenth-century Canadian growth.) These circumstances strongly favored Canada's bid to attract settlers. As Chapter 5 will discuss, the Laurier government organized a major immigrant recruitment program that transformed the prairie region within fifteen years. Like Argentina, Canada actively recruited immigrants, but the Laurier immigration policy went far beyond Argentina's, which simply opened the door to virtually all comers and did little for them once they arrived. The Canadian policy, based on Macdonald's Homestead Act, was to root the newcomers permanently on the soil and to tie them permanently to their new homeland by giving them land grants. By 1905 the prairies had grown so fast that the government finally created the two new provinces of Alberta and Saskatchewan, although neither these provinces nor Manitoba would receive jurisdiction over their public lands until 1930.

Prime Minister Laurier presided over a prairie development boom, but despite this economic prosperity, severe political tensions divided Canadians along numerous cleavages. For one thing, Laurier's labor policies, which not only did not support unions, but even saw the Mounted Police used as strikebreakers, earned the prime minister little favor with the working class. Although labor radicalism in the form of anarchism was not as widespread in Canada as in Argentina, labor policy was becoming a major Canadian political issue. For another thing, French and English-speaking Canadians divided bitterly over Canada's participation with Britain in the Boer War. But, at least for the prairies, the tariff remained the single-most-divisive issue in Canadian politics.

The National Policy had given direction and purpose to Canadian economic policy, but it also divided the nation politically along regional lines. Argentina had also suffered division over the tariff issue during the federalist-centralist conflicts of the early nineteenth century, but by the end of the nineteenth century, protectionism was no longer an important Argentine issue because most of the population lived in the pampas region, and these people, whether farmers, cattlemen, urban consumers, or union members, were all opposed to a protective tariff. There was still support for the tariff in some interior Argentine provinces, but the government had largely defused the issue by enacting selective tariff increases for sugar and wine, the two largest interior industries. But in Canada the prairie region hated the tariff with a vengeance. The failure of Laurier's Liberals to achieve significant tariff reform led to Western political disillusionment and a widespread belief that Eastern economic interests controlled both parties. Talk of a Western third party and even of secession began. To try to save his political position in the prairies, Laurier grasped at the hope of a reciprocity treaty with the United States, but the Tories defeated Laurier's Liberals in a 1911 election fought largely over the reciprocity issue as well as over Canada's relationship to Britain's military buildup.

Laurier had been unable to deliver to the prairies on tariff policy but, as noted above, he had made important railway rate concessions to the region. In a similar fashion the new Conservative administration of Robert Borden (prime minister, 1911–20) held to the main lines of the National Policy, but at the same time backed grain-trade reform legislation that materially improved the position of Western farmers. Although some Westerners would argue that the Borden government—and the Laurier government as well—acted to aid the prairies only when the reform would also benefit Eastern capitalists, it remains true that the federal government between 1896 and 1920 made a number of economic policy reforms that improved the profitability of Western farming. Although they only acted after intense prairie political pressure, both Laurier and Borden did promote prairie agriculture in various important ways, primarily through transportation and marketing reforms. These reforms were some compensation for the government's failure to reform the tariff decisively.

The First World War temporarily dampened the fires of prairie regionalism. High wartime wheat prices and preferential treatment by the British Food Ministry brought a wave of prosperity to the Canadian West, which, as we shall see, gave Borden strong support by 1917. But the labor and ethnic issues that had plagued the Laurier government continued during the war. Borden's generally antilabor stance (particularly with regard to the more radical Western unions) and his draft policy, which deeply angered Quebec, weakened the Conservative Party's support. After the war prairie farmers also turned against the high-tariff Borden government, and the Western third-party movement that had begun before the war now reappeared.

Exhausted by the war, Borden retired in 1920 and turned the party leadership over to an arch-protectionist from Ontario, Arthur Meighen, who called an election in 1921. The Liberal Party, which had been devastated by the draft issue during the war, now regrouped under its new leader (Laurier had died in 1919), William Lyon Mackenzie King. This shrewd and able politician would dominate Canadian politics until 1930 and indeed for most of the time until he retired in 1948. Later chapters will examine King's political and economic policies toward the prairie West, but we can introduce him here by emphasizing that he had to deal with a full-scale prairie political rebellion during the 1920's. A new party, the Progressive Party, which was based primarily on prairie farmers but also enjoyed Western labor support, held the balance of power in the federal House of Commons between 1921 and 1926. To maintain this Progressive support behind his minority government and also to rebuild the Liberal Party in the prairies, King made several tariff, marketing, and transport policy reforms that benefited agriculture. As we shall see, King used a fortuitous political crisis in 1926 to regain his majority and to end his dependence on the prairie Progressives, but he continued to court the West and to rebuild his party, particularly in Saskatchewan between 1926 and 1930. The Depression, however, caught King unprepared, and his Liberals lost the 1930 election to the Conservatives under the ultra-protectionist R. B. Bennett.

Two Political Traditions: Analysis and Conclusions

Like Borden and Laurier, King acted to moderate the National Policy's negative effects on the prairie farming economy. Nonetheless, in 1930 the original economic policy structure that Macdonald had erected over 50 years before survived basically intact. The National Policy still pointed the Canadian economy along an East-West transcontinental and transatlantic axis rather than in the direction of integration with the United States. The central feature of the National Policy—the protective tariff—was still firmly in place in 1930.

Continuity rather than change also marked the history of economic policy in Argentina, Canada's great agrarian competitor at the opposite end of the hemisphere. As in Canada, where both major parties operated within the framework of the National Policy, in Argentina both the Conservatives and the Radicals, the nation's two largest parties, supported the free-trade economic order. Although the emergence of the Radicals signaled the rise of mass urban politics in Argentina, Argentine urbanization before 1930 was based not on industry, but on international trade, and consequently urban political support for a protectionist economic policy was hardly any stronger in 1929 than it had been over the previous half-century.

The different role of the state in the two economies had major consequences for the development of wheat agriculture in the two nations. In Canada the National Policy from the start aimed to make the prairies a wheat export region. Because both the Conservatives and the Liberals realized that Western prosperity was vital to the entire nation, both parties aimed to make prairie wheat a "super-staple" that would be second to none in the world market. This policy required Canadian governments from Macdonald to King to make reforms and concessions that would improve the farmers' financial returns as well as the technical base of the wheat economy. Wheat also became a great Argentine export—in fact, the leading export—but no government between 1880 and 1930 enacted any significant policy reform or concession to aid wheat agriculture. This was because the cattlemen who composed the political elite continued to regard agriculture's proper role in the Argentine economy as second to cattle raising. The fact that wheat production grew rapidly during the 1880–1930 period only served to confirm to the landed elite that their agricultural policy (or rather their lack of a policy) was correct. They fell into the dangerous assumption that the fertile soil of the pampas was enough to guarantee Argentina's major position in the world wheat market indefinitely.

Grain exports earned most of Argentina's foreign exchange and thus effectively supported the entire national economy, but the landed elite remained remarkably indifferent to agrarian reform and development. It is doubtful, however, that either the Conservative or the Radical governments could have maintained this indifference if the masses of Argentine immigrants had become politically assimilated. Here is one of the principal differences between Argentine and Canadian politics, and one that is of central importance to the political economy of Argentina and Canadian wheat. Although both Argentina and Canada were huge immigrant countries, in the Canadian prairies most of the foreigners became citizens, and in the Argentine pampas most of them did not. Immigration led to the rise of a mass democratic electorate in the prairies that added to the region's political strength and forced Ottawa to grant more concessions to the prairies than either the Liberals or the Conservatives would have done otherwise. This did not take place in Argentina, where the agrarian population remained disenfranchised and without much political voice. We will conclude this chapter by examining the immigrant naturalization question in the two countries more closely.

The contrast in the naturalization rates of the two countries was marked. In Argentina the 1914 census reported that 33,219 immigrants were naturalized. This was 1.4 percent of the total foreign-born population. In Canada the 1911 census reported a very different result. A total of 353,583 foreigners, or 46.98 percent of the foreign-born population, had become citizens by that date.* What is even more significant is that the naturalization rate in the three prairie provinces was higher than the national average. In 1911 55.3 percent of the prairies' foreign-born population had become citizens. But in Argentina most of the naturalized foreigners lived in the cities, not in the countryside. This sharp contrast in the political assimilation of the Argentine and Canadian immigrant population in the prewar period did not change substantially over the next few decades.[20]

The Canadian government strongly encouraged—indeed in many cases it virtually required—the foreign-born to become citizens. Anyone who applied for a grant of Homestead Act land in the West—and as we shall see, many thousands of immigrants homesteaded—had to become a citizen before he could receive title to the land. This requirement doubtless caused thousands of foreigners to become citizens as soon as possible. Canada's high rate of naturalization was closely linked to its liberal land policy. Foreigners wanted homestead land, and to get it, they were willing to take the big step of naturalization.[21]

The high naturalization rate also reflected the fact that one of the two major parties—the Liberals—attempted to build a political machine on the basis of the immigrant vote. The Liberal Party, in power from 1896 to 1911, promoted massive immigration into the prairies. In the prairie provinces local Liberal leaders provided help and advice to the immigrants and, before 1914, smoothed the path to quick naturalization. These Liberal leaders long delivered the Continental European immigrant vote by the use of patronage, by supporting the ethnic press, and

---

*Technically, naturalized immigrants became not Canadian citizens but British subjects. Immigrants from Britain were not counted among the "foreign-born." They were automatically citizens of Canada.

by backing legislation to permit public schools to provide at least some instruction in foreign languages. Liquor also helped. As one Danish immigrant recalled of the 1921 Alberta elections, "A vote could be bought for a few drinks," and Liberal politicians gave out whiskey "by the quart."[22]

Although the government tightened things up in 1914, acquiring Canadian citizenship was a straightforward and relatively simple process. Prior to 1914 the residence requirement was three years. Local-level federal judges administered the paperwork and delivered the citizenship oath. New legislation passed by the Conservative Borden government in 1914 was designed to end the sometimes flagrant abuse of the naturalization laws by prairie Liberals. This legislation raised the residence requirement to five years and required local judges to submit each application to the secretary of state in Ottawa, who would make the final decision.[23] This change was designed to prevent judges from abusing the citizenship procedure for political reasons, but it did not mean that Canada opposed naturalization of its foreign-born population.

The situation was very different in Argentina, where the foreigners found few inducements and many obstacles to naturalization. For one thing, Argentina had no homestead act or other large-scale land distribution plan that would root immigrants to the nation by granting them ownership of the soil. As we shall see in Chapter 4, the Argentine landed elites owned most of the good land, with the result that immigrants became tenants on short-term leases. By its very definition a transitory status, the tenancy system encouraged many immigrants to come to Argentina for only a few years, just long enough to make money to take back to Europe. Argentina acquired a sort of colonial status for many of its immigrants. It was a place to go to earn money but not necessarily to settle in. Even when an immigrant decided to remain permanently, retaining his European citizenship was to his advantage. Citizens of foreign countries not only enjoyed all the protection of the Argentine civil code, which applied equally to foreigners and nationals; they also could call on the diplomatic protection of the mother country, usually anxious to maintain its ties with the immigrants.[24]

Argentine immigrants who did want to become citizens faced a slow and complex bureaucratic procedure. Although the residence requirement was only two years, the aspiring citizen had to undergo a thorough police investigation, which tended to disqualify any immigrant active in a labor union. If he received police approval, the applicant had to spend days or even weeks in crowded reception rooms awaiting an interview with one of the federal judges, the only government officials authorized to grant citizenship papers. The judges, whose number did not keep pace with population growth, claimed that they were overworked and could allot only three or four hours a week to naturalization matters. Huge piles of paperwork accompanied each application, so delaying the procedure that officials in Buenos Aires processed only 30 or 40 cases each week. This procedure was particularly difficult for foreigners living in rural areas. They found it difficult, if not impossible, to make the frequently long and expensive journey to the cities where the federal judges resided. What a contrast with Canada, where judges visited rural districts to handle citizenship cases![25]

The two major Argentine political parties—the Conservative PAN and the Radicals—did not want to change this situation. Both feared the political consequences of mass immigrant naturalization, and neither attempted to build large-scale immigrant-based political machines. If Argentina's millions of immigrants held the vote, so the argument ran, the newcomers might be able to seize power from native-born Argentines. Political upheaval and possibly even social revolution would result. On several occasions, Radical and Conservative congressmen proposed tougher naturalization rules, but Congress did not act on these proposals because the impediments to naturalization contained in the application procedure were sufficient to keep the flow of new citizens to a trickle.[26] Two small parties, the urban-based Socialists and the Liga del Sur (later the Progressive Democrats), cultivated the immigrant vote and urged more liberal naturalization laws, but the Radicals and the Conservatives ignored such proposals.[27]

This survey of immigrant naturalization procedures in Argentina and Canada has shown the sharp contrast that existed between the two countries on this question of fundamental political importance. As far as immigrant citizenship was concerned, Argentine politics were exclusionary while Canadian politics were inclusionary. That is, Argentina's political elites, who feared that immigrants might upset the established order, aimed to make politics off limits to the foreign-born. The Canadian government did not harbor such fears, at least not while the Liberal Party was in office from 1896 to 1911. Secure in the knowledge that the European immigrant vote could be manipulated, few Canadian political leaders opposed the principle of mass naturalization.

Argentina's exclusionary policy and Canada's inclusionary policy had major consequences for the political economy of wheat in the two countries. Between the 1890's and the 1920's Canadian politicians faced a growing prairie political rebellion against Eastern political domination. This rebellion, which would reject both of the traditional parties as mouthpieces of Eastern business, was at least partly based on the immigrant vote, and particularly on the British, who automatically were citizens, and on the Americans, who quickly became citizens. This agrarian revolt was able to force successive governments in Ottawa to make major concessions to prairie agriculture that significantly strengthened the competitive position of Canadian wheat growers. In other words, Canada's tradition of inclusionary politics made agrarian reform possible and helped keep Canada in the forefront of wheat-exporting nations.

But Argentina's exclusionary political tradition prevented the rural political mobilization that was a necessary precondition of agrarian reform. While it may be true that many immigrants never wished to become Argentine citizens, it is also true that the government discouraged naturalization. In any case, most immigrant farmers did not become citizens and, unlike their counterparts in Canada, could not form political movements to advance their interests. Because no mass agrarian political base existed in the pampas (outside of southern Santa Fe province and some parts of Entre Ríos), agrarian reform in Argentina could come from only two directions. One would be the rent strike, when tenants would refuse to pay rents until the landlords and government granted reforms. As we shall see, this tactic usually failed. The other potential avenue of agrarian reform was from above—from the enlightened self-interest of the propertied classes who composed the elites of both the Radical and the Conservative Party. Unfortunately for the future of Argentine agriculture, these classes were not very enlightened. They took very little interest in agrarian reform or development. Bereft of a farsighted elite, exclusionary politics in Argentina ended in agricultural stagnation. The inclusionary democratic system practiced in Canada, with all its imperfections, led to a far more modern and dynamic agrarian sector than in Argentina.

3

◄•►

The Economic Structure of
Argentina and Canada, 1900-1930

Argentina and Canada emerged as organized states during the 1860's, and as we have seen, both countries devoted the rest of the nineteenth century to developing their vast grasslands. By the turn of the century, the agrarian economy was booming in both the prairies and the pampas. Wheat cultivation and production were growing phenomenally in both regions, and as a result Argentina and Canada had moved into the forefront of wheat-exporting nations by 1910. This chapter surveys and compares the Argentine and Canadian economies during the era of the great wheat boom from 1900 to 1930. My purpose here is to complete the background overview that Chapter 2 began by providing the reader with the economic context needed to analyze the Argentine and Canadian wheat policy.

Economic Parallels

At the end of the First World War, Argentina and Canada were both countries of vast area and small population. Canada was much larger—it covered about 9,800,000 square kilometers compared with Argentina's 2,600,000—but the habitable area of both countries was approximately equivalent. Their populations were almost exactly the same. In fact, the 1921 Canadian census reported a total of 8,750,000, against an Argentine estimate for that year of 8,900,000. Per capita incomes in both were relatively high for the period. In 1929 Argentina's Gross Domestic Product per capita was $540, which was more than twice as high as the figure for Chile, the next-highest Latin American country, and was five times as high as Brazil's.* Argentina ranked about equal to the Netherlands and well ahead of several European countries, including Italy and Austria. Canada's per capita income of $1,030 was less than that of the United

*These statistics are in U.S. dollars measured in 1955 prices. They provide only a rough approximation of comparative living standards because price levels differed greatly among countries.

TABLE 3.1

Five Leading Exports of Argentina and Canada, 1910–1929

(percent of total export value)

Export product	1910–14	1915–19	1920–24	1925–29
Argentina				
Wheat and wheat				
flour	20.5%	20.9%	27.3%	24.2%
Maize	16.8	8.4	13.4	18.5
Meat	15.3	22.9	15.8	15.3
Linseed oil	9.5	6.9	12.4	12.2
Wool	11.7	12.8	7.8	8.2
Percent of all				
exports	73.8%	71.9%	76.7%	78.4%
Canada				
Wheat and wheat				
flour	27.7%	24.9%	29.2%	33.0%
Newsprint	—	—	6.7	9.2
Lumber	10.7	4.1	6.2	3.7
Munitions	—	17.4	—	—
Meat	—	5.5	4.0	—
Fish	5.2	2.7	3.1	2.4
Silver	5.5	—	—	—
Gold	2.6	—	—	—
Wood pulp	—	—	—	3.7
Percent of all				
exports	51.7%	54.6%	49.2%	52.0%

SOURCES: Woltman, p. 8; Canada [3], vols. for 1914–30.

States but about the same as the United Kingdom's. Argentina and Canada had not only large but rapidly growing incomes: between 1899 and 1929 Argentina's GDP per capita rose 54 percent, and Canada's 62 percent.[1]

The wealth of both countries came primarily from their leading role as exporters of raw materials. Argentina and Canada concentrated on the production and export of a few basic staples—wheat, newsprint, lumber, woodpulp, and fish in the case of Canada, and maize, flaxseed, wool, beef, and wheat in the case of Argentina (see Table 3.1). Foreign trade consequently loomed large in Argentine and Canadian economic life. Statistics for 1910, which show that Argentina's export trade amounted to $58 per capita and Canada's to $40 (compared with $22 for the United States), underscore the critical impact that the production and export of staple products made on the economies of the two countries. (See Table 3.2, which presents per capita export data for 1900–1929.)

This foreign trade boom rested solidly on wheat, which by 1910 was the leading export of both Argentina and Canada. After the prairies and pampas came under cultivation in the 1880's, wheat production and ex-

TABLE 3.2
Total and Per Capita Exports of Argentina, Canada,
and the United States, 1900–1929
(current U.S. dollars)

Year	Population	Merchandise exports	Exports per capita
Canada			
1900	5,301,000	$168,972,000	$31.87
1910	6,988,000	279,247,000	39.96
1920	8,556,000	1,132,143,000	132.32
1929	10,029,000	1,138,034,000	113.47
Argentina			
1900	4,607,000	$159,381,000	$34.59
1910	6,586,000	384,536,000	58.39
1920	8,861,000	934,252,000	105.43
1929	11,592,000	907,113,000	78.25
United States			
1900	76,094,000	$1,623,000,000	$21.32
1910	92,407,000	1,995,000,000	21.59
1920	106,461,000	8,481,000,000	79.66
1929	121,467,000	5,347,000,000	43.91

SOURCES: *Canada.* Dollar exchange rate, Urquhart & Buckley, series G627; all other data, F. H. Leacy, series G381 and A1. *Argentina.* Peso exchange rate, Díaz Alejandro, *Essays,* p. 484; population, Argentina [4], p. 159; Díaz Alejandro, *Essays,* p. 421; exports, Tornquist & Cía., pp. 133–34, Díaz Alejandro, *Essays,* pp. 474–76, 484. *United States.* U.S. Dept. of Commerce, Census Bureau, 1: 8, 2: 864.

ports grew at spectacular rates, with the result that Argentina and Canada became significant wheat exporters soon after the turn of the century (see Table 3.3). But before the First World War, neither country exported as much wheat as Russia, which was then the world's leader. The 1917 Russian Revolution dramatically changed the world wheat trade, however, for Russia virtually ceased wheat exports. This development left the world market primarily in the hands of the "Big Four" countries—Canada, Argentina, the United States, and Australia—which until the Second World War produced the vast bulk of the world's exportable wheat. As Table 3.4 shows, the Big Four supplied between 86 percent and 94 percent of the world's wheat exports between 1922 and 1930.[2]

The wheat trade gave the Argentine and Canadian economies numerous similarities. Wheat agriculture in both countries absolutely relied on the export market. Table 3.5 demonstrates this export dependence. Between 1909 and 1932 Canada exported from 57 percent to 74 percent of its total annual wheat production, and Argentina from 46 percent to 79 percent. This reliance on foreign markets was also true of Australian wheat, and in this respect these three sparsely populated countries were analytically distinct from the United States, whose huge domestic market absorbed most of its wheat production.[3] In fact the United States ex-

TABLE 3.3
Argentine and Canadian Wheat Production and Exports, Five-Year Averages, 1884-1932
(millions of bushels)

Period	Production Argentina	Production Canada	Exports Argentina	Exports Canada
1884–1889	19.9	38.3	4.1	4.2
1889–1894	47.3	41.0	28.0	7.4
1894–1899	59.6	45.4	29.4	13.7
1899–1904	93.3	74.0	54.9	25.7
1904–1909	158.1	101.2	108.9	43.7
1909–1914	147.1	167.5	84.7	95.4
1914–1919	167.4	248.1	77.7	158.7
1919–1922	188.0	252.5	149.4	148.2
1922–1927	211.2	389.7	135.9	286.8
1927–1932	249.2	416.6	163.3	277.8

SOURCES: De Hevesy, pp. 344, 394–95, except for the data on Canadian exports in the 1884–1909 period, which are drawn from Canada [6], p. 24. De Hevesy relied on earlier estimates for this period.

TABLE 3.4
"Big Four" Wheat Exports as a Percentage of the World Wheat Trade, 1909–1930

Year[a]	Argentina	Canada	Australia	United States	Combined total
1909–14	12.3%	13.9%	8.0%	16.0% (2)	50.2%
1922–23	19.4	39.0 (1)	7.0	28.2 (2)	93.6
1923–24	20.7 (2)	41.2 (1)	10.3	15.7	88.0
1924–25	16.1	24.8 (2)	15.9	33.1 (1)	89.9
1925–26	13.8	46.0 (1)	11.1	15.3 (2)	86.2
1926–27	17.0	34.3 (1)	12.1	23.7 (2)	87.0
1927–28	21.8	39.6 (1)	8.7	22.7 (2)	92.8
1928–29	23.8 (2)	42.4 (1)	11.6	16.4	94.1
1929–30	24.4 (2)	31.0 (1)	10.1	23.2	88.7

SOURCE: De Hevesy, Appendix 10. No data are given for 1914–22.

NOTE: Figures in parentheses show rank in the world market; Russia was the largest exporter in 1909–14. Rows may not add up to the four-country totals because of rounding.

[a]All rows except the first are for August 1–July 31.

TABLE 3.5
Wheat and Wheat Flour Exports as a Percentage of Total Wheat Production, Five-Year Averages, Argentina and Canada, 1904–1932

Period	Argentina	Canada	Period	Argentina	Canada
1904–9	68.8%	23.6%	1919–22	79.5%	58.7%
1909–14	57.6	57.0	1922–27	64.3	73.6
1914–19	46.4	64.0	1927–32	65.5	66.4

SOURCE: De Hevesy, pp. 344, 394–95.

ported only 23 percent of its production in 1922–27, 16 percent in 1927–32, and a mere 0.3 percent in 1932–37.[4] The overseas wheat trade thus assumed far greater political and economic importance in Argentina and Canada than in the United States.

Foreign immigration was vital for the development of Argentine and Canadian wheat agriculture and indeed for the economic development of the two nations. Between 1901 and 1930 net immigration (the excess of arrivals over departures) reached a total of 1,240,000 in Canada and 2,710,000 in Argentina. Chapter 5 will analyze the agrarian impact of immigration closely, but here we may point out that while Argentine immigrants came primarily from Southern Europe, Canadian immigrants tended to originate in Great Britain, the United States, and Eastern Europe. But whatever their origins, the immigrants provided the bulk of the labor force in the prairies and the pampas; the native-born became a distinct minority among the farmers of both regions.

Although Argentina and Canada imported their agricultural labor from different areas, British investments and the British market tied both wheat export economies closely to the United Kingdom. Although not part of the "formal empire" like Canada, Argentina was very much a part of what H. S. Ferns has called "Britain's informal empire." Numerous observers have emphasized the particularly strong economic ties between Britain and Argentina, which E. J. Hobsbawm called a "British informal colony" and an "honorary dominion." Likewise, many Argentines realized that the relationship was close, perhaps uncomfortably so; the noted essayist Raúl Scalabrini Ortiz argued in 1940 that "our government has been more submissive to the will of the British than the Parliament of India." Sir Malcolm Robertson, British ambassador to Argentina in 1930, pointed out that Argentina, although "not in the Empire, should be regarded as of it" because of the "signal importance" of the "gigantic" British investments there.[5]

British investments were indeed "gigantic" in both Argentina and Canada. As Ferns points out, Argentina, like the United States, Canada, and Australia, was one of the "significant frontiers of British business enterprise during the century before World War I."[6] For generations British investors had poured in the capital to finance the railways, communications systems, and utilities networks that composed the infrastructure of these two rapidly growing export economies. Table 3.6 shows that by 1913 foreign money amounting to over three billion current U.S. dollars was invested in both Argentina and Canada, and that British capital accounted for nearly 60 percent of this investment in Argentina and 76 percent in Canada. Referring again to the table, we can see that foreign investment had grown by the mid-1920's, but that U.S. investment had dislodged Britain as the largest source in Canada by 1925. At that point,

TABLE 3.6
Foreign Investment in Argentina, 1913 and 1927, and Canada, 1913 and 1925
(millions of current U.S. dollars)

Category	Total foreign investment	U.K. investment		U.S. investment	
		Amount	Percent of total	Amount	Percent of total
Argentina, 1913	$3,146	$1,866	59.3%	$39	1.2%
Canada, 1913	3,700	2,800	75.7	800	21.6
Argentina, 1927	3,485	2,009	57.6	489	14.0
Canada, 1925	5,700	2,300	40.4	3,200	56.1

SOURCES: Phelps, 'p. 108; Aitken, pp. 28–54 passim.
NOTE: Argentine figures, given in gold pesos in Phelps, have been converted at the rate of 1 peso = U.S. $0.968.

British investment still reigned supreme (although not unchallenged) in Argentina.

These were enormous amounts of money for that period, but it is important to note certain qualitative differences in the pattern of foreign investment in the two countries. In Canada British investors had long been content to assume a conservative role—ordinarily they did not actually establish branches of British companies, nor did they incorporate the Canadian businesses in which they invested as British firms administered from London. British investment in Canada was overwhelmingly—about 90 percent—indirect rather than direct. In other words, British investors purchased securities of Canadian companies that Canadian management operated from head offices in Montreal or Toronto. But British investment in Argentina was primarily direct—in that country, the major British businesses were incorporated in London and administered by British boards of directors. Britons apparently believed that entrepreneurial and management skills were less developed in Argentina than in Canada, where the operation of British-financed companies could be confidently left to Canadian talent. They may also have had less faith in the vagaries of Argentine politics than in the apparently stable political order of the northern Dominion. In any case, British capital was much more visible in Argentina than in Canada—and thus more open to nationalistic attack. In contrast, U.S. investment was overwhelmingly direct in Canada as well as in Argentina. In Canada U.S. investment concentrated on the resource-extractive sectors (especially mining), but in addition U.S. investors acquired or built numerous manufacturing factories, which became branch plants of U.S. parent firms. In Argentina U.S. investment also entered the manufacturing sector as well as public utilities and banking.[7]

TABLE 3.7
World Net Imports of Wheat and Wheat Flour,
Five-Year Averages, 1909–1937
(millions of bushels)

Period	United Kingdom[a]	Continental Europe	Other countries	Total
1909–14	217.7	326.7	98.8	643.2
1922–27	224.4	373.1	142.7	740.2
1927–32	236.3	373.9	170.5	780.7
1932–37	225.8	170.8	147.0	543.6

SOURCE: Canada [6], p. 121.
[a]The Irish Free State is included in the data for 1922 and after.

Britain's impact on the Argentine and Canadian wheat economies extended into another critical area—the United Kingdom was their largest single market. Although by the 1920's both Argentina and Canada imported more goods from the United States than they did from Britain, only in exceptional periods did the United States purchase significant amounts of their foodstuffs. The British market was central to the Argentine and Canadian wheat economies because the United Kingdom imported more wheat by far than any other country during the first 40 years of the twentieth century.* In fact, as Table 3.7 shows, Britain imported 30–40 percent of the world's total wheat exports between 1909 and 1937. For good reason Liverpool had long been known as "the central market of the world" in the wheat trade.[8]

The United Kingdom's predominant position in the world wheat market reflected the British government's momentous decision of 1846 to repeal the Corn Laws, embrace a free-trade policy, and end nearly all tariff protection of domestic agriculture. When ocean transport rates declined rapidly after 1870, the new grasslands countries found that they could profitably export wheat to Britain and Europe. In response, British wheat production, most of which could not compete with imports, fell by two-thirds between 1873 and 1896, and by 1913 Britain depended on imports for 80 percent of its wheat consumption.[9]

The importance of the British market to the Argentine and Canadian wheat export trade can be appreciated when one considers that in every year between 1921 and 1939, Britain purchased from 19.2 percent to 67.6 percent of Canada's total wheat exports, and that in nine of those years, over 40 percent of Canadian wheat went to Britain. Argentina relied less

*Later in this chapter we will examine Britain's key market position for Argentine and Canadian meat exports. Britain purchased most of its maize in Argentina; its oats imports came from both countries.

on the British wheat market—Brazil and, in some years, Italy and Belgium were also major customers—but in most years the United Kingdom remained the largest importer of Argentine wheat.[10]

Economic Contrasts

Sharp contrasts also existed between the Argentine and Canadian economies, and these differences affected the divergent agrarian policies and international trade policies that the two countries followed in the interwar years. For one thing, the agricultural production and exports of the pampas were more diversified than those of the prairies. Wheat was by far the most important crop of Alberta, Saskatchewan, and Manitoba; in many prairie districts it was the only cash crop. "No region of the country," writes one Western historian, "has been as influenced by a single industry as the prairies have by grain."[11] In this respect the prairies were a monocultural export economy, whose specialization of wheat, in the judgment of one distinguished work on Canadian economic history, was "extreme." In fact, wheat was the source of from 47 percent to 72 percent of the total cash income of prairie farmers between 1926 and 1939.[12] Although by the eve of the Second World War farmers were beginning to produce poultry and dairy products for local markets, diversified agriculture in the prairies was still in its infancy.[13]

The cattle industry played an important role in the Canadian prairie economy, but cattle raising was a minor business there compared with its position in the pampas economy. By the twentieth century prairie ranching had declined from the powerful position it had once held. The Macdonald government, which argued that southeastern Alberta was too dry for agricultural settlement, encouraged the cattle industry to develop there by granting ranchers long-term (21 years) leases of prairie land. Cattle production grew rapidly, and Calgary emerged as a regional metropolis primarily because of the cattle trade. In most years between 1881 and 1900, Canada's cattle exports actually exceeded its wheat exports.[14]

As with wheat, Great Britain was the principal market for Canada's cattle. Isolated from the sea, the Alberta cattle industry could not develop chilled beef exports to compete with Argentina's chilled trade, but it did export live cattle to England. This trade, however, long suffered from British restrictions designed to protect domestic ranchers. Beginning in 1892, all Canadian cattle exported to the United Kingdom had to be slaughtered at the port of entry.* This restriction prevented fattening Canadian steers in England and thus kept steer weights low.[15]

---

*Although the United Kingdom justified the 1892 restrictions on sanitary grounds, in fact the British were aiming to eliminate imports of U.S. cattle, which came to Britain via Canada. See Simon M. Evans, "Canadian Beef," p. 756.

Prairie cattlemen were an important power bloc in the Conservative Party, but the ranchers' political position began to suffer after the Liberals took power in 1896 and particularly after 1905, when government policy openly shifted in favor of agriculture. Ottawa withdrew much of the cattle lease land and opened it to settlement. Nature dealt the prairie cattle industry another harsh blow. The winter of 1906–7 was so severe that, as Wallace Stegner has pointed out, it "changed the way of life of the region." The seemingly endless blizzards that year killed many thousands of animals and drove numerous ranchers out of business.[16]

The prairie cattle business picked up after 1911, when the Tories returned to power and modified the lease policy, but real prosperity came a few years later, after the United States passed the Underwood Tariff Act (1913) removing duties on Canadian cattle imports. Although Canadian exports to Britain ceased during the First World War for lack of shipping, the prairie cattle trade boomed because of the new U.S. market. This trade, however, proved short-lived; the Fordney-McCumber Tariff Act (1921) restored heavy duties on imported steers, a move that within a year cut Canada's cattle shipments southward by over 90 percent from the 1920 level. Meanwhile, the British removed their 1892 restrictions on imported steers, a move that averted total disaster for the prairie cattlemen. But the new British market did not compensate for the loss of the U.S. market, and as a result Canada's cattle exports earned less in 1924 than they had 20 years earlier. Canada's cattlemen had fought hard to save their industry, but they lacked influence in the Liberal Party, which was in power for most of the 1900–1930 period, they lacked access to their natural market in the United States, and they were in no position to match Argentina's massive meat exports to the United Kingdom.[17]

On the pampas cattle raising was the business of Argentina's most powerful economic and political group, the elite class of large landowners. The land and climate of the humid pampas, particularly in Buenos Aires province, made the region especially well suited for breeding and fattening beef cattle. Alfalfa grew well on the pampas and proved to be an excellent feed crop. Once technical problems regarding refrigerated ships and meatpacking plants were solved at the turn of the century, Argentina's beef exports began a period of "spectacular growth."[18] When the wheat boom began shortly after, the cattle industry, unlike Canada's, did not cede its place. Instead, wheat production concentrated on the outer fringes of the cattle belt, or else landowners—who had held secure title to the Buenos Aires humid pampas since the mid-nineteenth century—grew alternate crops of wheat and alfalfa. By the late 1920's, as Table 3.8 shows, about 15,000,000 head of beef cattle were grazed in Buenos Aires and Entre Ríos provinces, along with a similar number of sheep and about a million swine. A further indicator of the prominence of the

TABLE 3.8
Livestock Population of the Prairies and the Pampas, 1927

Stock	Prairies	Pampas[a]
Cattle	2,448,000[b]	14,850,050
Sheep	815,000	15,963,150
Swine	1,745,000	1,081,798

SOURCES: Murchie, pp. 28–29; Sociedad Rural Argentina, pp. 231–32.

NOTE: In this and subsequent tables, unless otherwise noted "the prairies" means the provinces of Manitoba, Alberta, and Saskatchewan (or their earlier equivalents) and "the pampas" means the provinces of Buenos Aires, Santa Fe, Córdoba, and Entre Ríos and La Pampa Territory.

[a]Data for Buenos Aires and Entre Ríos alone, the only provinces that carried out regular stock censuses.

[b]Excludes dairy cattle.

cattle industry in the pampas is that in 1930 about 30,000,000 hectares were devoted to grazing (along with another 14,000,000 planted in alfalfa and other feed crops)—or well over half of the 60,000,000 hectares in use for all agricultural and cattle pursuits on the pampas. (Only 2,700,000 hectares were devoted to grazing in Alberta and Saskatchewan that year.)[19]

The cattle industry, then, was central to Argentine economic life, and it was the showpiece of the republic's proud and aristocratic landowning class. Argentina led the world in meat exports—particularly chilled beef—and about 40 percent of total world meat exports came from the pampas. As in so many aspects of Argentine economic life, the role of Great Britain, which was the world's largest meat importer, was preponderant in the cattle trade. In 1920 83 percent of Argentina's 721,000 tons of exported chilled and frozen beef went to the United Kingdom. During the 1920's, other countries, especially Germany and Belgium, increased their purchases, but in 1927 Britain still purchased 68 percent of Argentina's 939,000 tons of meat exports, and it remained the predominant purchaser through the 1930's.[20] One other potential large market, the United States, was closed to Argentine meat in 1926, when the Coolidge government, alleging the danger of foot-and-mouth disease, prohibited imports of Argentine chilled and frozen meat.* Despite the enormous amount of land devoted to cattle raising the strong influence that cattlemen exercised in forming Argentine economic policy, in the mid- to late 1920's meat exports were in only third place in Argentina's export mix

---

*This prohibition was a serious blow to Argentine-U.S. relations. Argentina argued that the restrictions were unfair and politically motivated; to substantiate its case, it pointed to Britain's decision to continue Argentine meat imports under stricter inspection and certification procedures. See Dana R. Sweet, "History of United States–Argentine Commercial Relations," p. 108.

(see Table 3.1). But hides and other cattle by-products were also significant exports.

The Argentine pampas, unlike the Canadian prairies, also produced and exported huge volumes of maize and linseed oil. The benign climate and excellent soils of the pampas gave the Argentine a great natural advantage over most countries in these products. In fact, Argentina's exports of maize were never less than 63 percent of total world exports during the 1924–39 period, and its share of linseed exports was never less than 75 percent.[21] In the Argentine export mix, maize ranked second to wheat in 1925–29, and linseed held fourth place (see Table 3.1).

There was thus a great difference between the monocultural wheat production of the prairies and the rich diversity of cattle and agricultural production of the pampas. Indeed, in 1939 the Canadian economist W. A. Mackintosh was quoted as estimating that, in terms of population, wheat was almost three times as important in Canada as it was in the exporting countries of Argentina and Australia.[22] True, wheat was more important to Canada than to Argentina, but as Table 3.1 has shown, Argentina relied much more heavily on agricultural exports than Canada did. Indeed, all five of Argentina's leading exports—which together accounted for from 72 percent to 78 percent of total Argentine exports between 1910 and 1929—were the products of farms and ranches. But in Canada, wheat was the only agricultural export among the five leading commodities (except for meat in 1915–24). The Canadian export trade was far more diversified than Argentina's; Canadian exports included the products of mines, forests, and factories as well as farm products. Yet another contrast between the Argentine and Canadian export trade (also apparent in Table 3.1) was that the five leading products accounted for a much higher proportion of total exports in Argentina than in Canada. The Canadian export trade, in other words, was more diversified both by sector and by range of products than Argentina's. Indeed, by 1925–29 automobiles were Canada's sixth-largest export at a time when Argentina still imported all its motor vehicles.

These distinct differences in export composition reflect the fact that industrial development had not accompanied the agricultural cornucopia in Argentina. In this respect, there was another clear distinction between the Argentine and Canadian economies. As Table 3.9 shows, although Argentina had many more industrial firms than Canada in 1937 and almost as many industrial workers, the value of output from Canadian industries was over three times as great as in Argentina. The qualitative difference between the patterns of industrialization in the two countries becomes apparent in Table 3.10, where we see that Canadian heavy industry (metals, vehicles, and chemicals) accounted for a much larger share of total

TABLE 3.9

Industrial Production and Employment in Argentina and Canada, 1937

Category	Argentina	Canada
Number of firms	49,300	24,800
Number of workers	642,000	660,000
Value of output in current U.S. dollars	$530,000,000	$1,850,000,000

SOURCE: Dorfman, pp. 380–81.

TABLE 3.10

Principal Industries of Argentina and Canada, 1937

(value of output in millions of current U.S. dollars)

Sector	Argentina		Canada	
	Value	Percent of total	Value	Percent of total
Food products	$115	21.7%	$300	16.2%
Textiles	65	12.3	180	9.7
Metal products	35	6.6	290	16.0
Machinery and vehicles	45	8.5	175	9.5
Chemical products	15	2.8	85	4.6
Total output	$530	—	$1,850	—

SOURCE: Dorfman, p. 381.

production (by value) than in Argentina, where the food and beverage industries clearly predominated. Many of these Argentine "industries" in fact were nothing more than small artisan shops employing a handful of workers.

Argentine industry concentrated primarily on the processing and refining of the country's bountiful agricultural and cattle production—and the products of much of this industry, meat packing for example, were destined for export. Considerable import-substitution industrialization took place in Argentina, particularly after 1914, with the result that the republic did produce a wide variety of nondurable consumer goods. Nonetheless, imports still supplied the bulk of manufactured products. One notable example was cotton textiles, which according to the economic historian Adolfo Dorfman, were "still in a pronounced stage of infancy" as late as 1913. This basic industry did not grow much in the postwar years: in 1925–29 imports still supplied 92 percent of Argentine cotton textile consumption. (In contrast, imports supplied only 30.5 percent of Canadian cotton textile consumption in the same period.)[23] Argentine textile production grew greatly in the 1930's, but as H. S. Ferns points out, "What distinguished Argentine industry . . . from the industry of a country of recent settlement like Canada was the absence of a capital goods industry and a motor industry." Dorfman made the same point in 1942, when he emphasized "the importance of the metallurgical sector" of Canadian industry in comparison with Argentina's industrial output. The first Argentine steel mill did not go into production until 1961; Canada, in contrast, had become a large-scale steel producer well before the First World War. The level of technology in Argentine industry likewise was relatively low; in the 1930's Argentine industry consumed only a quarter of the amount of electricity per worker consumed by industry in Canada.[24]

By that decade, in fact, Canada was becoming a major industrial country. The motor vehicles industry was firmly established, as were the steel industry and the manufacturing of farm machinery, heavy railroad equipment, and other capital goods. Canadian industry, centered in Montreal as well as in Hamilton and Toronto, had been making rapid strides since the government enacted protectionist policies in the 1870's. In Toronto industrial employment doubled during the 1880's (to 26,000), and by the First World War Canada manufactured the bulk of the goods it consumed. Although British capital played a major role in financing railway construction in both countries, as early as the 1896–1900 period, Canada imported only £427,000 of British railway equipment, compared with £2,773,000 for Argentina. This major difference stems directly from the fact that Canada already had a large and advanced industrial structure capable of producing steel rail, passenger and freight cars, and some very impressive locomotives. Argentina produced railway equipment, but not nearly to the same extent.[25]

These contrasting rates of industrialization reflected a number of factors. Certainly one of them was Canada's relatively richer endowment in easily accessible industrial raw materials, including iron ore, lead, zinc, and lumber. The Canadian shield (an enormous and rugged rocky plateau that separates southern Ontario from the prairies) long had appeared to be nothing but a useless barrier, but by 1900 it proved to be a treasure house of industrial raw materials. Moreover, Ontario and Quebec enjoyed access to abundant supplies of hydroelectric power, a source of energy that Argentina was not able to tap on a major scale until the 1960's.[26] The resource differential was not all in Canada's favor, however. By the 1920's Argentina was already becoming a major petroleum producer while Canada still produced very little oil. (By the 1960's both countries had become essentially petroleum self-sufficient.)

Another reason why Canada industrialized earlier and faster than Argentina was its close economic relationship with the United States. U.S. industrial investors, as we have seen, poured vast amounts of capital into Canada. Much of this investment went into branch plants that would give the parent U.S. manufacturing companies access to the Canadian market. By 1926 foreign investors, primarily from the United States, controlled 35 percent of Canadian manufacturing capital, a total that rose to 38 percent in 1939.[27] Foreign investors controlled an even larger segment of Argentine industry—Dorfman estimates that 50 percent of industrial capital was foreign in 1935, but as we have seen, Argentina's industrial sector was much smaller than Canada's.[28]

The Canadian resource base and the U.S. investment in Canadian manufacturing both help explain Canada's lead over Argentina in industrialization, but even together they cannot fully account for that lead unless they are analyzed in the context of tariff policy. Argentina, after all, could produce cotton, and it produced very fine wool, but the Argentine textile industry remained stunted prior to the Depression. Yet Canada, which produced no cotton, had a major textile industry. The fundamental reason for this anomaly is that Canada had a long and consistent policy of industrial protectionism and Argentina did not. Indeed, one may argue that the main reason why so much U.S. industrial capital entered Canada was that the high protective tariff left U.S. industrialists who wished to sell there no other alternative than to establish branch plants.

In Argentina, as we have seen, the landed elite dominated politics for most of the period spanning independence from Spain in 1816 to the rise of Juan Perón in 1943. This elite consistently supported free trade and opposed protective tariffs except for such special cases as the sugar and wine industries. Although the Argentine government did tax most imported goods, this tariff was designed primarily to produce revenue for the state rather than to protect industry.

But in Canada, in sharp contrast, political power was centralized in the hands of financiers and industrialists from Ontario and Quebec by the 1860's. In fact, Quebec and Ontario already had begun to industrialize when Confederation took place in 1867; the Canadian colonial parliament had passed a protective tariff as early as 1859.[29] Unlike Argentina, Canada had no dominant class of large and elite landowners; most Canadian farmers gained title to their land through homesteading. The Ontario and Quebec businessmen devised a policy to support protective tariffs, and the Conservative Party under Macdonald put this protectionist policy into effect. Macdonald's support for the protective tariff would have made Argentine industrialists envious. During his 1878 election campaign he reportedly advised Canadian manufacturers: "I can not tell what protection you require. But let each manufacturer tell us what he wants, and we will try to give him what he needs."[30]

As we have seen, Macdonald's National Policy became an enduring feature of Canada's political economy. The Conservative Party legislated high tariffs, built the Canadian Pacific Railway across the continent against formidable odds, and promoted the settlement of the prairies to provide a new staple export and a market for the emerging industries of Ontario and Quebec. The opposition Liberal Party, whose social base was among the small farmers of Quebec, Ontario, and the West, opposed industrial protectionism, but when the Liberals held power between 1896 and 1911, they found that the government needed the revenue the tariff provided, and that Canadian industry had become too powerful to defeat politically.

The "full dinner pail" argument that the Canadian Conservatives used to promote working-class political support for tariff protection was not heard in Argentina. Indeed, as Ferns has put it, "In few rapidly developing countries did the state make so few and such feeble attempts at industrialization." Outside of industries processing rural products for export, no large industrial working class existed in Argentina prior to the mid-1930's; the Argentine working class primarily was occupied in transporting and processing the rich produce of the pampas for export. Indeed, Argentine labor leaders as late as 1929 were firmly opposed to protection; they advocated free trade in the name of higher standards of living for the workers.[31] Deprived of political support from the landed elite or from the working class, Argentine industrialists remained a weak and isolated group. They found some support among nationalist intellectuals and military officers, and from certain leaders of interior provinces excluded from the boom taking place on the pampas. But this disparate coalition seldom could exercise real political pressure on government industrial policy.[32] Even when the Depression of the 1930's brought a de facto policy of industrial protection to Argentina, higher import costs were not a major issue among Argentine farmers. Their imported farm machinery entered at a low rate of duty (10 percent after 1931), and they bought "little or nothing else except bare requirements in clothing." Therefore, as a perceptive Canadian observer concluded, "these tenants . . . know little about and interest themselves less in the tariff."[33]

Industrial development certainly gave Canada a more diversified economic structure than Argentina had, but industrialization was not necessarily a blessing for the development of prairie agriculture. Prairie farmers believed that Canadian industrialists were diabolical exploiters who used the protective tariff to raise the cost of farm production—most of which had to be sold on the highly competitive world market—or to reduce net farm incomes below what they would have been had the tariff been set lower. Canadian farmers would fight a long and ultimately unsuccessful battle against the industrial interests of the Quebec-Ontario heartland that determined government economic policy. The prairies were a region with distinctly different economic interests from the rest of Canada; in this sense, it was a colonial economic region subordinate to central Canada. In contrast, Argentine farmers were closely allied with the dominant landed elite and urban consumers over the issue of industrial policy, and consequently, with few exceptions before the rise of Perón, tariffs remained at moderate levels. Argentine farmers, in this respect, were more thoroughly integrated into the international economic system than their competitors in Canada; the absence of Argentine industry was one of the reasons why Argentine cereal agriculture remained cost competitive throughout the interwar period.

Part II

Foundations for Agricultural Development

4

Land Tenure and Settlement Patterns

One of the fundamental differences between Argentina and Canada is the role that the state played in agricultural development. As we have seen, the Canadian federal government acted early and decisively to open and settle the prairies, while the Argentine government relied on a laissez-faire policy to develop the pampas. The land-distribution policies of the two countries demonstrate this contrasting role of the state clearly. Canada at an early date adopted a homestead policy that made free prairie land widely available. As a result, small, individually owned farms characterized prairie land tenure. In Argentina, the federal government never devised this kind of massive land-distribution program. Indeed, no national land policy existed in Argentina. The result was that in the pampas, land remained concentrated primarily in the hands of a small group of wealthy and powerful owners. Most pampa farmers were renters.

Two very distinct kinds of rural societies emerged as a result of these contrasting land-tenure policies. In Canada the homestead system fostered the growth of well-organized rural communities, strong cooperative institutions, and powerful agrarian political movements. In Argentina the rental system did none of this. The tenants who composed most of the pampas' agrarian population moved about frequently, and as a result, rural communal and political associations were weaker than in Canada.

This chapter examines and compares the land policies that Argentina and Canada formulated during the half century that preceded the Great Depression. We will focus on relationships between the land-tenure system and agrarian development in each society. And we will analyze the advantages as well as the disadvantages that characterized Argentine tenancy on the one hand and Canadian ownership on the other. Because land tenure is so basic to agrarian development, this chapter will introduce numerous issues that later chapters will examine in greater detail.

The Prairies and the Pampas—Two Natural Treasures

When they became modern states in the 1860's, Argentina and Canada very soon had to confront land-policy decisions that would fundamentally affect the course of their agrarian development. The policy choice was so important because the stakes were so high. Argentina and Canada contained grasslands that were natural treasures equaled in only a few other favored regions of the planet. In a hungry world here were two new regions—the prairies and the pampas—that were exceptionally well suited to massive grain cultivation. We will begin our study of Argentine and Canadian land policy with a brief survey of the two regions that became the heartlands of wheat agriculture in Argentina and Canada.

In the prairies the best wheat territory lies within a huge belt that extends between the Red River in Manitoba to the east and the Rocky Mountain foothills to the west (see Map 1). The northern limit of this belt stretches roughly diagonally for about 700 miles from Winnipeg through Prince Albert, Saskatchewan, to Edmonton; north of this line the short growing season, wetter climate, and consequently poorer-quality product rule out concentration on wheat (although one important wheat zone, the Peace River country of northwestern Alberta, is an exception). Nor are the southernmost regions of the prairie, near the U.S.

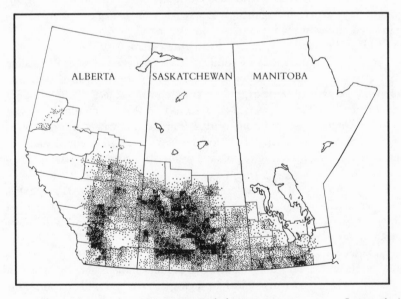

Map 1. Wheat acreage in the prairies, 1926. Each dot represents 3,000 acres. Source: *Agriculture, Climate and Population of the Prairie Provinces of Canada* (Ottawa: Dominion Bureau of Statistics, 1932).

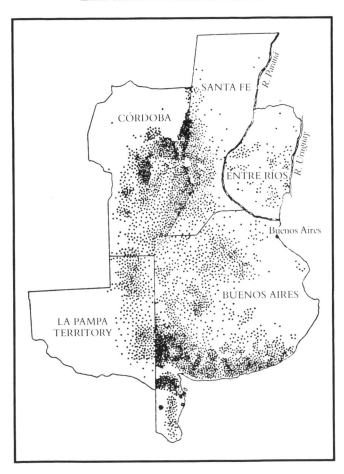

Map 2. Wheat acreage in the pampas, 1933–1934. Each dot represents 5.54 acres. Source: J. A. Strong, "Grain Farming in Argentina," *Commercial Intelligence Journal* (Ottawa), May 16, 1936, p. 899.

border (which is at 49° north latitude), particularly suited for wheat. Especially in southern Alberta and southwestern Saskatchewan, the extreme variations in rainfall have discouraged agriculture in favor of ranching (although irrigation today enables some exceptions). But in the wheat belt the soil is deep and rich, the rainfall is generally adequate, and the sunshine is abundant. Newcomers to Canada who were familiar with agriculture in the Old World were particularly impressed with the fertility of the prairies. One Scottish visitor, for example, noted that "the soil is very rich, and stands more cropping, without manure, than any other land in the world."[1]

This fertility, however, could not entirely compensate for a growing season that was much shorter than in Argentina or in most of the United States. The threat of frost, which can strike in early September or even in August, "hangs as a sword over the prairie," as one British settler put it.[2] As we shall see, the Canadian government's agricultural research stations eventually developed quick-ripening wheat varieties that matured in the short prairie growing seasons. But even when frost did not strike crops, the prairie's cold climate presented serious obstacles to wheat agriculture. The long winters required farmers to invest substantial sums of money in housing, clothing, and fuel.

None of this was a problem in the pampas, which lie much closer to the equator; here frosts and snowfall are rare, except in the far southern regions of Buenos Aires province. The fertile sections of the Argentine great plains extend roughly in a huge arc 300 to 400 miles north, west, and south of Buenos Aires, which lies at 35° south latitude. As Map 2 shows, wheat agriculture is concentrated most heavily on the outer fringes of this arc—in the western part of Santa Fe province, the southern and central regions of neighboring Córdoba, the eastern part of La Pampa territory and the southwestern sections of Buenos Aires province. This concentration in part reflects environmental limitations—land to the west of this arc is too dry for wheat while that part of Entre Ríos province lying north of the arc lacks optimal soils. The lush central regions of the pampas traditionally have been devoted to the cattle industry, as well as to maize and flaxseed.

Although there are numerous regional variations, the pampas possess superb soil and climatic conditions for cereal agriculture. Rainfall is generally adequate and well distributed. The soil, composed of finely compacted dust deposited by water and wind over the ages, is remarkably deep and free from stones. The geographer Mark Jefferson notes that "in the cliffs forming the walls of gullies," this loess could often be seen "50 or 60 feet deep without a single pebble." This combination of soil and climate led W. J. Jackman, one of the first Canadian agricultural experts to visit Argentina, to conclude that the pampas were "an ideal farming region."[3]

Land Policy in the Prairies

The powerful role the Canadian federal government would play in settling the prairies was clear soon after Canada acquired the region in 1869. Three years later, Sir John A. Macdonald, Canada's first prime minister, formulated the homestead policy that was to be the cornerstone of prairie settlement, and that established the small individually owned family farm as the predominant form of land tenure on the prairies (see Table 4.1). This free-land policy was a major event in prairie development, but as we have seen, it was only part of a much more ambitious development strat-

TABLE 4.1
Land Tenure in the Prairies and the Pampas, 1911–1933

Category	Total cultivators	Owners		Renters		Other	
		Number	Percent of total	Number	Percent of total	Number	Percent of total
Prairies							
1911	201,178	180,018	89.5%	10,493[a]	5.2%	10,667[b]	5.3%
1921	255,657	202,947	79.4	27,067[a]	10.6	25,643[b]	10.0
1931	287,125	198,770	69.2	42,709[a]	14.9	45,646[b]	15.9
Pampas							
1912–13	84,076	27,428	32.6%	46,325	55.1%	10,323[c]	12.3%
1923–24	117,766	42,096	35.7	65,183	55.3	10,487[c]	8.9
1932–33	132,479	47,847	36.1	80,809	61.0	4,174[c]	3.2

SOURCES: Murchie, p. 102; Argentina [12], p. 487.
NOTE: Percentages may not total 100 because of rounding.
[a]Both cash renters and sharecroppers.
[b]Part-owners, part-renters.
[c]Sharecroppers.

egy, the famous National Policy, that Macdonald and the dominant Conservative Party put into effect in the 1880's. To achieve its aim of settling the prairies with a massive farming population, the federal government administered the region until Alberta and Saskatchewan became provinces in 1905. (Manitoba had become a province in 1870). And until 1930 the federal government controlled the public lands in all three prairie provinces.

Federal control of the police and law enforcement in the prairie West was an integral part of Ottawa's settlement policy. Macdonald was determined that the rapid settlement of the prairies would take place in an orderly fashion, and that law should prevail in the West. The prime minister was convinced that U.S. precedents proved that leaving law enforcement in the hands of local officials was "an invitation to disaster," and would lead to another "wild west" in Canada. Accordingly, he placed the Northwest Territories under the jurisdiction of the Royal Northwest Mounted Police, which he established in 1873. Directly responsible to the government in Ottawa, the "Mounties" became a highly professional and successful law enforcement unit. Charged with creating a new and better version of Eastern Canadian society, they strove to avoid conflict between immigrants and native peoples and to integrate the immigrant masses into Canadian ideas of order and justice. To ease the settlement process, the Mounted Police advised newcomers about desirable homestead locations and gave advice on crops and weather. Most important, Mountie patrols tried to visit every settler at regular intervals. For these reasons, as R. C. Macleod points out, "their popularity among the citizenry of Western Canada never faded." As we shall see, the order that the

Mounties brought to the prairies contrasted sharply with the arbitrary and ineffective system of law enforcement on the pampas.[4]

When the settlement process began, Ottawa already had given away— or was in the process of arranging to give away—huge blocks of prairie land. Under the terms of the transfer of Rupert's Land to Canada, the Hudson's Bay Company received one-twentieth of the prairies as part compensation. The government of Macdonald granted 25,000,000 acres to the Canadian Pacific Railway, and later administrations made grants to other railroad companies. Other lands were set aside for the métis (who soon lost most of them to speculators), for Indian reserves, and for the support of public education. But the federal government still controlled the vast bulk of prairie land, and it acted vigorously to promote settlement. After the United States passed its Homestead Act in 1862, Macdonald had realized that no less generous a policy would attract population to the Canadian prairies.[5]

The Canadian homestead legislation stipulated that on payment of a ten dollar filing fee, any person who was sole head of a family or any male who was twenty-one years old could apply for a quarter-section (160 acres) of land. At the end of three years he could obtain title if he had prepared 35 acres for cultivation, had erected certain buildings, and had lived there for at least six months a year.* Later amendments (in force during the years 1879–89 and after 1904) enabled farmers to "pre-empt" and then purchase an adjoining quarter-section for a dollar an acre.[6]

Despite this offer of land, before the mid-1890's the prairie population grew slowly and agriculture made little headway (see Table 4.2). Political unrest, which culminated in the 1885 Northwest Rebellion, along with a series of poor crop years, the attraction of neighboring U.S. states for immigrants, high transport costs, and low wheat prices, all combined to discourage settlement in the Canadian prairies. But after 1896 wheat prices rose and railway freight rates fell. Moreover, Dominion agricultural experiment stations developed and publicized new dry-farming techniques as well as new quick-maturing strains of wheat, an important factor given the sometimes early prairie frosts. Once the economic and technical conditions for profitable wheat production were present, the "greatest rush for farm lands in the world's history" took place as immigrants flocked into the prairies. By 1911 the population was 1,300,000 and 9,200,000 hectares were under cultivation.† By 1928, a total of 56,400,000 acres (22,830,000 hectares) in the prairie provinces was being homesteaded.[7]

---

*Women were not eligible to apply for homestead land unless they were the sole head of a family with dependent children under the age of eighteen. See Georgina Binnie-Clark, *Wheat and Women*, p. xxi.

†One hectare equals 2.471 acres.

TABLE 4.2
*Growth of Population and Area Under Cultivation
in the Prairies and the Pampas, 1890–1931*

		Area under cultivation (000 hectares)	
Category	Population	Cereals and flaxseed[a]	Alfalfa
1890's			
Prairies	251,473 (1891)[b]	564 (1890)	—
Pampas	1,987,512 (1895)	1,738 (1890)	390 (1890)
1910's			
Prairies	1,328,121 (1911)	9,168 (1915–16)	—
Pampas	3,227,988 (1914)	13,020 (1915–16)	6,670 (1915–16)
1920's–30's			
Prairies	2,353,529 (1931)	13,954 (1926–27)	—
Pampas	5,761,586 (1931)[c]	15,969 (1926–27)	4,820 (1926–27)

SOURCES: *Population.* Argentina [4], p. 151; Argentina [6], 2:109; Kubat & Thornton, pp. 12–16. *Area.* Tornquist & Cía., p. 22; Sociedad Rural Argentina, pp. 119–24; Canada [1], 4: 408; Canada [2], 8: 540, 592, 664.
[a]For a breakdown of the area cultivated in wheat, see Table A.1.
[b]Includes Yukon and the Northwest Territories.
[c]Estimate.

As the settlers rushed in, so did the land speculators. The 1872 Homestead Act had prepared the ground for the wave of speculation that soon swept the prairies, for it had provided for the purchase of land at a dollar an acre up to 640 acres. As one scholar put it, these favorable terms were "an open invitation to land speculators."[8] Thousands of homesteaders had no intention of becoming farmers. Instead they "proved up," gained title, and then sold out. They were able to do this because, as a German traveler observed in 1909, "in the Canadian West more or less everybody speculates in land." Prices shot up rapidly. Rural land prices rose on the average 123 percent in Manitoba, 185 percent in Saskatchewan, and 201 percent in Alberta between 1900 and 1910.[9] Land speculation became even more rampant between 1910 and 1913, and not only farm lands were affected, but also properties in urban centers like Saskatoon, whose backers were proclaiming it "The fastest growing city in the world."[10]

Georgina Binnie-Clark, a doughty British farmer in Saskatchewan, described this runaway inflation:

The adjoining section of land, consisting of the orthodox six hundred and forty acres, was at that time the property of the Canadian Pacific Railway Company. I had always hoped it would become my own. In 1906 one could have bought it at five dollars an acre, in 1907 it was offered to me at seven dollars, in 1908 the price had risen to ten dollars, and at last in 1910, when I wanted the south-west quarter for a special purpose, the entire section had been purchased by its present owner at about eleven dollars an acre. It . . . owed this increase solely to the rising tide of Canadian land values, and without a solitary effort in the matter of contribution from the toil of man.[11]

Such rapid rises in prices were serious obstacles to established farmers who wished to buy more land and newcomers in districts where homesteads were no longer available. But as serious as the problem of prairie land inflation was, land prices were considerably lower there than they were in the Argentine pampas.

Inflation led to debt. By the early 1920's it was apparent that the original 160- or 320-acre homestead plots were too small for profitable grain farming, especially where environmental conditions were not optimal. Farmers who wished to expand had to resort to the public land market, which generally meant taking out a mortgage. This was particularly true of the people entering farming for the first time in the 1920's, when the best lands were already allocated. As a result, mortgages and other long-term debts became a major burden throughout the prairie provinces.[12] As early as 1915 a Scottish traveler wrote of the prairies that "the whole country is shadowed by debt." When Depression and drought hit in the 1930's, mortgage debt as a percentage of the value of farm property was 49 percent in Manitoba and 38 percent in Saskatchewan and Alberta. Saskatchewan, wrote C. E. Britnell, was essentially a "debtor community," an extremely perilous situation when wheat prices fell.[13]

The high price of land and frequent foreclosures stimulated a shift from ownership to renting among prairie farmers in the 1920's (see Table 4.1). Sharecrop tenancy on contracts of five years or less was the most common system used in Western Canada.[14] Many observers of the Canadian agricultural scene viewed the rise of tenancy with deep misgiving. One scholar quoted John Stuart Mill's contention that "if a man securely possesses a bleak rock he will turn it into a garden, but if he leases a garden he will turn it into a desert." It was generally agreed that tenancy, especially on short-term leases, resulted in soil destruction.[15]

The Prairie Cooperative Tradition

A vigorous and dynamic group spirit and cooperative ethic emerged among the homesteaders. British, Scandinavian, and German immigrants, as well as migrants from Ontario, brought with them a heritage of experience in the consumer cooperative movement, and the prairie land-tenure system of family-owned farms enabled the co-op tradition to become firmly rooted. Farmers were loyal to the cooperatives out of economic necessity, for they could reduce costs by pooling their efforts in joint concerns. This was an important consideration for prairie farmers, who were chronically short of capital and who were faced with a large initial investment when they started farming. By the 1920's the cooperative movement expanded to include a giant farmers' wheat-marketing association, the Canadian wheat pool.[16] Prairie settlers formed not only economic co-ops, but also social, recreational, and mutual-aid associations of all kinds. The result was, as one rural sociologist pointed out, that Saskatchewan "may in fact have the highest saturation of rural organizations of any region in the world."[17]

The abundant local histories of the West, along with surviving autobiographies, memoirs, and letters of Canadian farmers, provide eloquent testimony of the region's deeply imbedded cooperative ethic. One young Scottish farmer, John Stokes of Wood Bay, Manitoba, wrote his father in 1907: "We have got the telephone in the settlement now. Twelve of the people got together and formed a company, dug their own post holes, bought the poles and put them up, bought wire and strung it. . . . We had a great old time putting up the line." Four months later, he wrote that "we have formed a local society in Wood Bay this winter, . . . Wood Bay Mutual Improvement Society. It is divided into four branches: Christian Endeavor, Literary, Missionary, and Social."[18] Building a school was an especially important community activity; the school was not merely a place to educate, but a center of local life; schools were used for church services, lodge meetings, dances, political rallies, Wheat Pool meetings, and funerals. One settler's memoir relates that "when things settled down after Christmas, pa started talking about a school—there were four or five of us of school age. Pa called a meeting of the neighbors at our place. It was surprising the interest shown by all, even some of the bachelors." After the school came roads. Under a provision of Alberta law that divided the province into Local Improvement Districts, the inhabitants could elect a council to build roads. In many cases taxes paid by nonresident landowners financed the districts, while the settlers worked out their taxes by labor on the roads.[19] Throughout the prairies women participated actively and frequently as both leaders and members of cooperative groups.[20]

The cooperative movement provided much of the economic framework for the material and cultural standard of prairie farm life. The farmer's standard of living grew rapidly, and by the early twentieth century far exceeded that of their Argentine counterparts. Although there were many exceptions, especially among newcomers and settlers in fringe areas, housing was decent if plain. In stable, established farm communities the original sod or log house gave way to wood-frame structures.[21] One convenience that prairie farmers considered essential in this often isolated environment was the telephone. An astonishing number of farms—over 50 percent in Saskatchewan in 1931—had them. To facilitate the growth of telephone networks, all three provincial governments had assumed ownership of the telephone systems before the First World War. Another widespread convenience on prairie farms was the automobile—almost half the farmers in the Canadian West owned one in 1931.[22]

TABLE 4.3
Social Indicators for the Prairies and the Pampas, 1930's

Indicator	Prairies		Pampas	
	Date	Rate	Date	Rate
Illegitimacy (per 1,000 live births)	1936	36[a]	1937	203[b]
Illiteracy (percentage age 10 and over, Canada; age 14 and over, Argentina)	1931	4.0%	1943	11.8%–21.5%[c]
Infant mortality (deaths per 1,000 live births)	1934	55[a]	1931–1935	83.8; 111.7[d]

SOURCES: *Illegitimacy*. Britnell, p. 19; Taylor, p. 334. *Illiteracy*. Canada [2], 1: 1068; Taylor, p. 316. *Infant mortality*. Canada [3], *1936*, p. 176; Alejandro Bunge, *Una nueva Argentina*, p. 85; Argentina [4], p. 174.
[a]Saskatchewan only.
[b]Santa Fe province only.
[c]11.8%, Buenos Aires province; 14.9%, Santa Fe; 16.5%, La Pampa; 16.7%, Córdoba; 21.5%, Entre Ríos.
[d]Santa Fe (83.8) and Córdoba (111.7) only; average over the period.

Although most educators thought that the quality of prairie education left much to be desired, the literacy rate was over 95 percent in all three provinces by 1931 (see Table 4.3). Dropout rates were high, but the average number of years of school attendance in Saskatchewan rose from 4.96 in 1911 to 8.39 in 1931. Many farmers were voracious readers in the off-season; the Saskatchewan government had a traveling library service with over 100,000 volumes ("thank goodness," exclaimed one settler), and the *Grain Growers' Guide*, the premier farmers' newspaper and one of high quality, had a circulation of 120,000 by 1926.[23] Health care was also making notable strides. In Saskatchewan services improved rapidly after the introduction of "municipal doctors," supported by local taxes in 1916; in all three provinces the infant mortality rate fell steadily. Illegitimacy rates in the prairie provinces were lower than the national average (although very high among Ukrainian immigrants and Indians).[24]

By the late 1930's, then, prairie farm society had become securely established around a solid cooperative framework. Living standards were improving. But meanwhile mortgages and taxes were high, the Canadian tariff kept the cost of living higher than in the United States, and debts were rising.

Land Tenure in the Pampas

The land-tenure system of Argentina contrasted dramatically with that of the Canadian prairies. In the pampas over 60 percent of farmers were sharecrop or cash tenants (see Table 4.1), and tenancy was the central feature of Argentine farming. Unlike Canada, Argentina never had any national land policy that effectively made pampa land available to the public at nominal prices. There was, in other words, no Argentine equivalent of Canada's Homestead Law.

Unlike the Canadian government, the Argentine government lacked the power to develop a national land policy. Although the 1853 Argentine Constitution gave the provinces authority over the public lands, the central government had jurisdiction over land distribution only in the national territories. But with the exception of parts of La Pampa territory, the good farm lands of the pampas were all within Buenos Aires, Córdoba, Santa Fe, and Entre Ríos provinces. The result, as one Argentine scholar put it, was a "pulverization of authority." Despite the desire of nineteenth-century statesmen like Domingo F. Sarmiento and Nicolás Avellaneda to create a smallholder society, the Argentine federal government never was able to carry out a coordinated land settlement program.[25] The federal government did desire to promote agricultural production and exports, and to this end, it encouraged massive immigration and the construction of a large railway network. But the central government had to leave land policy to the provinces.

The pampa land-tenure system was not only a product of these political and constitutional circumstances; it also reflected the enormous power that the Argentine rural oligarchy held and used. Unlike the Canadian prairies, the pampa provinces had been controlled by a well-established landed elite since the late colonial period. During the early and mid-nineteenth century these landowners exploited little of their land intensively and instead made the rational economic decision to specialize in cattle and sheep raising. These estancieros (owners of large estates) became interested in promoting agricultural development and European immigration when world demand for grain increased late in the nineteenth century. Although interested in agriculture, the Argentine rural oligarchy viewed it as strictly secondary to cattle raising. From the viewpoint of the cattlemen, agricultural immigrants should remain tenants and not become owners, for the tenancy system gave landlords maximum flexibility to use their lands according to market circumstances.

Much new land came on the market in the 1860's and 1870's, when the provinces rolled back the Indian frontier and in the process acquired vast expanses of new public land. As we shall see, the provinces dedicated some of this land to settlement by immigrant smallholders. But the bulk of the newly acquired frontier ended up in the hands of the rural oligarchy. The provincial governments were in dire financial straits, partly because of the expense of the Indian wars. To raise funds, both Santa Fe and Córdoba adopted a policy of selling the public lands, often in large chunks of a square league or more. The cattlemen and entrepreneurs, many of them from the landed elite, who purchased this frontier land then often divided their purchases and rented plots to immigrant farmers.[26] Sometimes they formed so-called colonization companies. Some of these enterprises did facilitate purchase by farmers, but many "colonization companies" became only another type of latifundio (large estate) that rented land for cash or crop shares.[27]

Railway companies also acquired public lands, although to a much smaller extent than in Canada. The Central Argentine Railway received the only really large Argentine railway land grant in 1863. It was about 350,000 hectares (864,500 acres) in extent—in contrast to the 25,000,000 acres the Canadian Pacific later received. At first the railway's land company experimented with colonizing its properties, but by 1880 it merely sold or leased land without bothering to recruit settlers.[28]

Exceptions to this pattern of great estates and rented farms were found mainly in parts of Santa Fe, Córdoba, and Entre Ríos. As early as the 1850's the government of Santa Fe, motivated by a desire to populate the frontier as a buffer against Indians, began to provide public lands for immigrant colonization. These early colonization plans, which Córdoba and Entre Ríos also adopted, were successful. The availability of small plots of pampa land at reasonable prices and payment terms gave European immigrants opportunity and hope for the future. As one Italian colonist put it, "We worked hard in both Argentina and Italy, . . . but in Italy we never got any place while in Argentina we moved ahead." By the 1880's thousands of the early settlers in these immigrant colonies were becoming owners on the favorable terms that the colonization organizers had established to attract these people. This fortunate minority of the Argentine farming population eventually prospered and became a rural middle class. And it was in the well-established rural communities that grew up around these immigrant colonies that the Argentine rural cooperative movement emerged.[29]

But these provincial land-distribution plans were too limited in scope and duration to establish ownership as the dominant tenure pattern on the pampas. Faced with the continuing penury of their treasuries—a penury that made the sale of public land at high market prices a tempting policy—Santa Fe, Córdoba, and Entre Ríos virtually abandoned these land settlement programs by 1895.[30]

Buenos Aires, largest and richest of the pampa provinces, never did effectively promote small farm ownership. An 1887 law whose ostensible purpose was to promote the growth of a smallholder class did little more than aid speculators and become a public scandal.[31] After the turn of the century, land prices in Buenos Aires quadrupled—while cereal prices rose only 28 percent. Georges Clemenceau, who visited Argentina in 1910, noted that "the form of gambling which is special to Buenos Aires is unbridled speculation in land."[32]

By the 1920's pampa farms were much more expensive than equivalent size farms in the Canadian prairies. The *New York Journal of Commerce* noted in 1921 that "a small farm just two hours distant from Buenos Aires is just as expensive as a small farm in England, France, or Germany." Although some tenants eventually were able to buy, the average immigrant in Argentina found that landownership was a dream, an unattainable goal.[33]

The concentration of landownership in the pampas was extreme. As late as 1924 60 percent of the total area of the four provinces and one territory in the cereal zone was in the hands of 12,673 landowners who possessed 1,000 hectares or more; 33.4 percent was held in properties of 5,000 hectares or more. In the province of Buenos Aires, 14 families possessed 100,000 hectares or more (one family held 412,000 hectares).[34]

The wealth and power of this landed elite were legendary in Argentina. Even the richest Calgary cattleman could not have matched the lavish life-style that the Argentine landed elite led for generations. Prominent landowners not only had palatial homes on their estates but also maintained luxurious residences in Buenos Aires and sometimes in Europe. *The Review of the River Plate*'s obituary for Miguel Alfredo Martínez de Hoz, whose family owned 101,000 hectares in Buenos Aires province,* illustrates the rarified social atmosphere in which the landed elite moved:

We record with sincere regret the death [in 1935] of Don Miguel Alfredo Martínez de Hoz, the well-known "estanciero" sportsman, breeder of fine pedigree cattle and race horses and owner of the world-famous estancia "Chapadmalal." By the death of this distinguished Argentine gentleman England loses an old and loyal friend. Señor Martínez de Hoz had practically life-long vinculations with England, where he received part of his education, first at St. Peters School, Woburn Park, Weybridge, and at a later stage, we believe, at Christ Church, Oxford. For a number of years he leased a house in London and sent his sons to Eton. He was a member of the Newmarket Jockey Club, the Four-in-Hand Club at Ranelagh, the St. James' and the Wellington. Don Miguel Alfredo was for some time a familiar figure to the London public when driving his famous Four-in-hand coach along the Brighton Road. . . . The "Chapadmalal" estate was and is one of the "show" places of the country and has been visited by many eminent visitors from abroad, including H.R.H. the Prince of Wales. Señor Martínez de Hoz served for some years as a Director of the Banco de la Nación and also as a member of the Council of the Argentine Rural Society. He had also filled with distinction the Presidency of the Jockey Club.[35]

Despite this kind of concentration of landownership, Argentina offered economic opportunity, and immigrants flocked into the pampas by the 1880's. Among the Southern Europeans who composed most of Argentina's immigrants were numerous poor but ambitious individuals

---

*The Martínez de Hoz family was the fifteenth-largest landowner in the province of Buenos Aires in 1928 (Jacinto Oddone, *La burguesía terrateniente argentina*, p. 186).

who were willing to enter agriculture as sharecroppers or renters. These people were aware that the tenancy system—at least if good luck and high prices prevailed—offered farmers the opportunity to make money, which many of them intended to use to move to the city or back to Europe.[36] This abundant supply of immigrant labor enabled the pampas to grow spectacularly. Indeed, as Table 4.2 shows, both population and area under cultivation in the pampas were larger than in the Canadian prairies through the 1920's, although it should be noted that the urban population of the pampas was much larger than in the prairies. As we shall see, this extraordinary growth was due in part to the Argentine government's liberal immigration policy, which welcomed all able-bodied Europeans and which actively recruited Spaniards and Italians just at the time when Southern European emigration was reaching its height. In contrast, Canadian immigration policy was much more selective and divided Europeans into "preferred" and "non-preferred races." Spaniards and Italians fell in the latter category.

The pampas' relatively large population and cultivated area were also due to the earlier development of agriculture in the region than in the prairies, where as we have seen the settlement boom did not get underway until the mid-1890's. During the 1880's improved transportation, abundant labor, and the introduction of mechanization brought the onset of an agricultural export boom to the pampas. With few exceptions, wheat production and exports continued to expand during the 1890's, a trend that was due to the low price of Argentine wheat abroad. The relative cheapness of pampa wheat resulted in part from the depreciation of the peso during the 1890's.* But it was also due in part to Argentina's low cost of production. Indeed, the *Financial News* of London noted in 1894 that the cost of wheat production in Argentina was so low that it would "blanch the hair of American farmers."[37]

Throughout this era of rapid expansion in the pampas, the land-tenure system subordinated agriculture to the extremely profitable business of cattle raising. "No business in Argentina of the same importance has shown such good returns as cattle breeding, and these results have been chiefly brought about by the introduction of alfalfa," wrote the cattleman Campbell Ogilvie in 1910. Alfalfa, as he suggested, was a lucrative plant—steers fed on it could be sent to market sooner than those fed on native pampa grass. Indeed, the expansion of area under alfalfa from 400,000 hectares in 1890 to 6,700,000 by 1915 (Table 4.2) enabled the spectacular growth of Argentine cattle raising to occur. Cattlemen found that the land-tenure system was ideal, for it enabled them, with little cost or risk, to rent land to tenant farmers on condition that they sow alfalfa

*Argentina left the gold standard in 1885, and the peso depreciated rapidly through the early 1890's. The government returned to gold during the years 1899–1914, and again briefly between 1927 and 1929.

during the last year of their contract.[38] Alfalfa is a long-lasting crop—up to sixteen years in Argentina—but when it began to run out, estancieros would bring tenant farmers in again to sow grain crops and to prepare the land anew for alfalfa. As one Argentine writer pithily summarized this system, "At one moment it may be more convenient to have more 'cows than gringos' [a common Argentine term for Italians], and the next year to have more 'gringos than cows.' "[39] In this respect, the Argentine land-tenure system was a highly efficient mechanism that enabled the flexible use of the pampas without requiring fertilizers.[40]

Although many Argentine tenants had begun as *medianeros* (sharecroppers), by the end of the First World War, the vast majority had accumulated enough capital to become *arrendatarios* (renters). The *arrendatario* was a rural capitalist who often owned a variety of machinery and rented large expanses (500 hectares or more). One result of the extensive nature of pampa farming was the abuse of the soil. Rolf Sternberg, who farmed in Entre Ríos before moving to the United States to become a geography professor, noted that "tenant farmers viewed land as something to be exploited and then to be left discarded."[41] But extensive farming also meant that if prices were high, if the grasshoppers stayed away, and if the weather was favorable, *arrendatarios* could reap substantial profits. As early as 1904 one agricultural expert noted the "fever of rapid enrichment" that encouraged land speculation among renters. Some of them even sublet part of their land to *medianeros*. "These Italian peasants," concluded one English observer in 1911, "living though they do in a very rough and uncomfortable way, appear to handle and to make a good deal of money."[42]

This type of farming resulted in low crop yields. Argentine tenant farmers rented as much land as possible in the hope that good weather would combine with high market prices to produce a large profit. Farming on the pampas thus became, as the acute observer Juan Bialet Massé noted, a "game of chance." Particularly in the first 30 years of pampa farming, few tenant farmers bothered to give the soil or the crops the basic care required to raise yields substantially. Instead they trusted in nature and hoped for the best—always aware that in any case they would be moving along to a new plot in a few years. To be sure, Argentina's low yields cannot be ascribed solely to the tenant-farming system; as we shall see in Chapter 6, the government's neglect of agricultural research and rural education was a major factor in Argentina's low productivity. But the superficial and careless farming methods associated with extensive agriculture contributed in an important way to the low wheat yields shown below (in bushels per acre):

Period	Argentina	Canada
1909/10–1918/19	9.4	18.7
1919/20–1928/29	12.5	17.2

Argentina's wheat yields slowly improved. Nonetheless, they remained not only much lower than Canadian yields, but also substantially below yields in the United States and Australia. Argentine yields in 1923 were only marginally better than yields in Russia.[43]

Tenant farmers paid rents that varied with prevailing land price trends. These in turn reflected market circumstances for Argentine produce. During periods of prosperity rents rose steadily—often faster than crop prices—but when markets showed a long-term downward trend, rents fell, although usually more slowly than crop prices. Rents did drop rapidly, however, during the early 1930's, a downturn that forced numerous landowners to the verge of bankruptcy, for as in Canada, mortgages burdened Argentine rural property heavily.[44] The difference was that in Argentina the group that held these mortgages was not the mass of small farmers, since most of them were renters, but the big landowners and estancieros.

Total rural mortgage debt reached 2.3 billion pesos by the early 1930's, or 35–40 percent of the value of pampa rural property. In 1933 the government saved much of the landed class from what the *Review of the River Plate* called a "carnival of foreclosures" by enacting a national moratorium on amortizations over the strong protests of banks and insurance companies. This measure, which lasted from 1933 to 1938, enabled rents to fall still more, to 50 percent of the 1930 level in 1934 and 1935.[45] In other words, the cost of land—one of the principal determinants of the cost of crop production—was not a fixed cost as it was for so many Canadian farmers. Although at least one prairie province—Alberta—did enact a moratorium during the 1930's, the prairies suffered a wave of foreclosures. Of course, Argentine farmers also suffered, and often severely, when crop prices fell during the period of rental contracts.

Numerous observers have concluded that many of the ills that still afflict Argentine agriculture flow from this land system. They emphasize that short-term tenancy led to soil abuse, extensive and irrational cultivation, low yields, and the continuing impoverishment of the rural population.[46] Unquestionably many farmers were impoverished, and they did desire longer-term contracts, compensation for physical improvements to the property, and rent reductions when crop prices fell suddenly. As we shall see, a 1921 tenancy reform law did provide for minimum-term contracts and for compensation for improvements, but it contained numerous loopholes and was finally amended in 1932. Certainly many tenants wanted to buy their land, and thousands of them did so, as an analysis of cadastral records for the 1920's demonstrates and as Table 4.1 indicates. Nonetheless, most renters found it extremely difficult to accumulate the capital needed to become owners.[47]

Still, there is evidence that renters did not necessarily object to their temporary status. Many did not plan to remain in the pampas or even in

Argentina. In a highly revealing comment, Esteban Piacenza, president of the Federación Agraria Argentina, the principal farmers' organization, confirmed this tendency. Farmers, he said, "have migrated to this country . . . to work a few years, as few as possible, to make some money and then to return. . . . No one has come here from his homeland with the thought of remaining." This was an exaggeration, for numerous immigrants did desire to remain, but it does appear that during the era of international labor mobility prior to 1930, the land-rental system suited the economic interests of numerous immigrants.[48]

The Nomadic Society of the Pampas

The economic benefits of the tenancy system, then, were not necessarily limited to the landlords, but as Ferns notes, while the system "did not necessarily impoverish the tenants, . . . it tended to impoverish rural life."[49] The prevalence of renters blocked the formation of a close-knit rural society that could foster a sense of community and strong rural organizations. And as time went by, the term of rental contracts grew shorter. In the nineteenth century they often were for five years, but by 1914, as Francisco Delich's study of Córdoba shows, the term of most contracts was three years or less. Because of the instability of tenure, "the prevailing system of land leasing," as Rolf Sternberg argues, "reduced tenant farmers to agricultural nomads." Contemporary observers fully agreed; one likened the farmer to the "wandering Jew"; another called him an "agricultural Arab."[50] The novelist Alcides Greca expressed this situation in *La pampa gringa*, in which a rural schoolteacher, meditating on the local farming population, laments that "there was an enormous tragedy surrounding the pampas: it was the tragedy of the farmer who continues to think of himself as a foreigner . . . because the threat of expulsion is always hanging over his home."[51]

With the farm population constantly in flux, cooperative organizations were much weaker in the pampas than in the prairies. The exceptions were in the regions settled in the early colonization experiments, where the bulk of the farmers had become owners. Co-ops, as well as social and cultural associations, emerged in these regions, but elsewhere on the pampas they were rare. Indeed, as late as 1937 there were only 106 rural commercial co-ops in all Argentina. Their total membership was 32,000. The first co-op grain elevator in Argentina began to operate in 1930; by that time there were well over 1,000 co-op elevators in the prairies, and the Canadian wheat pools had 143,000 members.[52] Certainly, the isolation of rural life played a role in the weakness of these organizations on the pampas, but in the prairies organized farmers were able to overcome that problem—and in much worse weather conditions.[53]

Community life could hardly be expected to emerge on the pampas when few tenant farmers bothered to build houses. Small landowners in the early immigrant colonies eventually built substantial dwellings, but this was not true of the tenant masses. "They have no real houses, but merely uncomfortable adobe and mud hovels," wrote one English observer in 1911, "yet there is plenty of money to be made." As late as 1914 the landowner Herbert Gibson wrote that the farmer's house "is, at best, an enlarged sardine tin."[54] In fact, when farmers did build houses they tended to be mere shelters of mud or corrugated iron with a flat metal roof that produced great heat in the summertime. The 1937 rural census showed that 46.4 percent of the rural houses in Buenos Aires had mud walls and zinc roofs. Dirt floors were common, and unlike Canadian farmhouses, these pampa dwellings seldom had electricity.[55] Only the rural gentry had telephones. Because landowners paid no reimbursements for improvements, few renters bothered to plant shade trees (despite the heat), a condition that shocked numerous observers of the Argentine rural scene. Even vegetable gardens were rare.[56] Country life was "poor, mean, and joyless," wrote one observer in 1896, and this situation changed little for renters over the next 40 years.[57] A poem by José Padroni expressed the bleakness of tenant farms:[58]

> No one planted a tree
> And on the farms there were no sheep.
> Birds did not sing, the soft murmur
> Of bleating lambs was absent.
> Mothers raised sad children.

Although the Argentine educational system improved greatly in the late nineteenth century, it remained inadequate to the task, and the illiteracy rate was still high in the pampas in the interwar period (see Table 4.3). In part, the malaise of education was due to inadequate financing: in 1937 the Argentine economist Alejandro Bunge calculated that total government expenditures on education were 19.6 paper pesos per person in Argentina and the equivalent of 41.7 in Canada. Although these figures do not take regional variations into account, all evidence points to the generally poor situation of education on the pampas relative to the prairies.[59] The 1937 Argentine agricultural census reported that 81 percent of "producers" in the cereal belt were literate, but numerous Argentine educators claimed that things were much worse. Women were frequently illiterate, and the dropout rate for children approached 80 of 100 by the end of the third grade. After surveying this grim situation, the noted educator Ramón Cárcano concluded in his book *800,000 Illiterates* (published in 1933) that "our common rural school system lives in stagnation. It has not evolved in fifty years, defrauding the great law that created it."[60]

Moreover, unlike their counterparts in the prairies, the farmers of the pampas did not read much. Antonio Diecidue, a Santa Fe agrarian leader, recalled that in 1912, "with very few exceptions, it was difficult to find a book, a newspaper, or a magazine in the farmers' houses." *La Tierra*, the newspaper of the Federación Agraria Argentina, had only about one-sixth the circulation of the *Grain Growers' Guide* in the 1920's.[61]

If the standard of culture was not high, neither was public health particularly advanced for this period. The infant mortality rate, as Table 4.3 shows, was much higher than on the prairies, and although rural towns had doctors, local *curanderos* (folk healers) practiced among a large clientele.[62] Yet another consequence of the rootless and insecure life of the rural population was a very high illegitimacy rate. The severity of this social problem in the countryside was portrayed by the novelist Luis Gudiño Kramer. "On the farms," he wrote in 1943, "young girls barely entering puberty are violated by the peons, by their own brothers, or even by their fathers."[63] Nationally, the illegitimacy rate rose steadily between 1914 and 1950 (from 221 to 273 per 1,000 live births); by 1937 it had reached 192 per 1,000 in Buenos Aires, 203 in Santa Fe, 171 in Córdoba, 267 in La Pampa, and 390 in Entre Ríos. Illegitimacy appears to have been higher among the native Argentine poor than among immigrants (this was also the trend in the prairies, where the illegitimacy rate was highest among métis and Indians). In any case, Carl Taylor concluded of the pampas that "thousands of farm people are living in a cultural no man's land," and numerous children grew up bearing what Argentine culture considered a serious social stigma.[64]

In sharp contrast with Canada, the police and justice systems of the pampas were poorly organized, which only deepened the malaise of rural Argentine social life. In Argentina the provinces were responsible for the administration of justice—there was no equivalent of the Mounties—and these provincial police and local judges were notoriously venal, arbitrary, and ineffective. The testimony of Campbell Ogilvie illustrates this situation: he discusses one case in which the local police chief, who doubled as a grain merchant and whose greed was "insatiable," forced farmers to sell him wheat at 25 percent less than the prevailing price. With police like this, life and property remained very insecure in many parts of the pampas well into the twentieth century.[65]

Canada and Argentina: Early Encounters of a Comparative Kind

Prior to the early-twentieth-century wheat boom, Canadians and Argentines knew very little about each others' countries. But with the emergence of the prairies and pampas as great wheat exporters, this ignorance changed to curiosity, followed, in the 1910's and 1920's, by several serious attempts to compare Canadian and Argentine development. Some of the authors of these early studies did not write from firsthand knowledge, but others—including both Argentines and Canadians—made the long sea voyage back and forth between North and South America to study rural immigration or land tenure or marketing or cooperatives in the two big wheat producers at opposite ends of the hemisphere. We shall refer frequently to these early comparative studies throughout this book, for they provide interesting and sometimes unique perspectives on prairie and pampa agrarian development. Here we shall focus on the closely re-lated questions of land-tenure policy and immigration policy as they were viewed through this comparative perspective.

"Many thoughtful Argentines have recently been making comparisons between conditions in Canada and in Argentina," noted the respected *Review of the River Plate* in 1927, and "the conclusions have been most gall-ing to men possessed of sincere feelings of national pride and patrio-tism."[66] The distinguished newspaper *La Prensa*, then dean of the Argentine press, agreed. "We cannot exaggerate the note of pessimism," it observed, in face of the comparisons of Argentine and Canadian de-velopment that were appearing in the 1920's. The main lesson of these studies, *La Prensa* continued, was that "the fruits of our agricultural wealth are distributed with manifest injustice," adding that the concen-tration of land ownership was a prime cause of this maldistribution. The editorialist did not hesitate to blame the government's general inaction on agrarian questions for this situation.[67]

These negative conclusions were not new and had in fact been common currency in a number of earlier comparative studies. The pampa land-tenure system, argued one 1913 writer in the *Grain Growers' Guide*, led to "nomadic farming" that "never can build up a nation." Other Canadian authors thought that Argentina's ethnic composition was at least as neg-ative a factor as the land problem. For example, in 1919 a writer in the *Guide* predicted that Argentina would not become a major competitor of Canada, for the "genius of its Spanish [sic] population is all against grain growing."[68] Probably the most influential Canadian observer of Argen-tina in the 1920's was W. J. Jackman, the Buenos Aires agent of the Ca-nadian wheat pool. Although Jackman believed that Argentina had the potential of becoming an important competitor of Canada, he argued that structural and ethnic considerations made this outcome unlikely. Jackman emphasized that Argentina's insecure land-tenure system, as well as its primitive marketing system, led to poor quality wheat and a lack of standards that would unfavorably impress European buyers. Moreover, the population of Argentina lacked "that initiative and ag-gressiveness which distinguishes the northern races. Southern European blood transplanted to a soft climate is handicapped in competing with

Northern European blood transplanted to a hard and invigorating climate such as that of Canada."[69]

Few if any Argentines would accept the notion of racial inferiority that was implicit in these Canadian evaluations. But like these Canadian writers, Argentines who studied Canada often concluded that the social structure of the pampas needed major reforms if Argentina was to retain its position as a leading grain exporter. For an early example, we can turn to Domingo Bórea, one of the pioneers of the Argentine cooperative movement. "Let us turn our eyes to Canada," Bórea wrote in 1917. "With a less favorable climate and less fertile lands than ours, Canada is a stupendous model of energetic, farsighted, and wisely organized agriculture." One of the bases of this well-organized system, Bórea continued, was Canada's homestead legislation.[70]

The Buenos Aires *Herald*, one of the nation's most respected newspapers, agreed that the land system was a critical factor in Canada's success, enabling it to attract high-quality agricultural immigrants. "Argentina has yet to learn," the *Herald* noted in 1923, "that what she needs more than anything else is a subdivision of opportunities."[71] The same theme was emphasized by Tomás Le Bretón, whom we will meet later as Argentina's dynamic and able minister of agriculture between 1922 and 1924. Le Bretón had become familiar with North American agriculture while serving as ambassador to the United States between 1916 and 1922, and he was convinced, as he told an Argentine congressional committee, that Canada was "a country in a situation similar to ours. [It] goes in for cereals in the same manner that we do, . . . that is to say, in an extensive form." But, Le Bretón continued, Canada was progressing much faster than Argentina, and this was because of its "system of colonization (land settlement). . . . The decisive factor in attracting the best class of immigrant is easy settlement on the land."[72]

Conclusions

These early attempts at Canadian-Argentine comparisons were rather crudely and hastily done. While they correctly identified the key role land-tenure patterns play in agrarian social and economic development, these studies tended to ignore the strengths of the Argentine land-tenure system and to overlook the flaws of the Canadian system. In fact, the land-tenure systems of both the prairies and the pampas had strengths as well as weaknesses. In this concluding section, we shall present a summary analysis of the Argentine tenancy system on the one hand and the Canadian owner-operator system on the other hand.

Flexibility was the keynote of the pampa land system. Argentine land rents moved in the same direction as world wheat prices. In other words, the cost of land, one of the key determinants of the cost of production, was a flexible rather than a fixed cost in the pampas. Flexibility also characterized land use in the Argentine system, for landowners could shift their property from cattle to cereal crops as market circumstances justified.

The Argentine land system thus had certain strengths, but it also had major disadvantages. Aside from the problem of low yields, perhaps the key weakness of the land-rental system was that it produced a nomadic agrarian population. This agrarian rootlessness delayed the cooperative movement, prevented the emergence of a modern marketing system, and kept rural social and cultural institutions weak. The farmers' voice in Argentine politics remained faint as long as so many remained tenants. Unlike the farmers of the prairies, the tenants of the pampas had little reason to take out Argentine citizenship.

In Canada, on the other hand, the owner-operator system fostered more concern for yields. It also fostered strong rural communities and a thriving cooperative movement. And these communities became the social base for powerful agrarian political movements that were able to induce a succession of federal governments to make substantial rural reforms.

The worm in the apple of the Canadian land system was the high fixed cost of land that plagued so many prairie farmers. We have seen that homesteaders who wished to expand, as well as farmers who entered agriculture after the initial homesteads were taken up, had to take out mortgages, and that, as a result, a very high debt hung over Canadian agriculture. This meant that whereas land costs tended to be flexible in Argentina, they tended to be fixed in Canada, a serious disadvantage when world crop prices fell sharply.

Canadians could—and did—compensate for high fixed costs by increasing crop yields. But Argentine farmers were never entirely able to compensate for their lack of community, a factor that weakened the entire structure of pampa agriculture. The pampa land system functioned effectively, but only as long as Argentina had access to massive immigration. The millions of poor Southern European immigrants who streamed into Argentina between 1880 and 1930 provided agriculture with farmers willing to bear the hardships and risks of tenancy. And this constant supply of cheap farming labor willing to live at low standards helped keep the cost of wheat production in Argentina low. But this massive immigration also gave Argentine governments the opportunity to avoid agrarian reform and development legislation. Why bother with messy and costly reforms when more *brazos* (arms) were always willing to take the plunge in pampa agriculture?

Massive immigration was central to the history of pampa agriculture, and although immigration took a substantially different path in the prairies, it was no less important as a factor of agrarian development there. The spectacular economic development of both regions attracted millions of European workers. It is to an analysis of these people whose labor built Argentine and Canadian agriculture that we now turn.

5

Peopling the Prairies and the Pampas

In the brief period of 60 years, millions of immigrants transformed the virtually empty prairie and pampa grasslands into thriving agricultural regions. Whether in Argentina or in Canada, these newcomers shared a common aspiration—to find in New World agriculture the material prosperity and personal freedom that were so difficult for the poor to achieve in Europe. Many failed, but around the turn of the century enough succeeded to give the prairies and the pampas the reputation of golden opportunity. Although there were numerous similarities between the two regions in this respect, there were sharp contrasts as well, for Argentine and Canadian immigration came from different parts of Europe (and America) and from countries at different stages of economic and political development. This chapter first examines and compares the efforts the Argentine and Canadian governments made to populate their grassland regions and then focuses on the diverse currents of internal and international migration that provided the labor force essential for agricultural development in the prairies and the pampas.

To Govern Is to Populate

Argentine and Canadian statesmen long had realized that their vast and largely empty grasslands would be unable to satisfy Europe's growing demand for food without a large population, and that massive immigration was essential to the rapid population growth they desired. As early as 1852 the Argentine intellectual Juan Bautista Alberdi expressed this attitude in a famous dictum, "To Govern Is to Populate." Although both governments, particularly between 1880 and 1914, made immigration a high national priority, Argentina attracted and retained a much larger immigrant population. Between 1870 and 1930, as Table 5.1 shows, more than four times as many immigrants arrived and remained in Argentina as in Canada. The increase in the Argentine population after 1870, a direct

TABLE 5.1

Net Immigration, Ten-Year Totals, Argentina and Canada, 1871–1930

(excess of arrivals over departures)

Period	Argentina	Canada
1871–1880	104,095	-40,000
1881–1890	675,942	-154,000
1891–1900	462,318	-115,000
1901–1910	1,249,505	794,000
1911–1920	495,450	306,000
1921–1930	969,986	142,000
TOTAL	3,957,376	933,000

SOURCES: Argentine Republic, Consejo Federal de Inversiones, *Aspectos jurídicos, económicos y sociales de la colonización con inmigrantes* (Buenos Aires, 1963), pp. 205–6, 209, 214; Duncan M. McDougall, "Immigration into Canada, 1851–1920," *Canadian Journal of Economics and Political Science*, 27 (May 1961), p. 172. Despite the title of this article, the author analyzes Canadian population growth through 1931. His figures are for persons 10 years or older.

NOTE: The Canadian government did not keep a close count of departures, most of which went to the United States. For this reason, the figures for Canada represent estimates made by the Dominion Bureau of Statistics after an analysis of births, deaths, and immigrant arrivals during each ten-year period. The Argentine figures are the official statistics of the government's Dirección Nacional de Migraciones. They do not count clandestine immigration, which frequently took place in Argentina.

TABLE 5.2

Growth of Population in Argentina and Canada, National and Regional, 1869–1914

Category	Population in 1869, Argentina; in 1871, Canada	Population in 1914, Argentina; in 1911, Canada	Average number people yearly
Argentina, national total	1,856,490	7,885,237	133,972
Pampas			
Number of people	772,261	4,227,988	76,795
Percent of national total	41.6%	53.6%	—
Canada, national total	4,324,810	7,206,643	72,045
Prairies			
Number of people	73,228	1,328,121	31,372
Percent of national total	2.0%	18.4%	—

SOURCES: Argentine Republic, Superintendente del Censo, *Primer censo de la República Argentina verificada en los días 15, 16, y 17 de septiembre de 1869* (Buenos Aires, 1872), pp. 632–33; Argentina [6], 2: 109; Kubat & Thornton, pp. 14–15; Canada [2], 2: 141, 440.

NOTE: "Pampas" here excludes the capital city of Buenos Aires but is otherwise as previously defined: Córdoba, Santa Fe, Entre Ríos, and Buenos Aires provinces and La Pampa Territory.

reflection of massive immigration, was spectacular compared with the sluggish performance of Canada (Table 5.2).

Why did Canadian immigration lag so far behind the Argentine? Part of the explanation lies in the magnetic attraction the United States held for immigrants to North America, including those who originally sailed for Canada. Between 1873 and 1896 economic stagnation plagued Canada, and as a result, emigration to the United States reached massive proportions during this period. Indeed, more people left Canada (primarily bound for the United States) then entered it between 1871 and 1901 (Table 5.1). As Sir Richard Cartwright put it, "The Dominion which began in Lamentations seemed to be ending in Exodus." During the next two decades (1901–21), the period of the great prairie land rush, net immigration (the excess of arrivals over departures) totaled 1,100,000 but this number was still only 44 percent of immigrant arrivals.[1] Argentina also had a very high rate of transient immigration, although for different reasons. Before the First World War, as we shall see, tens of thousands of migrants came each year to work as migrant harvest laborers; thousands more came only to work for a few years and to return to Europe with their savings. Nonetheless, Argentina's net immigration was higher than Canada's. Between 1901 and 1921 3,100,000 foreigners entered Argentina, and 1,600,000 (52 percent) remained.[2] The lack of a competing immigrant country on Argentina's borders to serve as a magnet for the disillusioned (Brazil or Chile could hardly exercise the pull that the United States did in Canada) may account for some of its success in attracting and holding more immigrants, but this success resulted primarily from the country's immigration policy.

Although both governments mounted vigorous and expensive campaigns to attract immigrants, Argentina spent more on advertising than did Canada. And this campaign presented the opportunities the republic offered in terms that were not only highly favorable but even misleading. Prospective immigrants read that it was "easy" for newcomers "without capital" to acquire their own land, and that "virgin land" was "subdivided and sold by the owner on very easy terms."[3] To attract immigrants, the Argentine government also occasionally employed recruiting agents and at one period of particularly feverish growth, 1888 to 1890, paid ocean passages for 132,000 Europeans.[4]

Most important, however, the Argentine government prior to the Great Depression maintained a generous open-door policy that admitted all able-bodied Europeans regardless of ethnic background. Southern and Eastern Europeans found that the doors to the United States and Canada were swinging shut after the First World War, but Argentina imposed no ethnic or national quotas or other restrictions, and continued to admit all Europeans regardless of their birthplace. The only exception, after 1910, was in the case of "anarchist agitators." This policy reflected the position of the republic's leading interest groups, the most influential of which was the Sociedad Rural, the association of large landowners and cattlemen. Before 1914 the Sociedad never wavered from its position that unrestricted European immigration was vital to Argentine growth and prosperity.[5]

The Canadian government had begun to pay some subsidies and bounties to immigrant recruiters as early as the 1870's, but it accelerated its promotion campaign after 1896.[6] When the Liberal Party headed by Wilfrid Laurier came to power that year, world demand for wheat was rising, and news of a gold strike in the Klondike promised to usher in a period of prosperity. Laurier, who believed that the twentieth century would be "Canada's century," took advantage of the newly favorable economic circumstances to launch a major immigration program. His able minister of the interior, Clifford Sifton of Manitoba, organized and directed an immigration bureau that issued mountains of propaganda and paid bonuses to recruiters, who (like their Argentine counterparts) presented Canadian prospects in the most roseate terms. According to Brian McCutcheon, Canadian immigration agents spread "half truths, distortions, exaggerations, and outright lies" about facets of prairie life ranging from the climate to railway rates. Misrepresentation also appeared in the advertisements of private land companies. As one disgruntled English settler wrote home in 1914, "Different companys [sic] advertise for Canada, but you hear only the good side."[7] Sifton, who remained in the ministry until 1906, cast his net widely. In the United States, a major focus of the government's efforts, he carried out "an organized sales campaign." And, as we shall see, he recruited many thousands of immigrant farmers in the Ukraine.[8]

Unlike Argentina, where political support for continued massive immigration remained strong through the 1920's, Canada turned rather sharply against large-scale immigration after the First World War. The postwar Mackenzie King government believed that prairie agriculture needed few new farmers, and as a result, King gave at best lukewarm support to British emigration schemes and "Empire Settlement" plans of the 1920's. King was well aware of the fears in the prairie provinces that renewed massive immigration would result in more agricultural production and thus in lower prices. Always ready to believe in Tory plots, King thought that the Baldwin government in Britain and the Conservative Party in Canada had hatched a scheme under the high-sounding title of "Empire Settlement" to dispose of tens of thousands of the unemployed by shipping them to Canada. The prime minister was determined not to cooperate with what he regarded as a nefarious plot, and as a result, during the 1920's the Canadian federal government remained aloof from British emigration policy and indeed took a generally hostile stance toward immigration promotion of any kind.[9]

The generally more restrictive nature of Canadian immigration policy becomes clear when one compares the politics of ethnicity in Canada and Argentina. By the late 1890's the predominant flow of European emigrants came from Southern and Eastern Europe, and while Argentina welcomed these people, Canada (with some exceptions during the Sifton ministry) did not. Certainly Northern Europeans enjoyed the highest prestige as desirable immigrants in Argentina; a noted contemporary writer, for example, stated that one would have "to be blind or consciously close one's eyes" to fail to perceive this attitude. Social Darwinist notions of "superior races," after all, were well entrenched in Argentina (and Canada).[10] But Argentine statesmen were well aware that the republic would attract few people from Northern and Western Europe, given prevailing demographic trends and the competition of the United States and Canada. Consequently, the Argentine government actively encouraged Spanish and Italian immigration, and people from these countries, attracted by economic opportunity as well as by religious, linguistic, and cultural similarities, came by the millions.

Few Argentine intellectuals regarded these Southern Europeans as ethnic undesirables. In fact, the predominant tone of Argentine literature (at least until about 1905) was to praise the Italians as bearers of a distinguished cultural tradition, and the Spanish as thrifty and hard workers. Argentines did, however, regard Basques more favorably than other Spaniards and Northern Italians as preferable to Sicilians and Neapolitans. Numerous and influential Argentine voices objected to Syrian, Lebanese, and Asiatic immigrants as people who would ruin their nation's racial "homogeneity," and opposition also emerged against Jewish immigration.[11] Despite these rumblings of protest, the Argentine government kept the doors wide open to virtually all comers precisely when Southern and Eastern European migration reached its zenith shortly before the First World War. This policy peopled the Argentine pampas with thrifty and hard-working migrants who were not accustomed to standards of living as high as those common in Northern Europe (or Canada), and who were willing to enter agriculture as sharecroppers or renters rather than as landowners.

Canada, in contrast, strongly favored Northern and Western Europeans—and U.S. citizens. The government classified immigrants from Britain, Scandinavia, the Low Countries, France, Germany, Switzerland, and the United States (but not blacks) as people from "preferred countries." To encourage them to come to Canada, Ottawa advertised in these countries, and in some years also subsidized steamship fares.[12] Immigrants from Southern and Eastern Europe were "non-preferred," although Sifton temporarily challenged this categorization by recruiting Ukrainians, a policy that caused deep misgivings among many British Canadians, who accused him of bringing in the "scum of Europe."[13]

The classification of people into preferred and non-preferred groups reflected a deep-rooted Canadian conviction, fortified by Social Darwinist theories, that Slavs and Latins were racially undesirable, if not inferior, and should not be encouraged to migrate to Canada. Indeed, some of Canada's leading intellectuals, including the national humorist Stephen

Leacock, legitimized the idea that Southern and Eastern Europeans were of subordinate intelligence and would "pollute" the Canadian population through race mixture.[14] The Canadian government made no attempt to encourage Italian immigration, nor did it recruit Ukrainians after Sifton's ten-year ministry. Ottawa did not actually prohibit Slavic and Latin immigrants. These people were simply "non-preferred," and if they came to Canada, they faced public attitudes of prejudice or outright hostility. By the late 1920's the Ku Klux Klan was active on the prairies, and in the Canadian case the Klan sowed hatred not against blacks but against Roman Catholics, whether they were from Quebec or from Continental Europe.[15]

Although relatively few Latins might have come to Canada in any case, Canadian policy, based on rigid ethnic conceptions, certainly discouraged entrance to the massive Southern European migrations of the early twentieth century. This was one reason why the Dominion's population grew more slowly than Argentina's and why the prairies remained sparsely populated compared with the pampas. Canadian policy populated the prairies predominantly with British and Northern European immigrants who, like the Argentine immigrants, were thrifty and hard-working. But Canadian immigrants differed from their Argentine counterparts in several ways. With the exception of the Eastern Europeans, they were accustomed to a higher standard of living. Moreover, immigrants in the prairies aimed to start farming as independent landowners and to build solid rural communities. Most of them also came from countries where popular participation in democratic politics was well established, and they brought this tradition of populist reformism with them.

Immigration and Demographic Change

The first agricultural migrants who ventured to the newly opened Canadian prairies shortly after 1870 found these vast grasslands virtually deserted. A small cluster of métis lived near today's Winnipeg, and others gathered farther west, along the shores of the North Saskatchewan River. These people, along with the plains Indians and a few whites, totaled only about 73,000 at the time of the 1871 census. Forty years later, as Table 5.2 shows, 1,300,000 people inhabited the prairies.

The population growth of the Argentine pampas was even greater. There were 772,000 inhabitants of the pampa provinces in 1869, most of whom lived around the Paraná River port towns. The great pampa grasslands were still almost entirely empty except for a scattering of sheep and cattle ranches and a few early immigrant colonies in Santa Fe and Entre Ríos. Much of the pampa region away from the rivers and main towns was still Indian territory. But by 1914, when the government took its

TABLE 5.3

Native-Born and Foreign-Born Populations of the Prairies, 1911, and the Pampas, 1914

Area	Total population	Native-born		Foreign-born	
		Number	Percent of total	Number	Percent of total
Pampas					
Buenos Aires	2,066,165	1,362,234	65.9%	703,931	34.1%
Santa Fe	899,640	583,699	64.9	315,941	35.1
Córdoba	735,472	585,052	79.5	150,420	20.5
Entre Ríos	425,373	352,872	83.0	72,501	17.0
La Pampa	101,338	64,406	63.6	36,932	36.4
TOTAL	4,227,988	2,948,263	69.7%	1,279,725	30.3%
Prairies					
Manitoba	461,394	270,554	58.6%	190,840	41.4%
Saskatchewan	492,432	248,751	50.5	243,681	49.5
Alberta	374,295	161,869	43.2	212,426	56.8
TOTAL	1,328,121	681,174	51.3%	646,947	48.7%

SOURCES: Argentina [6], 1: 202; Canada [2], 3: 518–23.

NOTE: The Alberta and Manitoba totals given in the source were later revised slightly. Percentages in this and subsequent tables in this chapter may not total 100 because of rounding.

third national census,* the Indians were gone and immigrants were transforming the pampa grasslands into one of the world's richest agricultural regions. The population of the pampas region in 1914 was over 4,200,000. The region had grown by an average of 76,795 people per year between the 1869 and 1914 Argentine censuses; the prairie population increased at the much slower average pace of 31,372 between the 1871 and 1911 Canadian censuses (Table 5.2).

This contrasting rate of population growth had major political and economic implications for the two countries. The prairies remained a minor region, with only 18.4 percent of Canada's population in 1911. Although occasionally the federal government granted reforms to the region, the prairies were unable to exercise a permanent decisive influence on federal economic policy, for seats in the House of Commons were apportioned on the basis of provincial population. The rapidly industrializing provinces of Ontario and Quebec together held nearly 60 percent of the seats in Commons, and these provinces generally supported Canada's high tariff policy of industrial protection that the prairies so bitterly resented. But in Argentina the pampas contained over half the national population by 1914, and the pampa provinces (along with the federal capital, which was primarily a commercial outlet for the region) clearly dominated Argentine economic policy making, keeping tariffs low not to

*Argentina did not take another national census until 1947, so it is impossible to make exact comparisons between the prairie and pampa population in 1921 and 1931, when Canada took its regular ten-year censuses

TABLE 5.4
Native-Born and Foreign-Born Cereal Farmers in the Prairies, 1911,
and the Pampas, 1914

Area	Total cereal farmers	Native-born		Foreign-born	
		Number	Percent of total	Number	Percent of total
Pampas					
Buenos Aires	42,148	13,533	32.1%	28,615	67.8%
Santa Fe	28,496	5,723	20.1	22,773	79.7
Córdoba	18,496	6,461	34.9	12,035	64.9
Entre Ríos	11,060	4,569	41.3	6,491	58.5
La Pampa	4,317	541	12.5	3,776	87.6
TOTAL	104,517	30,827	29.5%	73,690	70.5%
Prairies					
Manitoba	44,969	21,924	48.8%	23,045	51.2%
Saskatchewan	101,101	36,458	36.1	64,643	63.9
Alberta	62,761	17,290	27.5	45,471	72.5
TOTAL	208,831	75,612	36.2%	133,159	63.8%

SOURCES: Argentina [6], 5: 309–16; Canada [1], 6: 52–229 passim.
NOTE: "Cereal farmers" includes alfalfa farmers in Argentina.

protect infant industries, but simply to provide government revenue. Argentine protectionist sentiment concentrated in the interior provinces, an area that in the early nineteenth century had contained a light industrial base and the bulk of the nation's population. Massive immigration decisively shifted the nation's demographic and political balance to the pampas, and this shift made a policy of industrial tariff protection—which both farmers and cattlemen opposed—a political impossibility throughout the interwar period. Argentine farmers, as a consequence, could purchase imported goods relatively more cheaply than Canadian farmers, and this helped improve the competitive position of Argentine agriculture relative to Canada's.

Despite their different rates of population growth, the prairies and pampas both became predominantly immigrant regions. On the eve of the First World War, as Table 5.3 shows, 30.3 percent of the pampa and 48.7 percent of the prairie population was foreign born. In both cases, and particularly in the pampas, the children of these immigrants (the censuses counted them as native born but they often remained unassimilated) composed additional and very large groups. The farming populations of both regions, moreover, were overwhelmingly composed of immigrants. Table 5.4 shows that 63.8 percent of the prairie cereal farmers and 70.5 percent of the pampa cereal farmers were foreign born. Clearly, it was immigrant farm labor that made possible the sudden and spectacular rise of Argentina and Canada to the position of leading world wheat exporters.

Patterns of Immigration

Our focus now shifts to a closer analysis of the migratory currents and major ethnic groups that formed the farming populations of the two regions. The tables that follow (5.5–5.7) detail the countries of birth of the prairie and pampa farming populations and will provide a convenient point of reference for our discussion of the impact that massive immigration made on the rural labor force. The contrast between the national origins of the Canadian and Argentine farmers will stand out sharply in this section.

Internal Migration

When Canada acquired the prairies in 1869, the Dominion already contained a large and experienced farming population in the heartland provinces of Ontario and Quebec. The prairies provided a potential outlet for this crowded rural population, and migration began in the 1870's, primarily from the farming districts of Ontario. This quickly became a major migratory movement. As Table 5.5 shows, among the prairie farmers in 1921, more had been born in Ontario than in any other Canadian province or foreign country. "There was scarcely a family in rural Ontario," one observer commented, "that has not sent a son and the bay colts and the second-best buggy to Manitoba or Saskatchewan."[16]

Immigration from Ontario began earlier than large-scale European migration to the prairies, and these early Canadian migrants transplanted many of the social and cultural norms and institutions of Victorian-age Ontario to the West. The Ontarians labored to create a new society in the image of the one they had left. Under the impact of these migrants, the family farm, the rural community with its schools and churches, and a popular devotion to cultural and civic progress—fundamental characteristics of Ontario society—took deep root in the prairies. Later waves of immigrants from the United Kingdom reinforced the dominant British Canadian culture of the prairie West.[17]

Few French Canadians, however, joined the steady stream of Ontario farmers who packed up to claim a homestead in the West. Although rural Quebec suffered severe overcrowding, the press and the powerful Catholic Church did not portray the West as a favorable destination for emigration. Particularly after the defeat of the Northwest Rebellion of 1885—a rebellion, it will be recalled, in which government troops crushed the forces of the French-speaking métis—the Quebec press considered the West a place of "exile" for a French Canadian, where his language and his culture were threatened with assimilation. As a result,

French Canadian migration turned toward the northern regions of Quebec or the United States, and only a few French Canadian farming communities formed on the prairies.[18]

Argentina did not share the Canadian pattern of a large internal migration of native-born farmers to the new agricultural regions. As already mentioned, "native-born" Argentine farmers that the censuses reported were usually the sons of immigrants. The overwhelming predominance of European immigrants and their descendants among the Argentine farmers resulted from the fact that, unlike Canada, Argentina possessed no previously settled agricultural region to serve as a source of internal migration. There was no Argentine equivalent of Ontario with its pattern of individually owned small farms. The Argentine interior provinces, which had been settled since the late sixteenth century, were divided into large estates, were devoted to cattle raising or plantation crops, and were worked by mestizo sharecroppers who knew little about cereal agriculture.

Much like the métis horsemen of Western Canada, the existing population of the pampas—the famous gaucho horsemen—did not make the transition to agriculture easily. Their world had been the hunt, not the plow and furrow, and as a result many gauchos disdained agriculture and refused to do farm work. Those who did try agriculture usually failed, victims of avaricious merchants and landlords and of their own lack of business experience. Such is the theme of one of the agrarian novels of the period, *Los gauchos colonos*, by Mario César Gras. The central character of Gras's novel, a hard-working and altogether decent gaucho, finds himself driven out of tenant farming by the local grain dealer, who is actually a loan shark. Meanwhile, the son of the local landowner rapes the hapless gaucho's daughter.[19]

Exploited by merchants and landowners, gauchos who entered agriculture received no assistance from the government. There were almost no land grant programs to aid the former pampa horsemen, nor did the governments at any level provide the kind of education they needed. Given this kind of policy, or rather lack of policy, the gaucho population dwindled. Those who survived ended up as workers on the big cattle ranches. As Walter Larden was led to comment in 1911, "The population in the 'camp' is composed of two distinct classes, who are also of two distinct races. These are . . . the peons, 'native' by race, who deal with the stock, and the colonists, mainly North Italians, whose work is agriculture."[20] As a result of this state of affairs, Argentine agriculture long was more "foreign" than prairie agriculture. In contrast with Canada, most Argentine farmers came from outside the national social and cultural tradition.

TABLE 5.5
Country of Birth of Prairie Farmers, 1921

Country of birth	Manitoba		Saskatchewan	
	Number	Percent of total	Number	Percent of total
Canada	22,630	42.5%	38,259	32.0%
Ontario	10,469	19.7	23,851	20.0
Quebec	1,804	3.4	4,129	3.5
Manitoba	9,122	17.1	3,005	2.5
Saskatchewan	130	0.2	3,086	2.6
Alberta	18	0.0	105	0.1
United Kingdom	9,722	18.3	21,226	17.8
England, Wales	6,315	11.9	14,616	12.2
Ireland	791	1.5	1,682	1.5
Scotland	2,616	4.9	4,928	4.1
United States	2,157	4.1	18,713	15.7
Austria[a]	4,336	8.1	5,800	4.9
Denmark	165	0.3	600	0.5
France	669	1.3	1,095	0.9
Galicia[a]	3,925	7.4	3,023	2.5
Germany	449	0.8	2,269	1.9
Norway	339	0.6	4,082	3.4
Poland[a]	715	1.3	855	0.7
Rumania[a]	441	0.8	2,191	1.8
Russia[a]	2,421	4.5	8,441	7.1
Sweden	849	1.6	2,924	2.4
Ukraine[a]	848	1.6	553	0.4
Not specified	1,376	2.6	4,955	4.1
All other	2,210	4.1	4,465	3.7
TOTAL	53,252	99.9%	119,451	99.8%

Country of birth	Alberta		Combined total	
	Number	Percent of total	Number	Percent of total
Canada	21,305	25.7%	82,194	32.1%
Ontario	12,675	15.3	46,995	18.4
Quebec	3,174	3.8	9,107	3.6
Manitoba	683	0.8	12,810	5.0
Saskatchewan	116	0.1	3,332	1.3
Alberta	1,298	1.6	1,421	0.6
United Kingdom	14,348	17.3	45,296	17.7
England, Wales	9,741	11.7	30,672	12.0
Ireland	1,254	1.5	3,727	1.5
Scotland	3,353	4.1	10,897	4.2
United States	21,172	25.5	42,042	16.4
Austria[a]	2,720	3.3	12,856	5.0
Denmark	837	1.0	1,602	0.6
France	448	0.5	2,212	0.9
Galicia[a]	2,366	2.8	9,314	3.6
Germany	1,561	1.9	4,279	1.7
Norway	2,686	3.2	7,107	2.8
Poland[a]	611	0.7	2,181	0.8
Rumania[a]	1,123	1.3	3,755	1.5

Country of birth	Alberta		Combined total	
	Number	Percent of total	Number	Percent of total
Russia[a]	3,023	3.6	13,885	5.4
Sweden	2,517	3.0	6,290	2.5
Ukraine[a]	432	0.5	1,833	0.7
Not specified	5,051	6.1	11,382	4.5
All other	2,754	3.3	9,429	3.7
TOTAL	82,954	99.7%	255,657	99.9%

SOURCES: Canada, Dept. of Trade and Commerce, Dominion Bureau of Statistics, *Sixth Census of Canada, 1921*, 6 vols. (Ottawa, 1924–29), 5: 78–79.

NOTE: Because the 1911 census did not report county of birth of farmers, the 1921 census is used for comparison with the 1914 Argentine census.

[a]The welter web of empires and nation-states in Eastern Europe makes it difficult to ascertain how many people from that region's various ethnic groups actually lived in the prairies, and the Canadian census statistics on which this and later tables draw, did little to resolve this complex problem. Through 1921 census takers usually classified East Europeans by the country of their birth, but the 1931 census usually counted Ukrainians from the parts of the Austro-Hungarian empire absorbed by the new Polish state as Poles. However, in 1921 some Ukrainians were counted as "Galicians," and in both 1921 and 1931 a few were even classified as Ukrainians!

British Immigration

Despite the Canadian government's desire to attract British immigration and use of "every possible means of advertisement" to induce it, Canada hardly enjoyed a stellar reputation in the United Kingdom as an immigrant country. It was hard to live down the admonition of Rudyard Kipling, who had called Canada "Our Lady of the Snows." The general tone of British travel literature on Western Canada was that it was "the Siberia of the British Empire," a country where life was very hard and where success in farming came only with great difficulty.[21] Nor did all Western Canadians think British immigrants were particularly suited for prairie life. When Kipling visited the prairies in 1908 and asked about English immigrants, he got an earful. Canadians, he learned, disliked English immigrants "because the English do not work. Because we are sick of remittance-men and loafers sent over here. Because the English . . . kick at our way of doing things. They are always telling us how things are done in England." Many Westerners considered the British uppity and condescending; others believed that Britain was merely palming off its undesirable population on Canada.[22] Despite these unflattering perceptions on both sides of the Atlantic, British immigration increased rapidly after 1900; Canada replaced the United States as the major destination for British emigration. The British became the largest single immigrant group among prairie farmers (see Table 5.5), and they were especially numerous in Saskatchewan. Many of them, however, tried rural life briefly and then departed, pulled by the magnets of the Canadian cities or the United States.[23]

If the British hesitated to go to Canada to farm, they positively avoided Argentine agriculture. Thousands of the English, Irish, and Scotch had migrated to Argentina in the mid-nineteenth century, had obtained land, and had often prospered in cattle or sheep raising. But in the late nineteenth and early twentieth centuries, British migrants found far greater opportunities to become independent landowners in the Dominions or the United States than in Argentina, which was not, warned one Englishman on the scene, "a country suitable for the English emigrant of the usual 'emigrant class!' "[24] The threat, until the 1880's, of frontier Indian raids, along with the unfamiliar language, Argentina's low standard of living, and its reputation as a country where rural justice and the police were capricious and corrupt, all combined to discourage British migration. In fact, net British immigration during the entire 1857–1924 period was only 19,056.[25] So few Britons entered Argentine farming after the turn of the century that by 1914 they were statistically insignificant (see Table 5.6). Argentina did have a large and powerful British community, but it was centered in Buenos Aires and composed of businessmen or descendants of the nineteenth-century migrants.[26]

For Argentines, a particularly poignant aspect of British migration was the departure of a group of Welsh farmers for Canada in 1902. Although these people had not settled in the pampas region, but took up farming in the Chubut Valley of Northern Patagonia, their case demonstrated the difficulties that Argentina faced when it tried to encourage the immigration of Northern Europeans. The first Welsh had migrated to Argentina in the 1860's, attracted by the federal government's offer of land grants and by the vision of establishing a Welsh bastion in the New World. The obstacles to successful farming in Patagonia were formidable; to overcome them the settlers had to learn irrigation techniques and had to devote themselves to hard, unstinting labor. About 3,000 Welsh eventually migrated, and by the 1880's the colony achieved a modest prosperity, primarily in wheat farming. The Welsh, in fact, grew superb wheat—it won a gold medal at the Paris exposition of 1889—and they proved that at least some parts of Patagonia, which most Argentines long had considered a wasteland, could be turned to agriculture. But discontent grew with the Argentine government's determination to assimilate its immigrants through compulsory military training (instituted in 1901) and education in the Spanish language.[27] W. L. Griffith, a Canadian immigration agent in Wales, heard of this situation, visited Chubut, and began to make plans for the Welsh farmers' emigration to Canada. Powerful support came from David Lloyd George, then representing Wales in Parliament. Lloyd George, who had visited the Canadian West, helped raise a fund to pay for the Welsh re-migration. The Laurier government, anxious to encourage these hardy farmers with their knowledge of irrigation, agreed

TABLE 5.6
Country of Birth of Pampa Farmers, 1914
(cereal and alfalfa farmers only)

Country of birth	Buenos Aires		Santa Fe		Córdoba	
	Number	Percent of total	Number	Percent of total	Number	Percent of total
Argentina	13,533	32.1%	5,723	20.1%	6,461	34.9%
Italy	15,766	37.4	17,436	61.2	9,995	54.0
Spain	7,172	17.0	2,594	9.1	1,092	5.9
Russia	1,619	3.8	418	1.5	25	0.1
France	1,822	4.3	325	1.1	213	1.2
Austria-Hungary	466	1.1	1,175	4.1	295	1.6
Uruguay	574	1.4	69	0.2	38	0.2
Switzerland	216	0.5	297	1.0	115	0.6
Germany	224	0.5	187	0.6	70	0.3
United Kingdom	180	0.4	31	0.1	50	0.3
All other	576	1.4	241	0.8	142	0.7
TOTAL	42,148	99.9%	28,496	99.8%	18,496	99.8%

Country of birth	Entre Ríos		La Pampa		Combined pampas provinces	
	Number	Percent of total	Number	Percent of total	Number	Percent of total
Argentina	4,569	41.3%	541	12.5%	30,827	29.5%
Italy	1,951	17.6	1,459	33.8	46,607	44.6
Spain	241	2.2	746	17.3	11,845	11.3
Russia	2,592	23.4	984	22.8	5,638	5.4
France	285	2.6	145	3.4	2,790	2.7
Austria-Hungary	344	3.1	46	1.1	2,326	2.2
Uruguay	536	4.8	113	2.6	1,330	1.3
Switzerland	248	2.2	10	0.2	886	0.8
Germany	142	1.3	50	1.2	673	0.6
United Kingdom	14	0.1	12	0.3	287	0.3
All other	138	1.2	211	4.9	1,308	1.3
TOTAL	11,060	99.8%	4,317	100.1%	104,517	100.0%

SOURCE: Argentina [6], 5: 309–16.

to set aside a 36-square-mile township for the Patagonian Welsh in Saskatchewan.[28]

In May 1902 230 Welshmen left Argentina en route to Liverpool and then Quebec. They settled on good land in the Saltcoats district of Saskatchewan, where they prospered and founded the town of Llywelyn. This affair unquestionably damaged Argentina's reputation in Britain as a country for emigration. In Wales the press spoke of the "practical serfdom" of the Argentine colony; in London the *Daily Telegraph* emphasized that the Patagonian affair demonstrated that immigrants should fol-

TABLE 5.7
Leading Countries of Birth of Pampa and Prairie Farmers, 1914 and 1921

Country of birth	Pampa farmers (n = 104,517)		Prairie farmers (n = 255,657)	
	Number	Percent of farmers	Number	Percent of farmers
Argentina	30,827	29.5		
Canada			82,194	32.1
United Kingdom	287	0.3	45,296	17.7
United States			42,042	16.4
Italy	46,607	44.6		
Spain	11,845	11.3		
Russia[a]	5,638	5.4	13,885	5.4
Poland[a]			2,181	0.8
Austria[a]	2,326	2.2	12,856	5.0
Galicia[a]			9,314	3.6
Ukraine[a]			1,833	0.7
Rumania[a]			3,755	1.5
France	2,790	2.7	2,212	0.9
Germany	673	0.6	4,279	1.7
TOTAL	100,993	96.6%	219,847	85.8%

SOURCES: Tables 5.5 and 5.6.
[a]See Table 5.5 note on the nationalities of East European immigrants.

low the flag. In Buenos Aires, where *La Prensa* equated the Welsh exodus with the "failure of colonization in Patagonia," the mood was sadness. Nonetheless, Argentina's attitude toward immigrant assimilation continued to harden as the First World War approached. A few more Welsh did trickle in, but their migration ceased by 1914 and never resumed.[29]

U.S. Immigration

Canada also had access to another group of immigrants that Argentina could hardly hope to tap. These were people from the United States, and they were particularly important to agricultural development because many thousands of them had previous experience in dry-land cereal farming. The United States became the second-largest source of immigrant prairie farmers by 1921 (Table 5.7). Encouraged by the availability of good, cheap lands in the prairie provinces, hundreds of thousands of people, particularly from the Great Plains states, trooped north after 1896. The Canadian government counted 133,000 entries in 1911–12 alone. The total number of U.S. immigrants is unknown. Ottawa reported 785,000 between 1897 and 1912, but other estimates range as high as 1,500,000. (In contrast, only 4,153 immigrants from the United States arrived in Argentina between 1881 and 1909.) Many people moved north in prairie schooners and did not bother to report at stations along the border, which was easy to cross unnoticed. Hundreds of thousands also soon returned to the United States. In any case, the 1921 Canadian census reported that the number of U.S.-born residents rose from 81,000 in 1891 to 374,000—and the vast majority of these lived in the prairies.[30]

These "Americans" came from diverse origins; some were from the families of earlier Canadian emigrants to the United States, others were farmers' sons who wanted land, and still others were recently arrived Europeans—especially Scandinavians—who wanted better land. The United States, in fact, was the principal source of Norwegian and Swedish immigrants in Canada. Whatever their origin south of the border, most Western Canadians warmly welcomed these newcomers. After all, they shared the same basic aspirations and values as the Ontario migrants: they strove to obtain title to their land and to achieve a comfortable standard of living in a stable rural community. The U.S.-born became particularly numerous in Alberta (where they were the largest immigrant group among farmers), but wherever they settled, they established a reputation as competent farmers. Indeed, the distinguished Canadian historian Arthur S. Morton concluded in 1938 that Americans "contributed more than any other nationality to increase the productivity of the Canadian northwest."[31] When R. B. Bennett, the future prime minister, stood up in the House of Commons in 1913 and suggested that U.S. settlers might be disloyal, the *Grain Growers' Guide* quickly repudiated his contention and emphasized their contributions to Western agriculture. Much more alarm about the consequences of U.S. immigration appeared in the British press and government than in Canada.[32]

Southern European Immigration

In sharp contrast with Canada, immigrants to the Argentine pampas were primarily from Latin Europe, and Italians led the way. Although wages were lower than in the United States, Canada, or Australia, millions of people from all parts of Italy viewed Argentina as a land of golden promise where the poor newcomer found prosperity, the climate was benign, and the culture and language presented no formidable barriers to economic assimilation. Net Italian immigration to Argentina between 1857 and 1924 totaled 1,300,000. At least 40 percent of all Argentine immigrants were Italians, more Italians migrated to Argentina than to any country except the United States, and in no other immigrant country did Italians form such a large part of the total population.[33] Given these facts, it is hardly surprising that Italians formed by far the largest single group of pampa immigrant cereal farmers. Indeed, they far outnumbered the native-born Argentine farmers in the 1914 census (see Table 5.6). As early as 1895 wheat growing was said to be "really in Italian hands."[34] It was this abundant supply of Italian immigrant labor that enabled pampa agriculture to expand so dramatically around the turn of the century.

Some Anglo-Saxon writers assumed that Italian farmers in Argentina were "of a low order of intelligence" and employed "ignorant methods,"[35] but the reverse was more often the case. Most Italians who migrated to the pampas were Piedmontese, Lombards, and Venetians; although they usually lacked capital and previous agricultural experience, they were extremely ambitious, "always feverishly anxious to get rich," as Robert Forester noted.[36] They did not necessarily object to the cash-rental system, at least not as long as crop prices were high, and rents remained at reasonable levels. Indeed, Italian newcomers viewed the Argentine land-rental system as an opportunity to accumulate capital and move up the economic scale. When times were good, their thrifty habits enabled them to save remarkable sums, good parts of which they remitted to Italy. As one recent scholar has described this migration, "Argentina offered work and land and an upward social mobility that Italy could not have provided at that time."[37] And, as mentioned in the previous chapter, many Italians never intended to remain in Argentina, but meant only to work as tenant farmers for a few years and then return home with their savings. Because of this trend (and because of the large numbers of Italian harvest workers who migrated seasonally to Argentina before the First World War), the rate of Italian re-migration was high: 49.6 percent of all Italian immigrants to Argentina left the country between 1857 and 1924.[38]

These plucky Italian tenant farmers often had large families—twelve to sixteen children were not uncommon—and all pitched in. As William Goodwin noted in 1895, "When there is work to be done, the Italian will not spare himself or his family, but will plough by moonlight or by starlight, resting during the hot mid-day hours, and will not cease from his hard and continual work during favorable weather."[39] Similarly, Campbell Ogilvie, a shrewd Briton, praised these "industrious and kindly people," who were "well content with their surroundings" and "careful and frugal in their living." He concluded that Italians adapted themselves to Argentine circumstances far more successfully than the British. Another British observer, Walter Larden, thought that "the industry of these colonists [was] wonderful."[40]

The Italians did not enjoy this reputation in Canada, where they were "non-preferred" immigrants. Nonetheless, a substantial Italian community emerged; in 1931 there were 42,578 people of Italian birth in Canada. The vast majority lived in the eastern cities, and only 3,388 Italians inhabited the prairie provinces, where many worked as railway section hands. About 1,000 were prairie farmers.[41] As Catholics and Latins, Ital-

ians who ventured to the prairies had to confront prejudice: various jour-
nalists claimed (with little evidence) that they were particularly prone to
crime, drunkenness, and other vices, and one Saskatchewan politician
even termed them "garlic smelling mongrels." Some prairie economic
leaders, notably the Calgary businessman C. W. Peterson, challenged this
racial mythology and pointed out that Italian immigration might lower
the high labor costs that plagued prairie agriculture. But these arguments
produced no change in Canadian policy or popular attitudes.[42]

Like the Italians, Spanish emigrants headed for Argentina in massive
numbers and avoided Canada. Argentina, in fact, was by far the leading
destination of overseas Spanish migration; net Spanish immigration totaled about 1,000,000 between 1857 and 1924. Some Spaniards went to
Argentina to avoid compulsory service in the army, which was engaged
in long and futile wars in North Africa. But the majority came to find
their fortune. They streamed out of the poverty-stricken northwestern
region of the Iberian peninsula known as Galicia.[43] Many of these "Galle-
gos," as they were called, aimed to work in Argentina for a few years and
then to return with their savings, and thus the Spanish re-migration rate
(42.5 percent) was high.[44]

The huge mass of Spanish immigrants, most of whom were illiterate
and about 70 percent of whom were male, congregated in the cities and
entered unskilled jobs. Far fewer Spaniards than Italians ventured into the
countryside to become farmers. In 1914 Spaniards composed 11.3 percent
of the pampa farming population (Table 5.6). The one Spanish group that
often took up agriculture was the Basques, who were particularly numerous in La Pampa territory.[45]

The Spanish community in Canada was very small. Whereas Argentina was the destination of 403,000 Spanish emigrants between 1911 and
1915, only 80 sailed directly for Canada. After working on the Panama
Canal or in the United States, other Spaniards trickled into Canada to
look for railway or canal construction jobs. A Spanish community
emerged in Toronto by 1915, but as late as 1931 only 552 persons of
Spanish birth lived in Canada—and of these, 70 inhabited the prairie
provinces.[46]

Eastern European Immigration

The groups discussed to this point—Ontarians, British, and Americans in the case of Canada, and Italians and Spaniards in the case of Argentina—formed the bulk of the immigrant farming population of the
prairies and pampas (see Table 5.7). There were, of course, numerous
other European groups present in both regions. Icelanders, Hollanders,
and Germans were scattered through the prairies, often in ethnic settlement clusters; and Mennonites came from all over to form communities

TABLE 5.8

Four Leading Places of Birth of Prairie Farmers, 1921 and 1931

Place of birth	1921 (n = 255,657)		1931 (n = 288,129)	
	Number	Percent of farmers	Number	Percent of farmers
Canada	82,194	32.1%	88,586	30.7%
United Kingdom	45,296	17.7	42,575	14.8
Eastern Europe[a]	43,824	17.1	53,481	18.6
United States	42,042	16.4	38,890	13.5
TOTAL	213,356	83.5%	223,532	77.6%

SOURCES: Table 5.5; Canada [2], 8: ccxi–ccxiii.

[a]See Table 5.5 note on the nationalities of East European immigrants.

there. Similarly, the pampas contained pockets of Swiss and Danish set-
tlements as well as a scattering of French, Austrians, and Yugoslavs. On
the whole, however, the population of the prairies was overwhelmingly
Northern European and the population of the pampas just as overwhelmingly Southern European. Only Eastern Europe was a common source
for major groups of immigrant farmers in both Canada and Argentina.
Although this migration was far more numerous in Canada than in Argentina, Eastern Europe provided the third-largest group of rural immigrants and farm operators in both countries.

The Eastern Europeans contributed enormously to the opening and
development of prairie agriculture. As Table 5.8 shows, in 1921 they
were nearly as well represented among the immigrant farmers as the British or U.S. immigrants, and by 1931 they surpassed both, to form the
largest single immigrant group. The vast bulk of these people were
Ukrainians who had come from Galicia and Bukovina, territories that
were under Austrian rule before the First World War, and then passed,
respectively, to Poland and Rumania.[47] Ukrainians found the free homesteads and absence of compulsory military service in Canada a powerful
magnet, particularly in view of the high taxes, avaricious landlords, and
small-sized farm holdings they were accustomed to. An estimated
270,000 entered Canada between 1896 and 1930, although many remained only temporarily. Ethnic Poles took a distinct second place to
them among the East European groups on the prairies.[48]

As mentioned earlier in this chapter, the Laurier government desired to
populate the prairies as rapidly as possible, and to carry out this policy,
Clifford Sifton, the minister of interior between 1896 and 1906, developed a program to recruit Ukrainian immigration vigorously. Temporarily, in fact if not in theory, Ukrainians became "preferred" immigrants.
In an oft-quoted remark, Sifton told a Toronto audience that "a stalwart
peasant in a sheepskin coat, born on the soil, whose forefathers have been

farmers for ten generations, with a stout wife and a half-dozen children, is good quality," and to emphasize this conviction, between 1896 and 1900 his ministry paid agents five dollars for each family head that migrated (plus two dollars per family member).[49]

Ukrainian immigration ceased during the war, and during the 1920's vociferous opposition to resuming it appeared. This view was largely based on fear of the social and cultural consequences of admitting more Ukrainians; most of this concern was centered in the prairie provinces, where many established farmers argued that they were undesirable because they remained aloof from the agrarian cooperative movements. During the early 1920's Ottawa responded to these attitudes by erecting various barriers to keep Ukrainian immigration to a trickle. But the Canadian National and Canadian Pacific railways, along with some Western businessmen, carried on an unremitting campaign on behalf of resuming large-scale migration from Eastern Europe. In 1925 the Liberal government of Mackenzie King devised an arrangement with the railway companies that admitted Eastern European farmers when prairie farmers were willing to guarantee them jobs.[50] But the pro-Ukrainian lobby found this policy far too restrictive. One leader of this group, C. W. Peterson, the Calgary businessman who also backed Italian immigration, strongly promoted massive Eastern European immigration during the late 1920's. He emphasized that if these people were settled correctly on less desirable dry lands and in "groups of from 10 to 15 families," they would solve two problems at once: they would speed the occupation of the fringe areas of the prairies, and they would form a labor pool that would "reduce the cost of operation of all farms in the West."[51] The Eastern Europeans, in other words, would help solve one of prairie farming's endemic problems—a shortage of harvest labor.

Indeed, Ukrainians had provided cheap labor in the prairie provinces since the beginning of their migration. As Peterson pointed out, the men had to go to work off the homesteads for months at a time to make the cash essential to begin farming. Although the largest number of Ukrainians settled in Manitoba, wherever they went, they usually occupied marginal land that homesteaders hitherto had avoided—especially bush land on the northern fringes of the prairies. Often arriving penniless, the Ukrainians' early years were full of hardships and suffering; their houses were hovels of logs and mud over a three-foot excavation. "Nothing could be more depressing to live in than that small, damp hole," one pioneer recalled. In the absence of the men, Ukrainian women cleared the land. Everywhere these sturdy migrants built close-knit communities, and by the First World War many of the early arrivals had achieved a modest prosperity.[52]

The origins of the Eastern European migrants to Argentina and the circumstances under which they arrived and settled differed significantly from the East European migration to Canada. Few Ukrainians made the voyage to the pampas, and although a substantial Polish migration took place in the 1920's, most of these people went to the cities or to the frontier semitropical territory of Misiones, where the federal government was making land grants available. Eastern Europeans on the pampas came primarily from the Russian Empire; they formed a large part of the 98,000 "Russians" who arrived and remained in Argentina between 1857 and 1924.[53] Of those who ventured to the pampas to farm, most were not ethnic Russians. One group, which with their families totaled perhaps 10,000, were not Slavs but Russo-Germans, or "Volga Germans," who had lived within the Russian Empire since the time of Catherine the Great and who began to migrate to Argentina in 1877. They were scattered around the pampas in tiny agricultural colonies in Buenos Aires, La Pampa, and especially Entre Ríos, where they were most numerous. They were extremely industrious—indeed indefatigable—farmers, and they were known not only as a people who loved the soil, but also as a group who formed a deep attachment to their new homeland. Nonetheless, the Russo-Germans maintained their ethnic homogeneity. Unlike many Italian and Spanish farmers, they did not intend to return to Europe, and in Argentina they typically bought their own farms and lived in villages.[54]

The largest group of pampa farmers who migrated from Russia were Jews; they settled in rural Argentina as part of a colonization scheme that was unique in the history of the Western Hemisphere. Appalled by the growing persecution of Jews in Russia, the wealthy Dutch Jewish philanthropist Baron Maurice de Hirsch established the Jewish Colonization Association (JCA) in 1891 to transport and settle East European Jews on lands that the association acquired in Argentina. (In 1891 and 1892 the JCA also tried to establish agricultural colonies in the Canadian prairies. These early attempts failed, but by 1903 a small Jewish colony existed in Saskatchewan.) By 1925 the JCA had purchased 617,000 hectares and founded ten agricultural colonies, primarily in Entre Ríos, but also in Santa Fe, Buenos Aires, La Pampa, and Santiago del Estero. At the high point of their population, during the mid-1920's, the Jewish colonies comprised about 33,000 Jewish settlers, 20,000 of them farmers and their families and the 13,000 others artisans and tradespeople, and 5,000 non-Jews.[55] At this time about 20 percent of all Argentine Jews lived on farms, and as late as 1935 5.8 percent of employed Argentine Jews worked in agriculture. Despite the international interest that these colonies aroused, they were only partially successful. Most of the Entre Ríos land they occupied was only marginally suited for wheat growing; in these circumstances, the 75- to 150-hectare farm each colonist received was too small to be very profitable; as one saying went, these farms were "too little to live on, too much to die from." Colonists bought their plots at prices much below current market levels on long-term payments from the JCA, but the association's French Jewish administrators were often highly inflexible and expelled farmers in arrears. Nor would the JCA grant land from its reserves to the sons of established colonists. As a result, according to Judith Elkin, "the JCA colonies became part of Argentina's persistent latifundia-minifundia complex of problems."[56]

Despite all these obstacles, many Jewish farmers made substantial progress. Gauchos taught them how to ride, herd cattle, and shoot. They organized the first major system of cooperative credit and purchasing associations in rural Argentina. Unlike most pampa farmers, they had schools at an early date. Mark Jefferson remarked in 1926 that a Jewish farmhouse he visited "was the most civilized farmer's house in the Argentine." Nonetheless, agriculture in the colonies remained a risky business, and in the 1930's they experienced a steady loss of population as more and more Jews, especially the young, departed for the cities.[57]

Wage Labor and Seasonal Migration

Until the wheat boom collapsed in the Great Depression, labor was usually scarce in both the prairies and the pampas. The mechanization of agriculture was only beginning, leaving Argentine and Canadian farmers, whether they were renters or owners, constantly in need of a great deal of wage labor, particularly at harvest time. Indeed, attracting and recruiting hired hands for agriculture was one of the major problems that confronted Argentine and Canadian wheat growers. Complex cycles of migration emerged to satisfy the demand, and in this way wheat growing in the prairies and the pampas not only attracted an immigrant farming population, but also gave impetus to regular seasonal migrations, both international and internal.

In Argentina the famous *golondrinas* ("swallows") had traditionally fulfilled much of the demand for seasonal farm labor. These workers, who were primarily Italian, used cheap steerage-class steamship fares to travel back and forth between Italy and Argentina to work in the harvests of both countries, which took place at roughly opposite times of the year. At its height during the 1908–12 period, golondrina migration brought between 30,000 and 35,000 temporary laborers to Argentina annually. Mark Jefferson compared them with European opera singers, who also habitually sojourned in Argentina during the Northern Hemisphere summer. Some golondrinas, he found, had made the trip seventeen times or more. A diligent seasonal worker could easily take the respectable sum of 150 Argentine gold pesos back to Italy; some netted nearly double that amount.[58] While it was at its height, this labor was vital to Argentine agriculture, so vital in fact that when Rome temporarily banned all emigration to Argentina in 1911 and 1912 following a dispute over sanitary conditions on immigrant ships, Argentine agricultural interests became deeply alarmed. *La Prensa* editorialized that "it seems that we are living in a nightmare."[59]

Golondrina immigration revived only slightly after the First World War. First of all, pampa farmers began to mechanize. Second, nationalist economic policies in postwar Italy kept more labor at home. And third, Argentine internal migration increased during the war, when Argentina suffered severe unemployment, and this pattern persisted through the 1920's and on into the following decade. From the war years on, the bulk of the seasonal migrants were native-born Argentines of mestizo stock. Some were descendants of the gauchos. Others were inhabitants of the poverty-stricken interior provinces—places like Salta, Tucuman, and Catamarca—whose economies had not shared in the benefits of the export agriculture boom. They were a large group—the 1937 rural census counted 277,000 *transitorio*, or temporary, workers in the pampa provinces.[60] And, whatever their provenance, they remained a depressed group near the bottom of the Argentine social structure. While the golondrinas often had been able to earn substantial cash by working two major harvests yearly, the Argentine rural wage-earning population was generally employed only during the three or four months of harvests in the cereal belt. The rest of the year they were unemployed, and the evidence points towards a life of misery for these people. As early as 1905 the agronomist Hugo Miatello noted that the seasonal migrants were "our errant Jews who have neither a roof over their head nor a stable family nor relationships to link themselves to the rest of society."[61] Throughout the 1910's and 1920's the harvest workers remained an oppressed and poverty-stricken group at the bottom of pampa society.[62]

Part of the responsibility for the plight of these workers rested with the provincial and federal governments, which did little or nothing to organize their deployment to areas where labor was needed. Instead, as *El Capital* of Rosario observed, they wandered around looking for work, "riding on freight trains, being stopped at every station, and taking an eternity to arrive where they were needed." (Men who traveled this way often were called *crotos*, apparently after José Camilo Crotto, the governor of Buenos Aires, who decreed in 1920 that two migrant workers could ride in each empty freight car within his province.) This dismal way of life gave rise to considerable discontent; on a number of occasions (discussed later in this book) rural laborers went on strike against farmers, disrupted production, and were repressed by the police, citizens' "patriotic brigades," and even the army.[63]

This rural labor protest almost always failed, for reasons that a study of a 1928 farmworker strike makes clear. This labor dispute broke out late in the year (during the harvest season) in Santa Fe province. Various unions emerged among farm laborers, as well as the draymen who carted bagged wheat from farm to market. Their demands included higher wages, recognition of the unions, and an end to the farmers' increasing practice of trucking their produce to market. The Federación Agraria Argentina, through its president, Esteban Piacenza, categorically rejected any notion of dealing with farmworkers' unions and argued that the issue was not wages or hours, but the farmers' right to select the peons they wished and to reject any who spread "inflammatory" propaganda about the Russian Revolution and workers' rights.[64] Tension mounted rapidly. The Rosario Chamber of Commerce, noting the need for Argentina to cut costs of production in the current international context of rapidly falling wheat prices, demanded recognition of the right to work. In the countryside the wheat harvest was delayed, violent confrontations took place between strikers and farmers, and one farmer was reportedly killed.[65]

President Hipólito Yrigoyen responded swiftly and offered troops to the provincial governor, Pedro Gómez Cuello, a member of an opposition wing of the ruling Radical Party. When Gómez Cuello declined the offer, on the grounds that the situation was not yet critical, Yrigoyen sent two regiments anyway and ordered them to enforce the "right to work." Labor leaders were arrested throughout southern and central Santa Fe, and the strike was broken, to the fervent applause of farmers, exporters, and the Argentine establishment.[66]

These events of December 1928 illustrate one way that the cost of Argentine wheat production was kept down. Pampa farmers, themselves obligated to landlords and often squeezed by grain merchants, were an incipient rural middle class who viewed any attempt to raise the price of harvest labor as a direct threat to their interests and perhaps to their survival. On the rural labor issue the farmers aligned with the government, which at all costs was determined to prevent any obstacle to placing the Argentine crop on the world market at the lowest possible prices. In this economic context the government would not tolerate rural labor unions. Although most farmworkers were Argentine citizens and could vote, no government in this period was prepared to act in their interests if such action would raise the cost of farm production and thus jeopardize Argentina's position in the world market. Not until the regime of Juan Perón would the state support rural labor organizations.

Canadian prairie agriculture also depended on migrant harvest labor, but in contrast to the turbulence on the pampas, labor peace prevailed among prairie farmworkers. Unrest or strikes that threatened to halt production or exports were almost unknown. Farm labor was, however, an expensive item that placed a substantial burden on the cost of Canadian wheat production.

Seasonal farm labor was in perpetually short supply in Western Canada. During the 1920's between 41,000 and 75,000 extra workers were needed every year, and at least two-thirds of these had to migrate temporarily from outside the prairies to assure an adequate supply between August and October. Prior to the First World War Canada's labor situation in some respects resembled Argentina's; there was no overall planning, and the *Grain Growers' Guide* complained that men "meander about carelessly," which "creates a surplus of help in some localities and a dearth in others." Labor was so scarce in some seasons that farmers began to call for the migration of East Indians, a group sure to arouse hostility in the rest of Canada. A Dutch immigrant reported in 1912 that "there is plenty of work to find around here at $3 a day. They are always short of men and people with threshing outfits ride through the country practically begging you to come to work for them." In a haphazard fashion the railroads attempted to meet the demand for labor by organizing special low-rate "harvest excursion" trains from Eastern Canada, which disgorged thousands of men at Winnipeg or Moose Jaw to look for seasonal farm work.[67]

Intense labor shortages during the war forced the government to step in to organize labor distribution more rationally. In 1918 the federal government established a national network of employment offices that estimated labor supplies and demands and attempted to direct workers where they were needed. The railways continued to offer their "harvest excursion" fares, and every year thousands of workers trooped west, mainly from the Maritimes, Quebec, and Ontario, attracted by the relatively high wages (double the rate of agricultural labor in the East). The Canadian Pacific Railway actively recruited harvesters; in 1926 it published a pamphlet with the title *50,000 Harvesters Wanted* in which it promised to distribute "excursionists" as efficiently as possible.[68] Despite this kind of appeal, harvest labor was scarce during the boom years of the mid-1920's. Consequently, in 1928 the railway companies and the government attempted to tap a new source of labor—unemployed British coal miners. Although some 8,500 miners made the trip to Canada that year, this experiment in seasonal migration was unsuccessful. The inexperienced British miners found that they were often the last to be hired. Discontent grew, and the government rather summarily sent most of the miners, in locked and guarded trains, to ports of embarkation; fewer than 2,000 stayed in Canada.[69]

Despite such instances of poor planning, Canada's system of providing seasonal farm labor usually worked effectively. Wages were good (the daily pay averaged between $3.48 and $4.50 during the 1920's), and as one "excursionist" pointed out, "it is very easy to save money." To be sure, abuses existed: housing was often substandard; some farmers refused to sign written contracts and others cheated their workers; and laborers suspected of "agitation" were promptly fired.[70] On the whole, however, the system was an improvement over the prewar situation, to the point where, in 1933, George Haythorne could write that government planning had provided "an important check to the growth of the undesirable social conditions so frequently connected with harvesting in many countries."[71] Haythorne may very well have had Argentina in mind, for by this time the rural labor situation in the two countries differed sharply.

Both countries tried to combat the rural labor shortage by agricultural mechanization, but Argentina replaced harvest workers with machines much more rapidly than Canada. As we shall see in the next chapter, the rapid rate of mechanization on the pampas was in part a result of the government's policy of admitting imported agricultural machinery duty free. Canada, on the other hand, protected its agricultural machinery industry behind a tariff wall that forced Canadian farmers to pay inflated prices for their combines and other equipment. Argentina, in other words, dealt with the volatile pampa labor situation by striving to make harvest workers obsolete. Canada, enjoying rural labor peace as it did, was not under the same compulsion to mechanize its agriculture.

Summary and Conclusions

Between 1870 and 1930, when the Depression temporarily halted the tides of human migration that were altering the demographic face of the planet, the Canadian prairies and Argentine pampas changed dramatically. Millions of immigrants had swept into these two grassland regions, and although many soon moved on, enough remained to provide the labor force that propelled Argentina and Canada into world agricultural leadership. Yet the different origins of the two migrations—along with the contrasting land-tenure systems of Argentina and Canada—spelled the emergence of two very different rural societies. The Canadian government maintained a racially selective immigration policy that—with the sole major exception of the Ukrainians—populated the prairies with Northern Europeans, while Argentina welcomed nearly all comers and consequently became a haven for the poor of Southern Europe. The prairies thus became imbued with British values and institutions, and wedded to the ideal of the family farm and the rural community, while in the pampas immigrants were willing to gamble for fortune under the circumstances of sharecropping and tenant farming. Because Argentine immigrants were willing to accept the highly mobile status that pampa agriculture demanded, the pampas did not develop the sense of community so predominant in the prairies. But while the pampas lost community, the Latin farmers of Argentina did not have to bear the high fixed costs so common among Canadian farm owners. Argentine immigration policy, in other words, contributed to keeping the cost of pampa cereal production low.

Agriculture in the prairies and the pampas depended on another kind of migration as well—seasonal migration for the harvests. And here again Argentina enjoyed an advantage over Canada, although it was an advantage obtained at the cost of the severe exploitation of the farmworkers themselves, especially after the end of golondrina harvest migration at the outset of the First World War. Canadian harvest migration was far better organized than in Argentina, and the labor strife that plagued Argentina was almost unknown in the prairies. The Dominion's farmers, however, purchased this supply of harvest labor at the price of high wages that added a considerable economic burden to prairie wheat production.

6

Tariffs, Technology, and Transport

Argentina and Canada had vast amounts of land to devote to export agriculture, and both countries also had access to plentiful supplies of labor. But land and labor alone are not enough to ensure that a country's cereal export agriculture will prosper in the competitive world market. To achieve that degree of economic success, an exporting country must put into effect a policy of agricultural development that focuses on raising yields and lowering costs of production and transport. In other words, the role of the state and of government policy is critically important in the effort to increase agrarian productivity.

This chapter examines and compares some of the agricultural policies the Argentine and Canadian governments adopted during the formative years of the prairie and pampa economies. We will analyze several key policy areas. First, we will look at tariff policy, specifically with reference to the impact that import duties made on the prices of products that farmers needed. How did the tariff affect the agrarian standard of living? Did government import duties place a serious burden on the farmer's cost of production?

Second, we will examine government support for rural research, technology, and education. Did the state provide significant support for agricultural science? Did the Argentine and Canadian governments effectively foster improved farming methods?

Finally, we will consider the development of the roads and railways that together form the transportation system essential to the smooth functioning of a rural export economy. What transport costs did Argentine and Canadian farmers have to pay? Did the governments in Ottawa and Buenos Aires promote efficient road and railway networks to give farmers dependable and economic access to markets?

Our analysis of these questions will reveal a sharp contrast between the agrarian development policies of the two new nations. On the one hand, the Argentine government's agricultural policy was weak, and successive regimes neglected agricultural development or took it for granted. On the other hand, the Canadian government carried out a vigorous and, on the whole, successful policy of rural development in the prairie provinces. Although Ottawa's tariff policies did not satisfy prairie demands, the Canadian government did strive to find other ways to make prairie agriculture more efficient.

As a result of these contrasting government policies, Argentina forfeited the advantage that its flexible cost of land, combined with the mild climate and the seaboard location of the pampas, gave to wheat production. Many of the roots of Argentina's long-term decline and Canada's long-term success as agricultural exporters are found in the rural development policies the two governments followed in the decades preceding the Great Depression.

The Burden of the Protective Tariff

Farmers—especially farmers in countries that export large amounts of agricultural produce—have long viewed industry with deep suspicion. And with good reason, for industries that enjoy high tariff protection raise the cost of goods that farmers must buy to produce food, and when this food must be sold on a world market whose prices the farmer is powerless to control, the result of protected industrialization is to reduce the agrarian standard of living. For a good example of agrarian hostility to the tariff, we can take the cotton planters of the southern United States before the Civil War, who desired free trade with England, their major market, and viewed northern industrialization as an economic threat. Anti-protectionism continued to be a powerful political force in both the South and the Midwest during the late nineteenth century, but gradually, as the domestic market absorbed ever greater proportions of the country's agricultural production, a political compromise on the tariff issue emerged. Eventually the agrarian bloc in Congress accepted the U.S. industrial protectionism policy with the proviso that tariff and quota protection be extended to agriculture. This high tariff policy on meat and cereals produced in the United States became especially pronounced during the 1920's. If U.S. industries were to be protected, the farm states had the political muscle to secure tariff protection too.

This model of a protected economy—which the United States largely abandoned after the 1930's—did not apply to the great grain-exporting countries of Argentina and Canada. As we have seen in an earlier chapter, both countries had small populations that could consume only a fraction of the cereals that their farmers produced. Large-scale agriculture in both countries could not exist without massive exports to foreign markets, but if Argentine or Canadian wheat was to be sold on these markets for a price at which farmers could make a profit, the cost of production had to be significantly below the market price. And since commodity prices changed constantly—and sometimes fell sharply—in response to world supply and demand, Argentine and Canadian farmers needed to keep their costs of production low and their yield high. We have already seen that Canadian farmers had high fixed land costs, and the next chapter will discuss the high fixed marketing costs that Argentine farmers faced. Given this situation, agricultural producers in both countries viewed protected industrialization as a direct threat to their interests. The higher the governments in Ottawa and Buenos Aires set the tariff, the more the pampa and prairie farmers had to pay for their supplies, machinery, and tools—as well as their clothing, housing, and even some of their food, such as sugar. For this reason, tariff policy was of central importance to agricultural development in both countries, and cereal farmers viewed urban-based manufacturing industry—and the politicians that supported it—as enemies.

But because there was a sharp contrast between the level of industrialization in Argentina and the level in Canada—and between the tariff policies of the two countries—protectionism was a far more important issue in the prairies than in the pampas. Not only was the Canadian tariff higher, but Canadian farmers also consumed more manufactured goods than the pampa tenants. The cold winters, which made housing, fuel, and heavy clothing essential to survival, meant that the prairie farming population had to consume a large variety of manufactured or processed goods that pampa cultivators could do without. And the higher living standards of the prairies had the same effect. The cars, telephones, and radio sets that prairie farmers purchased in increasing amounts—and that were little used in rural Argentina—increased the dependence of agriculture on industry in Canada.

Because climate, culture, and taste combined to increase the consumption of manufactured goods in the prairies, the tariff issue was constantly brought home to the farming population. Particularly galling—and particularly noticeable—to prairie farmers was the fact that the price of virtually every item they purchased, whether agricultural tools or domestic items, was higher than in the United States. In 1910 agrarian leaders estimated that the tariff cost the average prairie farmer $200 a year, and reduced the purchasing power of a bushel of wheat by 25 percent.[1] But in Argentina where, as a perceptive Canadian observer pointed out, most farmers bought machinery and tools but "little or nothing else except bare requirements in clothing, [the farmers knew] little about and interest themselves less in the tariff."[2]

By the late nineteenth century the protective tariff had become a fact of political as well as economic life in Canada. As we have seen, Sir John

Macdonald's National Policy was based on the use of high tariffs to stimulate the industrialization of Central Canada and to construct a political alliance between manufacturers and their workers. Between 1878 and 1889 (years when Macdonald's Conservatives held power) the average ad valorem rate of duty rose from 21.4 percent to 31.9 percent, and specific rates on various articles that farmers consumed, such as woolens, nearly doubled.[3] This high-tariff policy did not change substantially after 1896, when the Liberal Party held power in Ottawa. Before the 1896 election, the Liberals had promised Western farmers tariff relief. Indeed, Wilfrid Laurier, the party leader, at an 1894 Winnipeg speech, had denounced "the policy of protection as bondage—yea, bondage, and I refer to bondage in the same manner in which American slavery was bondage." But Laurier found it much easier to denounce the tariff during a campaign than to reduce it once in office. Not only did the government need the revenue that the tariff produced; the Liberal Party also had a protectionist wing, particularly in Quebec, that demanded satisfaction. So while Laurier shaved the tariff here and there—particularly on coal, iron and steel, and farm machinery—prairie farmers remained unsatisfied.[4]

By 1910 the prairie West was virtually in rebellion against the Laurier government over the tariff issue. And Laurier took this problem seriously. As the population of the prairie provinces as a percentage of the national population rose, so did their representation in Parliament, and Laurier, who faced a major challenge from the Tories, needed prairie political support to stay in office. The prime minister thus conceived a daring plan to negotiate a treaty with the United States for reciprocal free admission of each country's raw materials. This policy represented a major new direction in Canadian political economy—the last time reciprocity had been in effect was during the 1854–64 decade—and it aroused a controversy so intense that Laurier decided to call a general election for 1911 to confirm his policy.[5]

In this bitterly fought contest "economic issues were almost lost sight of in the hysteria that accompanied the election." Protectionists mobilized for battle behind the Conservatives, who devised the catchy election slogan "No Trade or Truck with the Yankees" and raised the specter of annexation if reciprocity passed. Canadian big business viewed the reciprocity in raw materials as the opening wedge to complete free trade, and the financial-industrial elite vowed to defeat it. As William Van Horne of the Canadian Pacific Railway (who viewed reciprocity as a threat to the East-West transportation axis) emphasized, "I am out to do all I can to bust the damned thing."[6] On the other hand, prairie farm organizations supported reciprocity with religious fervor and an attitude of moral righteousness. This election, editorialized the Grain Growers' Guide, was "a struggle between the common people and the privileged classes," for Canadian manufacturers had used the tariff to make farmers a "conquered people." The tariff, said the *Guide*, was a system of "legalized robbery." Western farmers would not tolerate politicians who took "an active part in robbing them."[7]

Laurier swept the prairies but lost the 1911 election, and in many respects the defeat of reciprocity was also the death knell of the traditional two-party system in the prairies.[8] During the 1920's Western farmers deserted the Liberals and Conservatives en masse to form new prairie-based political movements. In those years the prairies were occasionally able to exert enough pressure on the federal government to influence tariff formulation. Prime Minister Mackenzie King, for example, restored Laurier's cautious policy of cutting the tariff here and there to try to placate the West. But when the Conservatives under R. B. Bennett took power again in 1930, they raised tariffs sharply. After half a century of political struggle against the tariff, prairie farmers had been unable to end industrial protectionism, a situation that directly affected the standard of living in the West and Canada's ability to compete in the world wheat market, where prices fell steadily after the mid-1920's.

Prior to 1930 there was no equivalent protectionist policy in Argentina, and on the whole its tariffs were lower than Canada's. In 1922, for example, import duties as a proportion of the value of imports subject to duty were 24.9 percent in Canada and only 15.9 percent in Argentina.[9] This disparity reflected the fact that, unlike Canada, Argentina had no clearly defined tariff policy. Argentina was not a free-trade country; since the early nineteenth century governments had imposed duties for a number of purposes. One aim was to raise revenue; 40–45 percent of the federal government's revenue prior to 1930 usually came from import duties.[10] Another aim, and one that came into play in the late nineteenth century, was to promote forward linkages, that is, the development of industries to process or refine Argentine raw materials. In practice, this meant high tariffs on woolen textiles as well as on by-products of the cattle industry like leather goods and shoes. This kind of protectionism received a good deal of political support from the landed elite, especially in the 1920's.[11] Particularly high tariffs were levied on two mass-consumption items—wine and sugar. These explicitly protectionist duties supported the wine-producing province of Mendoza and the sugar industries that dominated the life of the Argentine northwest, where the livelihood of over 50 percent of the population depended directly on sugar. The wine and sugar duties served to placate powerful regional political groups, whose leaders argued that without specific protection the Argentine interior would be totally excluded from the benefits of the agricultural export boom taking place on the pampas.[12]

Finally, Argentine tariffs protected a variety of industries that by one means or another had curried favor with the federal government. These included certain factories that produced essential agricultural supplies, such as jute bags used for shipping grain. The high tariff on bags provided work for 2,000 people—and dividends of 15–25 percent annually for shareholders. Another protected product was the galvanized iron that many farmers used for housing material.[13]

Although some Argentine industries did enjoy marked tariff protection, no government during the 1880–1930 period attempted a consistent protectionist policy to spur industrial growth through massive import substitution. Indeed, the list of goods that entered Argentina duty free was extensive and included products that Canada taxed heavily, such as farm machinery and railway equipment. As mentioned previously, the political base for a reorientation of Argentine economic policy toward protected industrialization did not exist, for the pampa landed elite agreed with farmers as well as urban consumers that protectionism in an export-oriented economy would raise the cost of rural production and reduce aggregate real incomes. Although Congress did approve modest tariff increases in 1920 and 1923, the dominant landed elite defeated the ambitious protectionist schemes put forth by industrialists, leaders from the interior, and a scattering of intellectuals and military leaders.[14] Prior to the Depression, the comparative perspective with Canada clearly shows a relative absence of concerted government support of industrialization in Argentina. As Oscar Cornblit has put it, the Argentine experience offers "eloquent proof of the lack of political weight of the local industrial groups."[15] Because tariffs were generally low, tariffs and industrial policy remained matters of minor concern among Argentine farmers until the 1930's. Farm leaders occasionally mentioned tariff reduction as a goal,[16] but in practice, as we shall see, the issues of land tenure and marketing reform far overshadowed the tariff question in Argentine agrarian politics.

From the viewpoint of agriculture, one of the most eloquent and significant contrasts between the tariff policies of the two governments was in regard to farm implements and agricultural machinery. Although, as we have seen, Argentina protected the jute bag industry, the government exempted agricultural implements (except for light tools classed as hardware and a few items like windmills and pumps) from the tariff. Farm machinery entered free, and as a result, the domestic farm machinery industry, unable to compete with foreign imports, remained small.[17] Canada, in contrast, had a large, powerful, and highly protected farm machinery industry. In order to build it up, Ottawa had imposed high tariffs on imported agricultural machines even before the National Policy began; the tariff level was 20 percent in 1858. The duty on farm machinery rose to 25 percent in 1879 and to 35 percent in 1883. Between 1894 and

TABLE 6.1

Mechanization of Agriculture in the Pampas and the Prairies, 1920–1939

Category	Tractors	Combines	Trucks
Argentina[a]			
1920	253	797	n.a.
1930	16,220	28,656	n.a.
1939	23,540	42,729	n.a.
Prairies			
1921	36,485	n.a.	n.a.
1931	81,659	8,897	21,517
1936	81,649	9,820	21,294

SOURCES: Gravil, "State Intervention," p. 150; Britnell, p. 41; Murchie, p. 293.

[a]National totals.

1907, it stood at 20 percent. It then fell to 6 percent to 10 percent in the late 1920's, only to be increased to 25 percent by the Bennett government in 1930.[18] Canada's policy of protecting its farm machinery industry made these products about 15 percent more expensive in the prairie provinces than across the border in the United States. What seemed even more unjust to prairie farmers was that the big Canadian implement companies exported agricultural machinery and sold it in countries like Argentina well below the price of the same product in Canada. The farm machinery companies were able to sell their exports at these cut prices (up to 20 percent below the Canadian price) because they received various kinds of government export subsidies. The net effect was that Canada was "subsidizing the Argentine or Australian wheat farmer to assist him in competing with the Canadian wheat farmer in world markets."[19]

The availability of agricultural machinery at world market prices was a major reason why pampa agriculture mechanized at a surprisingly rapid rate. Indeed, one study concludes that prior to the Second World War Argentina did not lag behind "the most advanced countries" in agricultural mechanization.[20] Especially notable was Argentina's rapid adoption of the combine harvester. Argentine farmers were particularly anxious to introduce labor-saving technologies to avoid the high expense of harvest labor and the turbulent unrest that sometimes disrupted harvests.

The combine, first introduced in Argentina in 1921, enabled a great reduction in labor costs as well as a speedier harvest. It was a blessing to pampa farmers. Its use caught on very rapidly, and by the end of the 1920's, Argentine farmers had over 28,000 combines in use (see Table 6.1). About one-third of these harvesting machines were imported from Canada. What was particularly interesting about the mechanization of the Argentine harvest was that the masses of tenant farmers began to buy combines. As one Canadian observer noted (in 1938), "Among these Southern and Southeastern European tenant farmers one finds the paradox of a mud hut with a bench, a table, and a bed for furniture but outside a combine harvester, a tractor, and a motor truck of the latest models." Farmers who could not afford their own combine sometimes banded together to purchase one, and in some areas these harvesting machines were available for rent.[21]

Even after the Depression hit and the government imposed a 10 percent duty on imported machinery, farmers continued to import combines steadily. By 1936 combines harvested at least 65 percent of the Argentine wheat crop (some estimates ranged as high as 80 percent). Tractors were not as widely used as combines, because horses were cheap on the pampas and so was their food. Nonetheless, the number of tractors grew rapidly in the 1920's and continued to increase during the Depression.[22]

Pampa farmers, in sum, invested their savings or profits in agricultural machines. These represented capital the tenant could take with him as he moved from one estate to another, and they enabled him to lower his production costs. At the same time, however, farmers who tied up much of their capital in machinery had less money available for the purchase of land. In this sense, as James Scobie notes, mechanization may have "reinforced the hold of tenant farming on the pampas." Nonetheless, the combine cut the cost of wheat harvesting in Argentina by about two-thirds, and this dramatic saving helped keep pampa farming alive during the great wheat price slump that lasted from the late 1920's to the mid-1930's.[23]

Canada's agricultural mechanization lagged behind Argentina's. As D. A. MacGibbon pointed out in 1932, the mechanization of prairie farming was "still in its infancy." Far more tractors were in use than in Argentina (see Table 6.1), but the majority of farms still used horse-drawn implements through the 1930's. The total value of implements in use on Saskatchewan farms actually fell between 1921 and 1926; it grew only slowly thereafter, rising by about 10 percent by 1929. Especially striking is Canada's slowness in adopting the combine harvester. In fact, Canadian manufacturers exported more combines to Argentina (a total of about 14,000 between 1921 and 1936) than they sold in their own prairie provinces.[24]

Canadian farmers hesitated to purchase combines for a number of reasons. Some believed that climatic conditions were unfavorable; they feared that relying on combines postponed the beginning of the harvest too long and raised the chance of crop destruction by insects or hailstones. By the late 1920's, however, the new technology of windrow harvesting overcame that problem. Another reason—and perhaps the principal one—for the slow mechanization of the harvest in Western Canada was that the high initial cost of farm machinery (a cost inflated by the protective tariff) was too much to bear among a prairie farming population that was already heavily in debt.[25]

As a result, the highly labor-intensive traditional harvest continued in the prairies throughout the 1920's and the 1930's as well. This was possible because of the continued reliable supply of harvest labor (the costs of which, of course, declined sharply once the Depression began). Not until the Second World War did higher prices and labor shortages finally bring the traditional labor-intensive homestead to an end—and tariffs on agricultural machinery did not end until 1943.[26]

Agricultural Research and Education

Although the Canadian government found it difficult to reconcile its industrial protection policy with the economic interests of the prairie provinces, Ottawa did move decisively to create an advanced agricultural research and education program to increase the productivity of prairie farming. Thus, a tariff policy that reduced the profitability of agriculture was in some sense balanced by programs in agronomy and plant genetics that raised crop yields. But plant genetics research and rural education programs, so vital to the productivity of modern agriculture, received little more than a snub in Argentina, where agricultural technology has been limited almost entirely (at least before the 1960's) to labor-saving mechanization. Indeed, agricultural science is one of the areas where government policy in the two countries diverged most sharply. In the short run, Argentine agricultural productivity benefited from rapid mechanization, but in the long run, the government's neglect of agronomy has prevented Argentine agriculture from sharing in the dramatic increases in yields that have characterized North American and Western European agriculture.

As soon as Canada acquired the West, the government in Ottawa showed a strong interest in developing prairie agriculture—and this interest in financing prairie agricultural research and technology became an enduring feature of Canadian policy, no matter who was in power. Canada's willingness to invest in the future of its agrarian sector becomes clear when one examines total government spending on agriculture at both the national and the provincial level. In 1921 this spending amounted to about 1.00 Argentine paper peso per capita in Argentina, 2.55 in the United States, and 3.22 in Canada. Whereas the Canadian Agriculture Ministry dated from the birth of the Dominion in 1867, the Argentine government had no Ministry of Agriculture until 1898, and even after the ministry was created, it remained disorganized and poorly funded, the perpetual beggar in the Argentine bureaucratic structure.[27]

Argentine governments took it for granted that the rich pampa soil would give the republic a favorable position in the agricultural export trade indefinitely. As one agricultural expert lamented in 1933, "The fertility of the soil is the greatest enemy of agricultural education in our country." Although occasionally a government would make an initiative in the rural technology and education field, what Argentina lacked, as Noemí Girbal puts it, was "a coherently planned agricultural policy, one with clear national objectives."[28] Carlos Díaz Alejandro is quite correct when he calls the rate of change in the field of rural technology and education a "snail's pace."[29] Perón inherited this attitude of neglect from his oligarchic predecessors and did nothing to change it. Thus, while public research and extension services revolutionized temperate-zone rural technologies in North America and Western Europe, Argentina fell farther and farther behind. This neglect has cost the Argentine economy dearly, for while Argentina has the resource base to compete in the world agricultural market, the productivity of its farm sector has fallen way behind Canada and the United States.

During the interwar years, some of Argentina's most distinguished agricultural experts warned about the consequences of the nation's neglect of science and technology. "The total absence of technical education prevents a rise in cereal productivity and as a result, the cost of production stays high," wrote Miguel Angel Cárcano in 1921. (Cárcano would serve as minister of agriculture during the 1930's and would learn from first-hand experience how perpetually underfinanced the ministry really was.) Guillermo Garbarini Islas, a professional agronomist, lamented that "while the country fills up with more and more university graduates, magnificently prepared to become this national calamity that we call public employees, . . . we have three university rural science faculties and a few experimental schools."[30]

A succession of governments failed to heed these pleas and gave agricultural research and education very low budget priority. The worst offender in this respect was the Radical Party government of 1916–22. President Yrigoyen's oligarchic predecessors, beginning in 1896, had set up two secondary-level agricultural schools in the pampas, six experimental stations that provided elementary-level agricultural education, and two university-level faculties. By 1909 Tomas Amadeo, the government's director general of rural education and research, could proudly report that Argentine rural education "would honor the most advanced countries of the world."[31] But this initial impetus was lost. Under the impact of the financial crisis that began in 1912, the federal government cut its agricultural education budget severely, from 2,900,000 pesos in 1912 to 2,100,000 in 1916. Yrigoyen then slashed it by over half, to 900,000 pesos in 1917, where it remained until 1922. The president made this cut despite the fact that the government's revenue improved significantly beginning in 1918. In fact, the federal government's total operating budget rose over 50 percent between 1916 and 1922.[32]

These cuts forced a reduction in enrollment in government agricultural schools from a high of 634 in 1914 to 276 in 1921. The approximately 40 field agronomists employed by the Ministry of Agriculture (of these about half worked in the pampa provinces) were forced into a state of "semi-activity" by the cuts that Yrigoyen dealt to the ministry.[33] After Yrigoyen left the presidency in 1922, his successor, Marcelo T. de Alvear, increased the rural research and education budget, and enrollments rose again, to 472 in 1928. Yrigoyen, who returned to the presidency that year, slashed the budget again in 1929. During the 1930's enrollment rose—to 799 in 1939—but this remained a very small use of resources in the field of rural education in a country that depended on agriculture for its livelihood.[34]

Argentina's competitor at the other end of the hemisphere never took agricultural research and education for granted; on the contrary, it was made a cornerstone of government agrarian policy. Indeed, this Canadian government-financed work was fundamental to the success of prairie wheat growing. One distinguished observer of Canadian agricultural policy has questioned the motives of the government's agricultural education program, implying that it was a relatively cheap way to divert the attention of the prairies away from Ottawa's tariff policy. But the evidence clearly shows that the Dominion's emphasis on science brought major and lasting benefits to Canadian wheat agriculture.[35]

In 1886 the Dominion government began to establish a series of experimental farms, one of which was located at Indian Head in the Northwest Territories (later Saskatchewan). The system's first director, Dr. William Saunders, focused his energies on methods to make wheat agriculture feasible on the prairies, and this policy led to two great technical advances. One, the result of crossbreeding experiments at the Ottawa and Indian Head stations, led to the development (by Saunders's son Charles) of a new variety of earlier ripening wheat, which was named Marquis. It came on the market in 1911 and was an immediate success. Marquis wheat matured, on the average, in eight fewer days than Red Fife wheat, previously the most widely used strain but one that was frequently caught in the frost. Marquis also yielded seven bushels more per acre and was 20 percent more productive, on the average, than Red Fife. The government stations developed still more advanced strains in the 1920's.[36]

The second big contribution that the experimental farms made was to popularize the techniques of summer-fallowing. The problem was to find a way to prepare the soil so it would hold adequate moisture for wheat growing through the frequently hot and dry prairie summers. The

British and Ontario farmers who predominated in the first wave of immigration were accustomed to intensive farming and found themselves unprepared to cope with the prairie climate. But the Mennonite immigrants who entered Manitoba in the 1870's and who brought summer-fallowing with them from their native Russia attracted the attention of the Indian Head agricultural station. In 1888 the government scientists conducted experiments that demonstrated the value of summer-fallowing, and they then began a successful campaign to publicize the dry-farming technique throughout the prairies.[37]

After taking a look at the government's agricultural research programs, a shrewd Scottish visitor to the prairies concluded, in 1903, that "it would be difficult to laud too highly the work that is being done at the Government Experimental Farms."[38] Nothing comparable took place in Argentina. Canada put science, technology, and education to work to overcome the obstacles that the harsh prairie climate presented to wheat growing. The result was increased yields and higher quality. Argentina, which enjoyed so many natural advantages for wheat growing, including a mild climate and rich soils, did practically nothing in the area of agricultural technology.* In the long run, this technological stagnation severely impeded Argentine grain exports.

The Dominion government's agricultural policy supported more than research and technology. There was, for example, a finely organized system of gathering and reporting crop and production statistics. The Ministry of Agriculture, on the whole, enjoyed the reputation of a professional and competent nonpartisan organization.[39] How different this was from the situation in Argentina, where the Agriculture Ministry was not only poorly financed, but also notoriously corrupt and incompetent! "If it [the ministry] were abolished tomorrow," wrote one farmer, "outside the employees actually affected . . . nobody would be the wiser." The Federación Agraria Argentina complained in 1921 that the ministry had become a "nullity" that "[did] not even serve as a source of information."[40] Although some presidents of the period, notably Alvear, organized the ministry better than others, it never enjoyed anywhere near the success or prestige of its Canadian counterpart.

Yet another difference between agricultural technology in Argentina and Canada resulted from the role of the provincial governments. The governments of the pampa provinces were poorly financed and in any case, since immigrants lacked the vote, the officials were not responsive to the needs of agriculture. As at the federal level, the landed oligarchy— whose real pride was the cattle industry and not agriculture—were in control. Thus, with the exception of Entre Ríos, which had a higher per-

---

*It should be noted that the British-owned railways in Argentina did carry out some significant agricultural research there (a point that is discussed in more detail in Chap. 9).

centage of citizen-farmers than other pampa provinces, the provincial governments did practically nothing in the field of agricultural research and education. But in Canada, where provincial officials were responsible to the mass of farmer voters, the governments supplemented the federal efforts with comprehensive and effective agrarian education programs of their own. While the Dominion agricultural stations concentrated their efforts on experimentation, the provincial governments assumed responsibility for agricultural education and demonstration techniques. Perhaps the most notable example was in Saskatchewan, whose Ministry of Agriculture made "a gigantic effort to instruct new settlers and old alike in the best possible farming techniques." The ministry's numerous and imaginative programs included "Better Farming Trains" and "Special Seed Trains" to carry new techniques throughout the province, the publication of bulletins in various immigrant languages (including Icelandic and German), and the establishment of a "Co-operative Organization Branch."[41] In Argentina there were no provincial universities (and the national universities neglected the rural sciences); in Canada all three prairies provinces not only had universities by the early 1900's, but also established agricultural colleges, which, as a British observer put it, "have fine equipment, competent staff, and . . . magnificent buildings." These institutions organized extension services that taught advanced technology and organizational practices throughout the prairies.[42]

Transportation Policy

A third major area of public policy that directly affected the development of export agriculture in both countries was transport. Adequate rural roads, as well as dependable and economical railway service, were vital to the ability of Argentine and Canadian wheat to compete on the world market—and good transport was also essential to the economic survival of pampa and prairie farmers. Transport policy was a government responsibility, and here again, as with tariffs and agricultural technology, we will examine how the Argentine and Canadian governments responded to the challenge of developing the agricultural sector.

Argentina enjoyed numerous inherent transport advantages over Canada. The pampas are flat, and present no significant barriers to the building of roads or railways. Even more important, the pampas are close to the major Argentine ports of Buenos Aires, Rosario, and Bahía Blanca. Very few parts of the pampas are more than 300 miles from a seaport, and most of the farming sections in our period were much closer—generally within 150 miles. Under these circumstances, one would expect Argentine farm-to-seaport transport to be much cheaper than in Canada, where vast distances and rugged topography lie between the agricultural West and the seaports of either the Atlantic or the Pacific. The Great Lakes enabled wheat exports to move east by water from Port Arthur–Fort William (today's Thunder Bay, a name I shall use for convenience hereafter), but wheat still had to move several hundred miles by rail from the prairies across the rugged terrain of the Canadian shield to reach Lake Superior. Wheat moving west had to move across the Canadian Rockies to reach Vancouver. Canada's one major transport advantage over Argentina was the much shorter distance between its Eastern ports and the European market. Liverpool lies about 6,200 miles from Buenos Aires and about 3,000 miles from Montreal. But given that land transport is much more expensive than ocean transport, this maritime differential by no means meant that Canadian agriculture held a decisive advantage over the Argentine.

Nonetheless, as we shall see, Canada's transport costs were indeed dramatically lower than Argentina's. This difference was a direct result of government policy in the two countries. As a result of skillful political compromise, Ottawa gradually devised a Western transport policy that kept freight rates low. But in Argentina no government addressed the problem of national transportation in a concerted or consistent way; because of political irresponsibility, the regimes in Buenos Aires effectively squandered the natural transport advantages that the pampas possessed. In a manner similar to agricultural research and education policy, the Argentine federal government was content to leave the transport sector alone. As a result, Argentine agriculture was condemned to suffer the consequences of an inefficient and expensive transport system that directly affected the income of the nation's farmers.

Roads

The first stage in the transport process that any farmer faced was to move his crop from farm to railway station. The rural roads were thus a vital link in the transport system, and here again conditions in the prairies and pampas were in sharp contrast. Canada had a road network that was not only much more extensive than Argentina's, but much more modern. In 1930 one Argentine author estimated that while Argentina had about 0.5 km of road for each km of railway, Canada had 6.5. Another estimate, of 1928, was that Argentina had 9 km of road per 1,000 sq. km of territory, compared with 70.5 for Canada.[43] By 1930 the construction of "main market roads" around railway station towns, as well as of highways linking the major population centers, was advancing rapidly in the Canadian prairies. But before the mid-1930's Argentina's road system barely existed except on paper. In the words of one British observer: "Frankly, though I know most of the new lands of the world, I know of no region where the country roads are as villainous as in this Republic."[44]

As a consequence, Argentine farmers constantly faced the problem of slow and expensive transport of their crops from farm to railway station. In other words, the advantage they enjoyed over Canadian producers— the short railway haul to deepwater ports—was partly eliminated by the wretched state of local transportation facilities.

The contrast between the road systems in the two countries was partly a result of their different land-tenure patterns as manifested in the distribution of political power. In Argentina tenants on short-term leases lacked the organizational capacity to press for good roads; the general absence of settled rural community life only exacerbated the problem. Moreover, lacking the vote and largely excluded from political participation, Argentine farmers were unable to influence provincial or national governments to develop this critical aspect of the rural infrastructure. But in Canada, as we have seen, the rural population was firmly rooted on the land and organized a vigorous community life. Prairie farmers wanted roads, and on this issue, like many others, they used political power to get them. Moreover, farmers took advantage of legislation that enabled them to pay part of their taxes by working on the roads.[45]

The road system of the prairie provinces developed slowly until 1919, when the federal government passed the Canada Highways Act. This legislation, which appropriated $20,000,000 over a five-year period, granted each province $80,000 a year plus additional sums allotted in proportion to population. The purpose was to stimulate the provinces, which had jurisdiction over roads, to launch construction and improvement programs. Provincial governments could raise additional funds through gasoline and motor vehicle taxes as well as drawing on general revenues.[46]

During the 1920's the provincial governments pushed road construction and improvement programs steadily. While municipalities, which were able to obtain grants from provincial governments, improved local market roads, the provinces began to lay the basis for a modern highway system. In Saskatchewan, which had the largest road network in Canada, 5,960 miles of new road were built between 1919 and 1931; in the same period 879 miles were reconstructed, and 2,225 received new gravel surfacing.[47] By 1934 the prairies had nearly 10,000 miles of gravel highways and 118,000 of improved earth roads. In fringe areas roads were still primitive, but in more settled regions all-weather gravel highways were permitting a great saving of time and money.[48]

While Canada moved ahead rapidly on highway development, Argentina did practically nothing. The highway network was so poorly developed that no one was really sure how large it was. The government in 1935 claimed the republic had 300,000 km of roads but admitted that 293,000 km were "earth." This type of "road," according to one rural expert, "in the majority of cases . . . is the space between two barbed wire fences without any improvements."[49] An observer from the United States agreed that these were "little more than tracks across the fields" and were "impassable for automobiles during the rainy season from May to September." As late as 1923 no decent roads connected Buenos Aires with such major cities as Rosario (only 150 miles away) or Bahía Blanca. Alejandro Bunge counted 1,273 km of all-weather roads in 1925; ten years later a government report found 7,230 km, probably an optimistic estimate.[50] In 1941 another report concluded that Argentina had 4,800 km of "superior" roads, 23,300 of "intermediate," and 390,000 of "earth."[51]

Whatever the extent of the network, the roads were certainly bad and imposed a major burden on farmers transporting wheat and other grains to railway stations or ports. Although trucks came into use in some regions near ports by the late 1920's, farmers typically hired teamsters who would haul grain in huge carts drawn by oxen or from eight to sixteen horses. When it rained, such a journey could become a nightmare. This primitive method of transport, which further damaged the earth roads, cost Argentine farmers from two to three times what North American farmers had to pay to haul their harvests to the station.[52] One observer estimated the annual cost to farmers of transporting crops to the station at 75,000,000 pesos; another estimate was 100,000,000. Better roads would have reduced this burden at least 40 percent.[53]

The lack of policy did not reflect a lack of awareness about the serious difficulties the road problem created for the agrarian economy. Congressmen—mostly from rural areas—proposed federal highway development legislation on at least eighteen separate occasions between 1912 and 1932, but all these proposals died in committee, the victims of Argentina's fractured politics in some cases and of presidential indifference in others.[54] The availability of potential sources of financing was not an insuperable obstacle to Argentine road development. Bunge calculated that in 1929 Argentina's 350,000 automobile owners, most of whom lived in cities, paid 90,000,000 paper pesos a year in gasoline tax, license fees, and import duties on their cars. Yet the government used none of this for road construction.[55]

Another source of financing was available through a 1907 law called the Ley Mitre, which enabled the federal government to levy a 3 percent tax on the net receipts of the railway companies (the impact of this tax on the railways will be discussed later). The proceeds were to be used to improve the roads, particularly access roads to railway stations. The Ley Mitre produced about 70,000,000 paper pesos between 1908 and 1930, but a chorus of critics, inside and outside Congress, charged that the government was flouting the intent of the law and doing very little road work with these funds. A congressional investigation in 1917 showed that 22 percent of the road fund was spent on "administration." Twelve years later, Bunge charged that half the fund was being spent on projects and salaries that had nothing to do with roads. Moreover, by law the commission that supervised the road projects included the managers of the four largest British railways, a provision that brought angry complaints that the government had conspired with the railways to prevent construction of a competing highway system. In fact, whatever road work the Ley Mitre accomplished scarcely amounted to more than repairing the wretched roads that already existed.[56] With all its deficiencies, the Ley Mitre was the only road legislation Argentina possessed until 1932.

Railways

The railways were the transport backbone of the two nations and the vital lifeline that linked pampas and prairies to port cities and foreign markets. Dependable and economical railway service was crucial to the export-oriented economies of Argentina and Canada. Because rail transport played such a central role in economic life, railway policy—and particularly freight rate policy—was a major public issue in both countries. But Canada's railway rate policy was far more favorable to wheat agriculture than Argentina's. The political victory that prairie agriculture secured in 1897 over the railway companies kept Canadian wheat competitive in world markets. It was an especially notable victory when compared with the Argentine, where the railroad companies enjoyed a very favorable political climate and where the government did not intervene on behalf of wheat farmers.

The railway networks of Argentina and Canada were similar in three respects. First, as Table 6.2 shows, both countries had extensive railway systems; in fact Canada had the world's sixth-largest system in 1916, and Argentina the ninth.[57] Second, a very large proportion of both networks (nearly two-thirds in Argentina) was concentrated in the prairie and pampa regions. Third, the railroads in both countries shipped more grain than any other commodity—and in Argentina grain amounted to nearly 40 percent of total freight tonnage. These facts make it clear how closely tied the railways were with Argentine and Canadian export agriculture. While both countries had encouraged railway building to promote political unity and military security, the primary reason for building them was to link the prairies and the pampas to foreign markets.[58]

Because Britain was by far the most important foreign market for the agricultural exports of both countries, it should not be surprising that British capital financed a great deal of their railway systems. For Canada this meant that, as of 1910, "practically the whole of the capital which [had] been spent upon railway construction [had] been provided by the investors of Great Britain." Indeed, the British provided 84.6 percent of

TABLE 6.2
Railway Networks in Argentina and Canada, 1920's

Category	Argentina	Canada
Kilometers of track (1924)	37,800	64,495
Kilometers of track in the pampas and prairies (1924)	24,376	26,663
Tons of freight hauled (1929)	44,300,000	115,200,000
Tons of grain hauled (1929)	17,300,000	14,500,000
Grain as percent of all freight	39.1%	12.6%

SOURCES: Argentina [4], p. 451; Alejandro Bunge, Economía argentina, 1:101; Canada [3], 1926, p. 589, 1931, pp. 663–64.

all new railway investment in the Dominion between 1900 and 1913. The largest and most profitable railway, the Canadian Pacific, was about 60 percent owned by British investors in 1913; only 11 percent of its common stock was held in Canada.[59] The British financial stake in Argentine railways was just as impressive: during the 1920's about 70 percent of the entire railway network was British-owned, and these lines included the most lucrative routes in the nation's transport market.[60]

Although before the First World War British investment in Canadian railways had been somewhat larger than in the Argentine lines (£224,000,000 in Canada and £186,000,000 in Argentina), it dropped after the war, when the Canadian government bought up many of the private companies. But British investors continued to put money into Argentine railways, with the result that total British investment rose 26.5 percent (to about £235,000,000) between 1913 and 1931. By then the British financial stake in Argentine railways was much larger than in Canadian lines.[61] This divergent trend of British railway investment in the two countries had important political implications, particularly because in Argentina the railway companies were British companies organized in the United Kingdom and with their head offices in London. The Canadian railway companies, by contrast, were incorporated in Canada and had their head offices there. Because the British railways in Argentina were not only huge and growing, but also foreign-owned and -operated, they were highly visible targets for economic nationalists. Nothing of this sort occurred in Canada, where British investment in the railways was declining and the companies were Canadian-operated.

In addition to these privately owned railways, both countries had government-owned railroad systems. In Canada the state system was relatively small until the First World War, when two of the three privately owned transcontinental railways, the Canadian Northern and the Grand Trunk Pacific, went bankrupt. To keep the lines operating, the government purchased the companies, along with several financially troubled Eastern lines, and combined them into the state-owned Canadian National Railways, a giant system that possessed over half the railway mileage in the country.[62] The Canadian National was organized to prevent political interference, but it did receive large-scale government financial assistance at the outset. These funds enabled the CNR's capable management to re-equip the system, to rebuild some of it, and to compete vigorously with the Canadian Pacific, especially in the prairie provinces. Nonetheless, the CNR had financial problems—after interest charges it lost $456,000,000 between 1922 and 1931.[63]

The CNR satisfied an old prairie aspiration—it constructed a new line to the port of Churchill on Hudson's Bay to provide an alternative export route for Western wheat. Prairie leaders long had argued that this line should be built as a much shorter route to shipping facilities than the traditional export route via Montreal (3,790 miles for the Saskatoon-Churchill-Liverpool route versus 4,870 miles for the Saskatoon-Montreal-Liverpool route). But what the prairie leaders too often overlooked was the short shipping season (an average of 71 days a year) at Churchill.[64]

The federal government had granted a charter for construction of a railway from the prairies to Hudson's Bay as early as 1885, but financial problems, the opposition of existing railways, and lack of interest in Ottawa kept the work sporadic. In 1908 the line had reached The Pas in northern Manitoba. This was the northern terminus until 1926, when Prime Minister Mackenzie King, who desired to attract Western political support, promised to complete the long-delayed project as part of the Canadian National system. The last spike was driven in April 1929, and the first wheat exports from Churchill (where the federal government built a 2,500,000-bushel terminal elevator) moved later that year. Although grain rates to Britain were cheaper on the new route than via Montreal, the line never transported more than a small fraction of Canada's wheat. The importance of the Hudson Bay railway is that it provided enough competition to keep freight rates down on other routes.[65]

Although the government-owned railway system in Canada confronted numerous financial problems, it not only operated a major and in many respects excellent transport network in the prairie provinces, but also provided strong competition for the privately owned Canadian Pacific. This was in sharp contrast with the Argentine State Railways system, most of which served the sparsely populated, poor, and underdeveloped northern and western regions of the republic that private railway investors had largely avoided. In 1929 the state railways (including one line owned by the province of Buenos Aires) had 7,000 km of track, or about 20 percent of the total Argentine network. Almost all the track in the state system was one-meter narrow gauge, while almost all the British track was 1.67-meter broad gauge, the gauge used in Britain. Thus, traffic interchange between the state lines and the foreign companies was impossible without the expensive unloading and reloading of cars (except for two small French-owned railways, built on the one-meter pattern). In general, the state railways provided no effective competition for the British lines.[66]

The level of railway freight rates for cereals was critically important to rural producers in both countries. Because the farmer had no control over the price he received for his crop, the higher the railway rate, the lower his income. There were basically three contenders in the political arena of freight rates: the farmers, the railway companies, and the government (which had the power to set rates). Because of their political strength, the Canadian farmers received much more equitable treatment on the rate question than their pampa counterparts, who had little influence—with the railway companies or the government.

The British-owned railways in Argentina were an extremely powerful group of companies both economically and politically. Their size alone gave them a great deal of weight in any negotiations with the government; in 1915 they accounted for about 10 percent of the entire national wealth, and during the First World War years their revenues were greater than the income of the federal government.[67] Because Argentine governments traditionally had attempted to attract foreign investment, the railroads long received favorable official treatment. The Conservative administrations that controlled Argentine politics until 1916 believed that the British railways were fundamental to Argentine economic prosperity and growth, and that their interests must be respected. Despite some rhetoric to the contrary, the Radical Party governments that held office between 1916 and 1930 continued this policy and gave the railways favorable treatment on the crucial question of freight rates.

Argentina's basic rate legislation was the Ley Mitre of 1907. It exempted the companies from all import duties, as well as all municipal and provincial taxes, for 40 years. In return the law allowed the federal government to impose an annual tax of 3 percent on the net income of the companies. This tax, as we have seen, was to be used for road construction, and it was the only tax the federal government could levy on the railways. The law also gave the government authority to reduce rates if the net profits of a company exceeded 6.8 percent on invested capital for three consecutive years. The government had the right to establish the exact amount of the railways' capital accounts.[68]

This legislation aroused a storm of controversy and angry charges from the opposition that the government had capitulated to what amounted to a transportation monopoly. Opposition deputies in 1907 muttered that bribes had changed hands to promote the Ley Mitre, and although these charges cannot be substantiated, critics of the legislation ever since have complained that the railways were treated much too leniently. For one thing, as the economic historian Ricardo Ortiz argues, the tax exemption cost the Argentine government 15,000,000 pesos a year—enough to mount an ambitious road-building program.[69] For another, numerous observers were convinced that the law failed to define the concept of capitalization adequately. In other words, the railways could avoid rate reductions by concealing their profits through such financial maneuvers as stock watering, which had the effect of making investment appear larger than it really was. This ambiguity in the law gave fuel to economic nationalists who, in the words of Raúl Scalabrini Ortiz, charged that "the railways can extract their profits in the manner that best suits them."[70]

In fact, far from suffering rate cuts, the railways were allowed to raise their freight rates on several occasions. Beset with wartime inflation and higher labor costs, the companies approached the Yrigoyen government in 1919 and asked for higher rates. Yrigoyen challenged the companies' alleged capitalization but allowed substantial increases. In January 1921, during the postwar recession, the railways petitioned for another increase. This time, however, Yrigoyen, who feared adverse political consequences in the upcoming provincial and national elections, refused to grant the increase and instead appointed a commission to study the rate problem. It eventually recommended a 10 percent rise in cereal and other rates, but no increase on animals or meat—a reflection of the political power of the ranching elite and the political weakness of the farmers. After Yrigoyen's party won the 1922 presidential elections, Yrigoyen granted the increase the commission had recommended. The president, after all, wanted to float a loan in London.[71]

The 1922 rates hit agriculture hard, for wheat prices were falling sharply. Prices recovered by 1925, but then fell nearly in half by 1930, while railway rates stayed high. Argentine agricultural organizations pleaded for relief, and in 1926 President Alvear asked the companies to make a temporary emergency reduction on wheat rates of from 5 percent to 10 percent, depending on distance. But when wheat prices slumped further in 1927 and 1928, the companies argued that they could not afford to continue the reduction and ended it. Alvear tried to force the companies to restore the lower rates, but his presidential term was ending and the railways employed legalities to stall for time. He forced only one of the four big companies to cut rates before 1928, when Yrigoyen returned to the presidency. The aged Radical leader, who was attempting to improve relations with the British, shelved Alvear's policy, and the other companies maintained their high 1922 rates. By 1930 even *The Review of the River Plate*, the voice of the British community in Argentina, was charging that railway rates were "enormous and unbearable" and a serious threat to the entire export economy.[72] But not until the mid-1930's did the British railways encounter any serious highway competition or pressure from the state to reduce rates. Until then pampa farmers were essentially powerless against the huge railway companies on which they depended.

"The Government on Wheels," as many Western Canadians called the Canadian Pacific Railway, was in many respects as powerful as the British railways in Argentina. During the nineteenth century it enjoyed a near monopoly over prairie transportation, and, as in Argentina, farmers viewed it as a foreign company, headquartered in distant Montreal, and allied to the industrial and financial elite of Ontario and Quebec. In the prairies the image of the CPR was one of an octopus that used its immense political power to keep competing railways out, that charged what the *Grain Growers' Guide* called "exorbitant freight rates," and that gave poor freight service, particularly at harvest time.[73] Faced with a long freight haul of 500 to 1,000 miles to lake ports, prairie farmers in the 1890's argued that high CPR rates, combined with the tariff, were not only impoverishing the prairies, but impeding settlement and discouraging agricultural production. But the Canadian farmers were not politically powerless like the tenants in Argentina. After the government managed to cut freight rates, agrarians fought successfully to keep them down. This victory proved to be a key element in the long-term development of Western agriculture and the economic viability of Canadian wheat on the world market.

These cuts came in 1897, when the Liberal government of Prime Minister Wilfrid Laurier, which had come to power the year before, negotiated the famous Crow's Nest Pass Agreement with the Canadian Pacific. The railroad, which wanted to construct a route across the Crow's Nest Pass in the southern Rockies to tap the rich mining regions of British Columbia, asked for a government subsidy to build the new line. Laurier, who desired to expand production and population in the prairies and who was free of the political ties to the CPR that had characterized his Conservative predecessors, granted the subsidy in return for a company agreement to cut its rates on grain and flour three cents per 100 pounds from all points in the prairies to the head of the Great Lakes. The CPR also agreed to reduce rates 10 percent on westbound prairie freight traffic. When Parliament passed the Crow's Nest Pass Act in 1897, it made the new wheat rate statutory—that is, subject to change only by act of Parliament—and applicable to all new railways built in the prairies.[74]

It was not agrarian political pressure that caused Laurier to cut rates in 1897. The prairies were still too empty, and the farmers too disorganized to be a significant political force. Laurier had acted because he was convinced that all of Canada would benefit if the West was rapidly developed. But once the Crow's Nest Pass rates became law and the West filled up, the prairie region made it clear that the Crow's Nest rates must not be repealed. During the First World War, when wheat prices were high, the West was willing to accept the policy of the Conservative Borden government, which suspended the Crow's Nest rates to enable the railroads to cope with wartime inflation. But in 1919, much to the disgust of prairie farmers, Ottawa extended the suspension for three years. Thus, in 1922 (when wheat prices were falling), the new Liberal government of Mackenzie King (elected in 1921) faced a major political issue. The railway companies claimed that they would be unable to reduce rates elsewhere in Canada if the Crow's Nest Pass rates were restored; this argument was particularly effective in the poverty-stricken Maritime Provinces, which long had complained of excessive rates. But Mackenzie King needed the votes of prairie members of Commons to stay in power, for the Liberals were one short of a majority in the House. The new agrarian-based Progressive Party, which had emerged in 1921 when the farmers deserted the Liberals, held 65 seats and might well have brought down the government on the rate issue. Ever the political realist, in 1922 King supported legislation restoring the Crow's Nest rates on grain and flour. His government then fought a long and ultimately successful court case (finally settled in 1925) to make the restored rates permanent. Also in 1925 the government extended the Crow's Nest rates to cover prairie grain moving west to Vancouver. The Maritime Provinces received only token rate cuts (although their rates were cut further in 1927).[75]

Farmers were jubilant over the 1922 legislation. They had been unable to eliminate the hated import tariff, but they had effectively employed their political muscle to reduce railway rates. The *Grain Growers' Guide*, which considered this event a "clear-cut Progressive victory" over the CPR and the traditional political parties, viewed the Crow's Nest affair as ample justification for mobilizing the farm vote behind the new third party. In effect the 1922 decision made Canadian grain freight rates 40–50 percent lower than those in the northern United States. This reduction, prairie leaders pointed out, enabled Canadian wheat growers to compete with the Argentine and Australian farmers, who had only a short land haul.[76] The Crow's Nest rates, which one scholar called "the best transportation bargain on the continent," became for the prairies a sort of inalienable right. Prairie schoolteachers taught their pupils to recite, " 'The Crow rate is half-a-cent-per-ton-per-mile FOREVER.' " The famous rate— unchanged since 1897—remained on the books until 1983, when Parliament abolished it as part of a long-term Western railway development plan.[77]

TABLE 6.3

*Average Transport Costs of Argentine and Canadian Wheat
to Liverpool, 1931*

(Canadian cents per bushel)

Item	Argentina	Canada
Ocean freight	12.50	6.00
Lake freight	—	9.50
Railway freight	11.74	13.50[a]
Hauling to railway station	6.50	2.00[b]
TOTAL	30.74	31.00

SOURCE: Alberta Wheat Pool, File 818 B, Glenbow-Alberta Archives.
NOTE: Canadian costs also included loading and unloading at Montreal; and the total ran higher in the winter after ice closed the waterways.
[a]Rate from "typical Saskatchewan point to Ft. William."
[b]Estimate.

TABLE 6.4

*Average Total Costs of Handling and Shipping Argentine
and Canadian Wheat to Liverpool, 1931*

(Canadian cents per bushel)

Item	Argentina	Canada
Transport	30.74	31.00
Marine insurance	1.50	1.75
Country elevator handling	—	3.50
Grain bags	5.00	—
Other storage, port, and handling	7.96	2.00
TOTAL	45.20	38.25

SOURCE: Same as Table 6.3.

The final stage in the process of transporting the wheat of the two countries to market was the ocean voyage to Europe. In the Canadian case the first leg of the shipment by water from the Lake Superior port of Fort William–Port Arthur (Thunder Bay) was to Montreal or some other Eastern port, where wheat was transshipped for Europe. Lake freight was expensive; the 1931 rate was 9.5 cents per bushel, or more than the Montreal–Liverpool ocean haul. Water transport from lake to Liverpool thus averaged 15.5 cents per bushel, considerably more than the Buenos Aires–Liverpool rate (see Table 6.3). Argentine farmers thus would appear to have enjoyed a favorable competitive position in transport costs to the British market. Their waterway freight rates were less than Canada's, and so were their railway rates because of the short haul to port, which averaged only 144 miles. Despite these apparent advantages, transport from the pampas was only marginally cheaper, for Argentine farmers faced high costs in the initial transport stage, from farm to railway station.

But this rough equivalence of transport costs was only an illusion, in part because of the extremely high operating cost of Argentina's seaports. The notoriously inefficient port of Buenos Aires, which the *Review of the River Plate* complained was the "dearest port in the world," charged a 6,000-ton ship 6.5 times the fees levied at Montreal.[78] And then Argentine farmers had to bear another large cost—the cost of transporting their wheat in jute bags, which had to be loaded and unloaded at railway stations and seaports. As will be seen, Canada had developed an efficient system of bulk-grain transport that utilized grain elevators (often owned by farmers' cooperatives) and the mechanized unloading of grain cars at the lake port. Argentina had none of this. As a result, as Table 6.4 shows, the total cost of handling and shipping from a prairie farm to Liverpool was substantially lower than the cost from a farm on the pampas.

The State and the Agrarian Infrastructure

Canadian governments during the late nineteenth and early twentieth centuries adopted a series of important policies to improve the competitive position of prairie wheat in the world market. Government-sponsored programs of scientific research and agricultural education made major advances in plant genetics and disseminated this research among the farmers. The federal government forced the railways to reduce wheat rates and to keep them down. And, as Chapter 7 will show, government programs not only improved the Canadian wheat-marketing system, but also supported the marketing plans of farmer-owned cooperatives. The one major agricultural policy area where no Canadian government took decisive action was the tariff, but even there rates on agricultural machinery fell, at least prior to 1930.

Argentine government policy was much less favorable to the development of the agricultural infrastructure. One important contrast with Canada was in the role of the provinces. The prairie provinces all mounted impressive programs in the area of road building and agricultural education; the pampa provinces (which were financially weaker and less politically autonomous than the Canadian provinces) took little action in the area of agricultural development. At the federal level Argentine governments were lethargic and displayed a complacent laissez-faire attitude. The tariff burden on farmers was light compared with Canada, but this policy resulted not so much from any desire to aid agriculture as from the power of the ranching elite and their urban political allies. In the transport sphere the government virtually let the British railways charge what they wished and did nothing about road or highway development. Next to nothing happened in the area of agricultural research and education.

The complacent attitude that the Argentine landed elite and federal government shared about pampa agricultural development reduced the income of the nation's farmers and, in the long run, discouraged Argentine agricultural development. Despite the advantages Argentina had as a grain exporter—a mild climate, a favorable seaboard location, a variable-cost land system, and low-cost agricultural machinery—the cost of wheat production in the pampas remained high. Although agricultural cost-of-production statistics are admittedly difficult to determine and are at best probably only estimates, various sources do agree that Canada produced wheat more cheaply than Argentina. Paul de Hevesy estimates that in the late 1920's the cost of wheat production in Argentina was 13.08 gold francs per quintal as compared with 12.24 in Canada.[79] And, of course, Argentine transport costs were higher than Canada's.

We have seen in this chapter that Argentine governments failed to support transport modernization and agricultural research; in Chapter 7 we will see this same story of neglect repeated in the area of storage and marketing. The Argentine ruling elite assumed that agriculture would somehow take care of itself and thrive without any consistent state development policy. This proved to be a fatal mistake, for while the Argentines were mired in inaction, the Canadians plunged full speed ahead.

The comparative perspective on agricultural policy employed in this chapter raises some important questions about agrarian political power in the two nations. Argentine farmers were politically weak in comparison with both the powerful and well-organized families who composed the cattle-ranching elite and the emerging mass of urban consumers in the rapidly growing Argentine cities. Canadian farmers, on the other hand, were able to make effective use of the political system to achieve at least some of their objectives. The restoration of the Crow's Nest Pass rates in 1922 was one good example of their political weight, and we will examine others in the next chapter. Yet one major theme in Canadian agrarian history is that the farmers were in fact politically "powerless" and their organizations "of negligible importance" in shaping government policy. The Saskatchewan economic historian Vernon Fowke, looking back from the 1950's, expressed this view with particular force. But the comparative perspective reveals that Fowke suffered from a bit of myopia. He emphasized that "political power in Canada lies within the Cabinet" rather than among voting blocs like the farmers. This may be true, but the evidence is clear that the Canadian government did act decisively to develop export agriculture at a time when the Argentine government was mired in lassitude and complacency. Fowke may be correct that prairie agriculture received help only when this was in the national economic interest. But in Argentina agriculture received little or no help at all.[80]

Part III

Patterns of Agricultural Policy, 1900–1930

7

Agrarian Struggle and Rural Organization, 1900–1917

By the first decade of the twentieth century, the bloom was beginning to fade from the rose of Canadian and Argentine agriculture. Immigrants still flocked into the prairies and the pampas, but by 1910 they found that the best land was occupied, and that whatever good land was available was becoming increasingly expensive. Equally ominous for the farming population, the feverish period of rising crop prices that had begun in the mid-1890's was ending. By 1910, when crop prices began to decline, prairie and pampa farmers had become acutely aware that their livelihood was inextricably linked to a world market over which they had no control.

Because the individual farmer could do nothing to change the world price for wheat or corn or any other cereal, he had only a few alternatives if he wished to increase his income. One was to raise yields by applying technology to agriculture. But this alternative was practical only so long as world market prices remained above his cost of production. If market prices fell, increasing yields would do the individual producer little good. If the market price fell below the cost of production, raising more bushels of wheat per acre would only increase the farmer's losses.

Higher yields thus would not necessarily resolve the farmer's plight, but the individual producer did have another alternative, and that was to cut costs. We already have seen how pampa farmers mechanized their harvests and how prairie farmers lobbied for reduced railway rates, but the rural producers, Argentine and Canadian alike, were determined to cut other costs as well. In Canada farmers aimed to force the grain trade to reduce the cost of storing and marketing wheat, while in Argentina farmers attempted to force landowners to cut rents. In both countries the agrarian population organized bitter and protracted struggles to achieve these ends.

Although the rhetoric of the agrarian protesters was sometimes radical, in fact these movements were basically conservative. Wheat farmers (whether they were landowners or tenants) were essentially small-scale capitalists who aimed to maximize their profits. One analyst has called them "independent commodity producers"; another has termed them a "petit bourgeois class." Like capitalists elsewhere, they employed wage labor, particularly at harvest time. Although the farmer usually also worked with his own hands, he could not be called a member of the working class.[1]

Because of the farmers' reliance on the world market, they often felt deprived and manipulated by outside economic forces. In Canada their response was to attempt to regulate the grain merchants or even to replace them with farmer-owned cooperative marketing networks. In their struggle against intermediaries, prairie farmers attacked what they called the unjust expropriation of a large part of their return by urban interests who were not themselves producers. Although the Canadian agrarian reform movement called for government intervention to regulate the grain trade and especially for state ownership of terminal grain elevators, on the whole the prairie agrarians were not socialists. They were populists who aimed, not to end capitalism, but to redistribute its fruits.[2]

Like the Canadian farmers, pampa agrarian leaders believed that the grain trade was utterly extortionate and the cause of many of the evils that beset the farming population. But unlike the Canadians, Argentine agrarians realized that they would be unable to reform grain marketing until they first changed the land-tenure system. Until the Argentine farmer owned his land or at least enjoyed a longer-term lease on the land he worked, he would be unable to organize the cooperatives or to apply the political pressure necessary to create the kind of farmer-owned grain-marketing system that was coming into use in Canada. For this reason— and also because land rents were becoming intolerable—pampa farmers concentrated their protests against the land-tenure system. Land reform, they believed, would not only cut rents and thus reduce production costs; it would also provide more stability of tenure and thus provide the key to the reform of Argentina's archaic grain-marketing system.

Whereas in Canada farmers used political and economic organization to reform the marketing system, in Argentina the farmers used more direct tactics. Political and economic organization was difficult for the rootless and unnaturalized farmers of the pampas, but they could use an action tactic, the rent strike, to hit the landlords and merchants where it hurt—in the pocketbook. The great rural strikes of 1912 and 1919 halted production, led to severe property damage, and brought violence and repression, but despite the atmosphere of conflict that surrounded Argentine agrarian organization, the pampa farmers' movement did not aim at social revolution in the countryside. Argentina does not fit Jeffrey Paige's paradigm that "agrarian revolution" will result from conflict between tenants and landowners.³ As in Canada, farmers in Argentina organized only to secure a larger part of the return from their labor, not to challenge the capitalist system itself.

The modern Argentine and Canadian agrarian movements were born during the struggle for reform that prairie and pampa farmers waged during the 1900–1914 period. This chapter will first analyze the causes of agrarian unrest in the two societies, and then turn to the events that occurred when organized farmers challenged the established power structure. We will examine why the agrarian protest movements of the two countries differed so greatly and why, in the long run, agrarian organization was more successful in Canada than in Argentina. This second point requires us to focus on how the Argentine and Canadian governments responded to the agrarian unrest that swept the prairies and pampas during the years immediately preceding the First World War. This chapter—and the chapters that follow it—will contrast Canadian government policy, which was prepared to accept and even to promote certain major agrarian reforms, with the Argentine government, which viewed agrarian unrest as a serious threat and refused to promote any meaningful changes in the land or marketing systems. Because the Canadian state supported the farmers' call for a reform of the grain trade, Canada's wheat-marketing system would become the most advanced in the world. But in Argentina, where the state did nothing, the grain-marketing system remained primitive and a serious burden on the farmers of the pampas.

Protest and Organization in the Prairies

By the beginning of the First World War, Canadian farmers could look back at fifteen years of organized struggle against grain merchants, elevator companies, and other intermediaries who had controlled the marketing of wheat. Convinced that they were being exploited as long as the storage and marketing system remained in the hands of "parasitical" urban middlemen, prairie farmers launched vigorous cooperative and political movements with the aim of creating a marketing structure that the farmers themselves would own and operate. The Canadian rural cooperative movement was a unique achievement—in no other grain-exporting country did farmers combine on such a large scale or so effectively to create a vertically integrated system of businesses to compete with the private marketing structure.

The roots of the cooperative marketing movement in the Canadian prairies were numerous and complex. For one thing, the heritage and tradition of British and continental European cooperatives that immigrants brought to the Canadian West provided a valuable reservoir of ideas and experiences for the organized farmers' movement. The prairies also received the Canadian heritage—through the migrants who poured in from Ontario—of using rural organization and political pressure to fight for agrarian reforms. Several Ontario farmers' organizations had existed since the 1870's and had gained experience, if not success, in dealing with provincial and federal governments. British radicals, socialists, and trade unionists who migrated to the prairies contributed their own traditions of political protest. And, finally, there were numerous U.S. influences on the prairie farmers' movement. As Paul Sharp has pointed out, "Every major farmers' movement in the United States before World War I moved into Canada in some form."[4]

In the prairies these traditions of protest, organization, and cooperation received powerful support and ideological justification from a new brand of North American Protestantism—the Social Gospel. Most prairie farmers were Protestants, and the bulk were members of the mainline churches: Baptists, Lutherans, Methodists, Presbyterians, and of course, Anglicans. The United Church of Canada, formed in 1925 when the Methodists united with most Presbyterian congregations, became the largest church in the prairies. Whatever the denomination, attendance at prairie churches was high, and the church building served not only as a religious center, but as a social and community institution.[5]

By the First World War these churches, particularly the Methodists, were beginning to provide what Richard Allen has called "a new social faith" to the organized farmers' movement. Indeed, the Social Gospel became "the religion of agrarian revolt." At church colleges throughout the prairies, future preachers learned that the central concern of Christianity ought to be the eradication of injustice from society and the promotion of cooperation rather than competition.[6] One of the most influential Social Gospelers, Dr. Salem Bland, a professor at Wesley College in Winnipeg, was the keynote speaker at the 1915 convention of the Manitoba Grain Growers' Association. In his address he endorsed the grain growers' decision to demand the taxation of unused land, and emphasized that the co-op movement was "part of the divine plan of human brotherhood." By the time Wesley dismissed Bland in 1917, he had influenced a generation of students. Bland then turned to the Chautauqua platform and to journalism; the *Grain Growers' Guide* employed him to write a regular column. The teachings of Bland and others fell on a receptive audience, for many prairie farmers read widely and were acquainted with the Social Gospel movement in Britain and the United States.[7]

The Social Gospel was in many respects a theology of revival and of protest, and as such it made a direct appeal to a movement that was itself born as a protest against the virtual monopoly power the private grain trade held over individual farmers. By 1900 five large companies owned about three-quarters of the grain elevators in Western Canada, and even in towns where two or more companies owned elevators, farmers found that effectively no price competition existed. This situation was the result of an early policy of the Canadian Pacific Railway, which had lacked the capital to build elevators itself and instead offered to lease land free to elevator companies. To induce them to build, the CPR agreed that farmers would not be able to load grain directly into cars or use warehouses at railway stations to store grain. Farmers had to accept the grade for their wheat assigned by the elevator companies, and this was a crucial question, for grade was the prime determinant of price. The only exception to the tyranny that the elevator companies held over the grading process was in a few cases where farmers could obtain separate elevator bins and then negotiate the grade with a commission firm in Winnipeg.[8]

At the center of the wheat-marketing system in the Canadian prairies was the Winnipeg Grain and Produce Exchange, which opened in 1887 and was incorporated by the Manitoba legislature in 1891. This "central market institution of the wheat economy" was not only a place for grain exporters to deal with the elevator companies; after 1904 it was also a center of the wheat futures market. Its power over the prairie wheat economy was immense.[9] Each day a committee of its grain dealers set the "spot," or daily price, and telegraphed this information to the country elevators that purchased grain from farmers. As the prairie wheat economy grew, so did the size and power of the Winnipeg Exchange. By the 1920's it had become "the largest cash grain market in the world," and together with the other chief exchanges in Chicago, Buenos Aires, and Liverpool, it formed a "closely integrated system of markets which together produced the constantly changing world price for grain."[10]

The individual farmer was helpless when he confronted the elevator companies and the grain merchants. As one British immigrant explained, whenever he went to sell his grain to the elevator operator, "I was practically in his hands, unless I was prepared to throw my grain by the roadside or to haul it home." Robert J. C. Stead expressed the same dilemma in his novel *Grain*, published in 1926. In it Gander Stake, a Manitoba farmer, confronts three buyers "who, for competitors, seemed to him to be on much too friendly terms with each other." All three judge his wheat to be "three Northern," although the young Manitoban, proud of his ability as a farmer, is sure it is "one Northern." Stake has no alternative. He accepts the lower price for the "three" grade but, as Stead puts it, "his gorge was boiling within him. He had the farmer's deep-rooted sense of injustice over the fact that whenever he bought he had to pay the seller's price, but whenever he sold, the buyer dictated the figure."[11]

This sense of injustice had been building up among prairie farmers since the 1880's, when the first farmers' unions appeared in Manitoba. These groups, including the Manitoba and North West Farmers' Union

and the Manitoba and North West Farmers' Co-operative and Protective
Union, suffered from inexperience, poor management, and political in-
volvement in the Northwest Rebellion, and soon disappeared. Following
them was the Patrons of Industry, which appeared in Manitoba in 1891.
This group utilized a political approach to the farmers' problems and in
1894 launched its own political party, which soon collapsed. In 1896 the
new Laurier government turned a sympathetic ear to the problems of the
prairies and, as we have seen, the Crow's Nest Pass Agreement was one
result. Economic prosperity in the late 1890's, along with a sympathetic
government in Ottawa, dampened the fires of agrarian unrest, and the
Patrons died a quiet death in 1899. Although the early farmers' organizations failed, they did awaken rural Manitoba to the advantages of producers' cooperation, and as early as 1890 isolated groups of farmers
around the province began to form farmer-owned municipal elevator
companies.[12]

One result of the agrarian protest of the 1890's was to demonstrate to
the Laurier government that the economic power of railway and commercial interests in the West was actually stifling expansion, settlement,
and wheat production. This was directly contrary to Laurier's determination to develop the West as rapidly as possible, and the government
moved quickly to enact reforms. In 1899 Laurier appointed a Royal Commission to investigate the grain trade and named three Manitoba farmers
to sit on it. The prime minister's purpose was to get rural protest on record and to awaken the Canadian public to the need for change. As a result
of the commission's work, the federal Parliament passed the Manitoba
Grain Act of 1900, which instituted government regulation and supervision of the grain trade, directed the railways to provide loading platforms
and warehouses at stations where farmers petitioned for them, and ordered the railways to provide boxcars equitably to these facilities, as well
as to the line elevators.[13]

Farmers, however, did not quickly perceive the benefits of the Manitoba Grain Act, for the Canadian Pacific continued to favor the established elevator companies when it distributed boxcars. This became an
acute problem during the 1901 harvest; a shortage of cars left 30,000,000
bushels of wheat in farmers' hands when the snow fell. Anger rose fast
on the prairies: as one farmer put it, "It was the hot anger of a slap in the
face."[14] Rather than wait for the government to move, farmers began to
organize, and the modern prairie farmers' movement was born.

These agrarian protestors were not from the poorest sectors of the
farming population. The bulk of the support for the Canadian farmers'
movement—and certainly most of the leadership—came from what
James M. McCrorie calls the "agrarian middle class—those operating
medium-size farm units."[15] These people had invested substantial capital

in their farming operations but found their aspirations for upward mobility frustrated. The agrarian organizations identified two principal culprits responsible for the plight of prairie agriculture. One was the group of Eastern industrialists and politicians who kept the hated customs tariff so high. The other was the private grain trade.

Prairie agrarian mobilization coincided with a period of stagnant or declining wheat prices (see Table A.2). Middlemen and especially the Canadian Pacific Railway were the targets of this agrarian movement. In December 1901 a group of angry farmers at Indian Head, Saskatchewan, formed the Territorial Grain Growers' Association (TGGA) and the new group quickly launched a successful lawsuit against the Canadian Pacific. This suit resulted, in 1902 and 1903, in amendments to the Manitoba Grain Act designed to force the company to distribute freight cars equitably. Farmers in Manitoba and Alberta watched these events with intense interest, and the organizational movement spread rapidly. In 1902 the Manitoba Grain Growers' Association (MGGA) formed, and in 1905, the year that Alberta and Saskatchewan became provinces, the Alberta Farmers' Association was born. The following year the TGGA changed its name to the Saskatchewan Grain Growers' Association (SGGA), and in 1909 the Alberta Farmers' Organization combined with another Alberta group, the American Society of Equity, to form the United Farmers of Alberta (UFA).[16] The organizational structure of the prairie farmers' movement was now complete.

A spirit of missionary zeal pervaded the grain growers' associations in these early years. Ambitious membership campaigns organized locals in practically every farm district in the prairies. The fervor of these early organizers was remarkable. One MGGA organizer named McCuish reported in 1915 that at Riverton, on the west shore of Lake Winnipeg, "I engaged an Indian to take me over land to Hodgson with a dog team—a distance of 42 miles. . . . The last 20 we had no snow, so were obliged to walk to Hodgson." There he signed up fourteen new members. The broad appeal of the movement reflected its many functions; the grain growers' association was, as one enthusiast put it, "the twentieth-century church of the prairie, a Y.M.C.A., literary society, debating club, mutual improvement society, and board of trade all rolled up in one." Educational work was a strong priority; as W. L. Morton has emphasized, the "greatest work" of the MGGA was persuading farmers "to be thoughtful and active citizens."[17] The associations encouraged farm women to join, and they responded enthusiastically. Indeed, women became one of the main strengths of the grain growers' movement and were especially effective in promoting rural education.[18]

By the end of 1909 the three associations had over 75,000 members and had become a potent political force that no prairie government could afford to neglect.[19] They moved into the federal political arena too. In 1910 the three prairie associations, along with the Dominion Grange (the United Farmers of Ontario after 1914), formed the Canadian Council of Agriculture (CCA). Its role was to coordinate the work of the provincial associations and to serve as lobbyist at Ottawa.[20] The CCA lost little time in making its presence felt in Canada's capital. In mid-December 1910 it organized a delegation of 800 farmers to meet with parliamentary leaders to press for the farmers' desires for reform. Although this event is sometimes called the "Siege of Ottawa," the atmosphere was one of restrained dignity. In the House of Commons, where Laurier himself presided over the meeting, the farmer-delegates were given the members' seats; Governor General Earl Grey and Lady Grey also entertained them at a formal reception. The farmers lost no opportunity to impress on their hosts the urgency of their requests, which included the reduction of the tariff, the building of Hudson's Bay Railway, easier credit, and the public ownership of terminal elevators.[21]

The prairie associations also entered journalism. They adopted a newspaper, the *Grain Growers' Guide* (founded in 1908) as their "joint official organ." This extremely influential paper, which had a circulation of 30,000 by 1914, 70,000 by 1921, and 120,000 by 1926, was, as one British writer put it, "the most potent political force in the West," with the possible exception of the *Winnipeg Free Press*. As the voice of the organized farmers' movements, the *Guide* strongly promoted the cooperative movement and "created a uniformity of opinion and a solidarity of purpose rare in farmers' associations in the United States."[22]

The first editor of the *Guide*, E. A. Partridge, was an extremely important figure in the prairie farm movement. He stayed with the *Guide* for only a short time, but in the meantime he had numerous other schemes afoot. Partridge's name became a household word in the prairies. To him, farmers "were participants in a gigantic class struggle," as one historian has put it.[23] A visionary and a rebel, Partridge was one of the original founders of the TGGA. He was a Ruskinian socialist who espoused the Social Gospel. Physically imposing and possessed of a vibrant personality, Partridge had lived in the West since 1883. He understood and appreciated ordinary prairie farmers. Like them, he knew the dangers and hazards of farm work—he farmed for many years at Sintaluta, Saskatchewan, where he lost a foot in a binder accident. And the farmers deeply respected Partridge. One letter writer to the *Guide* summed it up by saying that "no one in our association is more widely known, respected, and honored." The established political and economic elites, however, regarded him quite differently. To Premier Walter Scott of Saskatchewan, Partridge was a "blatant demagogue."[24]

Cooperative Capitalism in the Prairie Grain Trade

Partridge's major contribution to the prairie agrarian movement was to recognize that farmers could go into the grain-marketing business themselves. Although he was deeply suspicious of the motives of big Eastern capitalists, Partridge had a shrewd eye for business opportunities. He was convinced that the grain trade exploited the farmers, but he also realized that the capital required to enter the trade was small, that management problems were comparatively simple, and that the rapid expansion of the trade presented an excellent opportunity to launch a farmer-owned company. Partridge also realized that a farmer's company should not be run by the grain growers' associations—the political arm of the agrarian movement—but should be an entirely separate enterprise. After traveling to Winnipeg to make a personal investigation of the grain business in 1905, Partridge and a group of friends launched the Grain Growers' Grain Company (GGGC) in 1906 and bought a seat on the Winnipeg Grain Exchange.[25] The company was launched as a cooperative enterprise in which profits were to be distributed to all shareholders. Only farmers could buy shares (at $25 each), no one could buy more than four, and each shareholder had only one vote. But because its plan to pay dividends to its shareholders conflicted with the rules of the Winnipeg Exchange, the GGGC dropped the profit-distribution scheme.[26] Although the GGGC never was a true cooperative, its formation represented a most important step in the prairie agrarian movement.

The company was to become extremely successful, but its early years were surrounded with controversy. After only six weeks of membership, it was expelled by the Winnipeg Exchange, ostensibly because of the profit-distribution plan mentioned above. But there is evidence that some exchange members were less than satisfied with Partridge, who had called the exchange "a combine with a gambling hall thrown in." Under pressure from the Manitoba government, the exchange readmitted the GGGC, but the exchange evaded government attempts to force it to change its rules by dissolving and reorganizing as an "unincorporated voluntary association." The farmers' company realized it would have to play the game by the rules that the established grain trade set.[27]

But the GGGC played that game brilliantly. The company enjoyed excellent leadership under another farmer, Thomas A. Crerar, who was president between 1907 and 1930, a period in which he also served as minister of agriculture in one federal government (1917) and minister of railways and canals in another (1929–30).[28] Although Partridge had brought Crerar into the company, the two men later broke when Partridge discovered that the GGGC was speculating in wheat futures.

(Nevertheless, the company later gave Partridge a pension.) From 1907 to 1912 the number of shareholders increased from 1,853 to 27,321, the paid-up capital from $11,795 to $586,472, and the bushels of grain handled from 2,300,000 to 27,800,000. By 1913 it was the largest grain company in Canada.[29]

The next big step for the farmer-owned grain company came in 1917, when the GGGC absorbed a provincial elevator network in Alberta and changed its name to United Grain Growers Limited (UGG), its current name. The new company had 30,000 farmer-shareholders and owned or leased 343 country elevators in Manitoba and Alberta, as well as two terminal elevators on the lake front. By 1929 the UGG had 463 country elevators. It was a powerful and financially sound company, and it never would accept the panacea that radical farmers would adopt in the 1920's— that Canadian farmers should be compelled to sell to a wheat pool on the theory that this would raise prices. In these respects the UGG's stand was similar to that of the private elevator companies.[30]

After helping to launch the GGGC, Partridge had embarked on another scheme—government ownership of grain elevators. Although this plan foundered, the GGGC eventually was able to acquire a major network of elevators because of Partridge's plan, and these elevators became a profitable part of the company's business. Partridge, who believed that the farmer would continue to be at the mercy of the grain merchants as long as they owned the line and terminal elevators, unveiled a proposal in 1908. In what was known throughout the prairies as the "Partridge Plan," he suggested that each provincial government should own and operate a system of grain elevators both at country towns and at the lake terminals. Partridge assured the readers of the *Guide* that his plan would be "another nail in the coffin of the combine," and that "even your most fat-headed legislator must see the necessity of it now."[31]

The idea took firm root among prairie farmers. The Partridge Plan became a particularly hot political potato in Manitoba, where the MGGA threw its support behind the idea. The Conservative government of Rodmond P. Roblin adopted the plan in 1910 and passed the Manitoba Elevator Act that year. The results, however, were disastrous. As R. D. Colquette points out, "Here was a great chance to unload old or poorly built elevators on the government." The purchase price was too high, charges of corruption flew fast and furious, and the government elevators, which lacked the monopoly that Partridge had wanted, faced stiff competition from private elevators. The public elevators lost $110,000 in the first year of operation, and as a result, the Manitoba government quickly got out of the business and leased most of its elevators to the GGGC in 1912.[32] Roblin commented: "I took the voice of a demagogue for the voice of the public and consequently I made a mistake."[33]

Roblin was not quite correct. True, public ownership of line elevators was discredited, but throughout the prairies farmers deeply distrusted the private elevator companies. In Saskatchewan and Alberta governments learned from Manitoba's mistakes, rejected public ownership, and began to urge the establishment of farmer cooperative elevator companies. Under pressure from the SGGA, the provincial legislature in 1911 passed a bill to form the Saskatchewan Co-operative Elevator Company Limited, a joint-stock cooperative company that would distribute patronage profits and whose shares would be sold only to farmers (maximum ten shares each). The most interesting feature of the bill was that the Saskatchewan government loaned the new company 85 percent of the purchase price of its elevators and agreed to guarantee the company's credit. In 1913 the Alberta legislature passed a similar scheme, which also included the 85 percent advance.* The provincial governments took a first mortgage on the elevators they financed.[34] This solution, in which the government supplied funds to create a major cooperative enterprise was, in the words of Harald Patton, "without precedent in the history of government relations with agricultural producers."[35]

The Saskatchewan company was "tremendously successful" and ultimately ran the world's largest co-op elevator system. By 1920 the UGG and the Saskatchewan Co-op were the two largest companies operating on the Winnipeg Exchange. Together the two farmer-owned companies handled between 20 percent and 25 percent of prairie wheat, had elevators at 50 percent of all prairie shipping stations, and owned or leased 40 percent of all lake terminal capacity. In the 1917–22 period the Saskatchewan Co-op's 314 country elevators handled an average of 27,000,000 bushels of grain a year; the company also acquired a seat on the Winnipeg Exchange, where it marketed its wheat through its own commission department. This vertical-marketing system was further integrated in 1918, when the Saskatchewan Co-op opened a large terminal elevator at Port Arthur.[36] The company carefully maintained a policy of catering to the interests of the farmers who used its elevators. For example, it offered "special binning" facilities for farmers who preferred to ship their grain for grading at Winnipeg; most private elevators offered this service on a very limited basis. The Co-op also bought the farmers' "street wheat" (less-than-carload amounts) for two or three cents a bushel over the price of the private elevators' bids.[37]

The Saskatchewan Co-op's elevator company also made very handsome profits. Between 1917 and 1924 it paid its shareholders 8 percent dividends and an annual bonus of between $.50 and $4.50 a share. But farmers who were not shareholders did not receive patronage dividends,

---

*As noted above, the Alberta Co-operative Elevator Company amalgamated with the GGGC in 1917 to form the UGG.

and in this respect, like the UGG, the Saskatchewan Co-op was not a true cooperative. Instead, it plowed much of its profits into expanding its facilities. In the long run this policy alienated the company from the poorer farmers of Saskatchewan and caused political problems in the 1920's. As one farmer pointed out, "Inasmuch as most of the pioneer settlers are too poor to hold shares, it is doubtful if it [the co-op company] has helped them much, except as a powerful and keen competitor with other firms."[38] Like its counterpart the UGG, the Saskatchewan Co-op was suspect in the eyes of poorer farmers as just another grain company—with the difference that it was owned by relatively wealthy farmers.

Although public ownership of country elevators did not take root, the organized farmers' movement continued to press for government ownership of the terminal elevators, the huge storage facilities at certain inland points and at the ports where the exportable surplus of Canada was concentrated. During the "Siege of Ottawa," the Canadian Council of Agriculture emphasized two points Partridge had made earlier—that the terminals mixed grain grades so that the average export quality was inferior to what many farmers actually sold (dropping the price), and that the companies cheated on weight by not cleaning grains sufficiently. Farm leaders were convinced that these practices took place, and because the grain in store at the terminals basically commanded the prices that were set for country elevators, it seemed to them particularly important to control abuses by the terminal companies.[39]

The organized political pressure that the farmers were exerting began to show important results by 1912. Anxious to act to assuage rural discontent and to avoid creating the impression that the high-tariff Conservative Party was neglecting the prairies, the new Borden government in 1912 passed the Canada Grain Act, "the most thoroughgoing system of national grain trade regulation to be found in any country," in Patton's view.[40] This legislation, which incorporated the 1900 Grain Act and subsequent amendments, created a new Board of Grain Commissioners and charged it with regulating the terminal elevators according to a strict series of inspection, grading, weighing, and mixing rules. The mixing of grain was prohibited in public elevators, although private terminal elevators could still do it. The 1912 act also authorized the government to acquire terminal elevators at the lake ports and charged the board with operating them. The following year, the Borden government built a big (3,250,000-bushel) terminal elevator at Port Arthur. By 1930 six government-owned terminals were in operation in Canada.[41]

The Canadian grading system quickly became respected worldwide, to the point where the London Corn Trade Association (the major organization of world grain merchants) purchased Canadian grain sight unseen and on the basis of the inspection certificate alone. Argentina developed no grading system that was remotely comparable to Canada's, a circumstance that cost pampa farmers dearly. The United States was the only other major wheat exporter to standardize and grade its wheat, but the Canadian system was superior to the U.S. system.[42] Most prairie farmers hardly would have agreed with Duncan MacGibbon that the 1912 act was the "Magna Carta of the grain grower," since it did not attempt to fix prices, but there is no question that this federal legislation greatly improved the overall efficiency of Canadian wheat marketing.[43] The new grading and standardization system made Canadian wheat the standard of the world and opened new markets among quality-conscious buyers. Thus, the Canadian government had shrewdly devised a grain-marketing reform that benefited the wheat farmers while it aided the entire Canadian economy.

Indeed, the Canadian farmers had achieved some notable gains by the time the First World War broke out. Drawing on their cooperative legacy, as well as the Canadian tradition of agrarian organization and political protest, prairie farmers had forced not only the federal government, but the governments of all three prairie provinces, to enact major reforms in the marketing system. But impressive as the Canadian prewar agrarian legislation was, the fact remained that the Canadian farmers had been unsuccessful in combating the despised high tariff policy. And many prairie farmers, especially the smaller and poorer ones, received few if any benefits from the new marketing schemes. The prairie agrarian population, in other words, still seethed with restlessness, and as we shall see, when a severe cost-price squeeze hit after the war, the movement continued to grow—and to make more radical demands for reform.

Grain Marketing in the Pampas

Despite the bitter perception of social and economic injustice that rankled prairie farmers so strongly, the Western Canadian agrarian movement remained nonviolent. The political system was sufficiently open and democratic to allow farmers a reasonable hope of gaining redress through traditional political methods. Such was not the case in Argentina, where most farmers lacked the vote, where the provincial governments were weak and had little policy-making autonomy, and where the federal government, despite a system of formal political democracy between 1916 and 1930, remained beholden to the elite group of large landholders that had traditionally dominated Argentine politics. The fact that most of the pampa farmers did not own land and tended to move about frequently certainly constrained the political potential of agrarian unrest. For this reason, agrarian organization in Argentina proceeded along very different lines from the form it took in Canada. Although Argentine farmers pressed governments to grant reforms, they also employed direct-action techniques—especially the rent strike—which lent an atmosphere of conflict and even violence to the pampa agrarian movement. The use of the strike points out the difference in goals between Argentine and Canadian farmers: whereas the prairie farmers primarily aimed to control or dominate middlemen, the pampa tenants confronted not only intermediaries, but also landlords. Like their Canadian counterparts, Argentine farmers were small businessmen who needed to cut their costs of production when commodity prices slumped. The landowning Canadian farmers could do little to cut the cost of land, but the pampa farmers could and did attempt to force landowners to lower rents. This attack on the Argentine elite gave the pampa agrarian movement a radical image, but in fact, like prairie farmers, the pampa farmers wanted a larger share of the rewards of the capitalist system, not socialism.

The marketing system the Argentine farmers faced was far more primitive than that of Canada. Even before the major cooperative reforms of the early part of the century, the prairies had a system of grain elevators offering at least a minimum of storage facilities. But the tall, stately silhouette of the elevator was absent from the pampas. Grain storage facilities hardly existed in Argentina, and pampa farmers had to ship their wheat to market in clumsy and expensive jute bags.

The Argentine marketing system—or more correctly, the lack of system—made the nation an "unsupervised playground" for three or four private grain companies, which dominated the nation's cereal exports. The individual pampa farmer stood at the bottom of a financial pyramid over which he had very little if any control. It was, as one author puts it, "one of the most abject commercial relationships possible."[44] Despite all the abuses that Canadian farmers laid at the doors of the grain trade, the fact remains that, in Canada, wheat deposited in the country grain elevator became a negotiable instrument. The elevator company would issue the farmer a warrant—a certificate of deposit—against which he could borrow money anywhere. But in Argentina, which had almost no grain elevators or other rural storage facilities, wheat was not a commercial security. The federal government had passed a warrant law in 1914, but in the absence of storage facilities, it remained a dead letter. With almost no exceptions before the 1920's—and with few after that—the farmer had no alternative but to sell grain to local merchants called *acopiadores*. These people were usually also the only source of credit. What was worse was that *acopiadores* were often employees of local landowners, who made it a condition of their rental contracts that the tenant buy and sell only through these agents.[45]

The lack of rural credit in Argentina was notorious. As mentioned, farmers could not borrow against their wheat. Moreover, the banks would not make loans to tenant farmers without land—and the Banco de la Nación, which dominated Argentine finances and controlled 80 percent of Argentine banking transactions, operated on a very strict commercial basis. Its loans at best had a 180-day limit, which made them useless for all but the wealthiest farmers who also were landowners.[46] Year after year, as early as 1911, reformers in Congress proposed the creation of government-backed rural credit programs, but the farmers had no political power, the landed and merchant elite saw no need for reform, and the credit system remained unchanged. Among the provincial governments, only Entre Ríos (in 1909) had legislated a rural credit system—and its scope was narrow. Aside from the unworkable warrant law, the only other legislation that facilitated rural credit was a *prenda agraria* (rural pawn) law of 1914 that enabled farmers to borrow from banks against their tools and implements. The limitations and dangers of this procedure made it of little use in increasing the supply of rural credit.[47]

In this situation, the typical small farmer had no place to turn but the acopiador. These merchants would advance supplies (at interest) throughout the year and then receive the crop to liquidate the debt—and if it was a good crop year—would pay the farmer the balance in cash. In poor years farmers might receive no cash at all and have to carry a margin of unpaid debt into the next season. As a result, a kind of debt peonage appeared in the pampas. Complaints about the avarice of acopiadores, some of whom charged two to three times the market rate for goods advanced on credit, were numerous. As Joseph Tulchin shows, storekeepers who made loans to farmers charged 20–25 percent annual interest. These merchants obtained this capital at 12 percent from the big grain companies, who got it from the bank at 6 percent.[48] As even Emilio Coni, who pointed out how much of a risk these small-town merchants themselves ran in advancing credit for such a tricky business as farming, conceded, "It is certain that often the businessman makes his loans at usurious rates."[49]

The acopiadores served as the agents for the handful of big grain companies that controlled the Argentine grain trade. Three firms—Bunge y Born, Louis Dreyfus y Cía, and Luis de Ridder, exported over 80 percent of Argentine wheat in the 1930–31 crop year, and Bunge, the largest, shipped 43 percent of the total.[50] Although the three big export grain firms earned an average of only 2.2 percent on their total sales of Argentine wheat, their net earnings as a proportion of invested capital were high: at the close of the 1920's they ranged from 22.7 percent in 1927 to 12.0 percent in 1930. In 1922, a year of agrarian depression, Bunge y Born nonetheless earned 33 percent of its capital.[51] Bunge y Born possessed a far-flung business empire. It included not only the grain trade, but also flour and textile mills, as well as factories producing paint and grain bags.[52] Among pampa farmers, the company's image was that of an exploitative monopoly. It was widely known as "The Octopus."*

Not all farmers were in the grip of this chain of exploitation, to be sure. Tenants who enjoyed successive years of good crops when prices were high might avoid the need to borrow from local merchants. Tenants who bought land (and, as we have seen, their numbers were increasing), as well as the lucky descendants of Argentina's early immigrant colonists who bought land before the wheat boom began, could sometimes avoid acopiadores altogether and deal directly with the grain companies. This was especially true of the larger farmers. But even prosperous farmers suffered dearly from Argentina's lack of grain elevators and from the antiquated system of transporting wheat in jute bags.

A symbol of the Argentine wheat trade was the circle of mountains of bags—some of them 50 or 60 feet high—surrounding a pampa railway station at harvest time. The lowly jute bag also symbolized the faulty organization of Argentine agriculture and the exploitation of the farmer by the major grain companies. The cost of bags—which James Scobie estimated at 4 percent of the cost of wheat production—was a serious burden on all pampa farmers. Indeed, the Canadian expert W. J. Jackman estimated that Argentina's expenditure on the 100,000,000 bags needed for each crop was "large enough to build quite a line of elevators each year." The use of these sacks instead of the bulk storage and transport of grain meant that Argentina had no official standards of grading wheat for export.[53] Many bags were imported from India; the rest were produced in Argentine factories from Indian jute and burlap (hessian cloth) imported by Bunge y Born and other merchants. This was big business—one British trade counselor wrote that a shift to grain elevators in Argentina would deal "a serious blow to the valuable Calcutta trade in hessians." In 1913–14 alone, Argentina imported £115,795 of gunny bags and £1,847,947 of gunny cloth.[54]

Farmers purchased their bags from the same acopiadores who bought their wheat and sold them their other supplies. The bags were loaded with threshed wheat, hauled to the station, loaded onto freight cars, hauled to the ports, and (since most of the Argentine grain was exported in bulk), emptied into the holds of ships. Not only was this system clumsy and costly; it also exposed the wheat to rain, the cause of a great deal of damage to the harvested crop. The Argentine government exercised no supervision of quality or standards. Indeed, without elevators,

---

*The company gained a different kind of notoriety in 1974, when leftist terrorists kidnapped Juan and Jorge Born, the director and general manager, respectively, and confined them to a "people's prison." They were not released until the firm paid a ransom of *sixty million dollars.*

such quality control was virtually impossible. The only "quality control" existed at railway stations, where representatives of the buyers thrust a pointed tube into each bag as peons carried them past the inspector from the pile to the freight cars. At the ports the companies carried on a rudimentary system of inspection to sort the grain into various categories, but no effective method existed for cleaning grain prior to export.[55]

When Argentine grain arrived in England, the London Corn Trade Association took samples from each cargo, to determine the "fair average quality" (f.a.q.) of the wheat. This rating served as the basis for payments for wheat arriving the following month. Within Argentina grain exchange officials determined f.a.q. twice each crop season; final settlement to the farmer made deductions for grain below f.a.q.—but paid no premiums for superior grade. Export firms were in a position to mix wheat—thus raising f.a.q. and penalizing farmers.[56]

The need for an adequate system of grain elevators to allow bulk handling, as well as to save time and labor, was widely recognized, but as so often happened in Argentina, no government took decisive action on this problem despite the numerous congressional proposals to build country elevators that had been made over the years. The Ministry of Agriculture in 1929 estimated that Argentina's clumsy and inefficient system of handling wheat cost the country at least 42,500,000 paper pesos a year, in addition to handling charges and loss of time in transport. Two million man-hours a year were spent loading bags onto freight cars; this amounted to 11 percent of the Argentine railway system's rolling-stock productivity. The British Chamber of Commerce in Argentina agreed that a system of grain elevators would yield substantial dividends in savings—10,000,000 pesos a year would be saved in losses and shrinkage alone. Despite all these obvious advantages, the first cooperative elevator in the country was not inaugurated until 1930, and that year there were only 23 grain elevators in the entire republic, most of them at seaports.[57]

As we have seen, unlike Canada, Argentina had no flourishing cooperative movement on the pampas that could challenge the government's inertia on the grain marketing problem by developing a co-op elevator network and marketing system. Everywhere commerce, which Domingo Bórea called "the mortal enemy of true cooperatives," fought against them. As late as 1925 there were only 92 rural co-ops in all Argentina with a total membership of 22,775.[58] Only in the province of Entre Ríos, where the Jewish colonists sponsored by the JCA established a successful co-op movement, could the system be said to be flourishing. Entre Ríos also was the province with the highest proportion of landholding farmers, and the only one in which the government gave effective assistance to the co-op movement. By 1935 40 percent of the province's farmers were members. Most of these cooperatives were limited to providing insurance and selling merchandise to their members, but a few of the larger ones also marketed grain.[59]

Before 1926 Argentina had no legislation covering cooperatives, a situation that severely impeded their growth. In that year, however, the Alvear government pushed two pieces of legislation through Congress that gave cooperatives legal standing, authorized the Banco de la Nación and the National Mortgage Bank to make loans to them, and exempted them from certain taxes. Co-ops, however, were forbidden to grant credit or make loans. The legislation enabled the cooperatives to begin building their own grain elevators, the first of which went into operation at Leones, Córdoba, in 1930. By 1937 cooperatives marketed 197,000 tons of wheat and 519,000 tons of grain of all types.[60]

Almost all these cooperatives, however, existed in areas where the farmers owned their land. Indeed, all observers of the Argentine rural co-op movement agreed that the principal cause of its stunted state was the land-tenure system. Tenants and sharecroppers on short-term leases were difficult to organize and reluctant to join any organization that required a long-term financial commitment. Emilio Coni summed up the movement's dilemma when he wondered: "How can cooperation be successful among wandering farmers, who move from place to place every year or so, always poor, always migratory?"[61]

Strikes and Violence in the Pampas

Despite all the problems that the marketing system created for Argentine agriculture, immigrants poured into the pampas to become tenant farmers during the first decade of the century. Crop prices were rising steadily, and farmers gambled on high prices, favorable weather, and the absence of grasshoppers. Agriculture was, as one careful observer pointed out, "a game of chance."[62] Success stories of immigrant farmers who became rich, like Juan Fuentes, a Spaniard who had arrived with two pesetas in his pocket and who by 1912 had a fortune of 3,500,000 pesos, promoted the notion that farming was one road to wealth for the poor immigrant. Indeed, tenant farming required only a modest investment—about 1,200 paper pesos for animals and basic machinery—and given ideal conditions and high crop prices, a tenant farmer could net 400 to 500 pesos annually from 80 hectares.[63] Farmers typically gambled by renting as much land as possible, and during the first decades of the twentieth century, what Carl Taylor called "a frenzy of speculative wheat farming" was the result. Many of these agrarian wildcatters eventually hoped to buy land, but in the meantime their competition for land steadily drove rents up. (Other growers faced the same situation. In southern

Santa Fe the maize farmers' rents increased from 18–20 percent of the crop in 1904 to 35 percent in 1912.)[64]

But this speculative boom began to end by 1911. Several consecutive years of high and rising crop prices—accompanied by rapidly rising rents—ended. Wheat prices in 1908 were 49 percent higher than the 1902 prices, but by 1913 they had dropped by 13 percent.* Linseed oil rose 132 percent between 1903 and 1911; it then fell 38 percent by 1913.[65] But the worst crisis came in maize. First the weather dealt a harsh blow; in 1911 drought ruined the crop, which fell to 713,000 tons from 4,500,000 the previous year. In 1912 the harvest was superb, reaching 7,500,000 tons, but this quantity was so great that it depressed the world market price, with the result that farmers were unable to recover from the disaster of 1911.[66]

Two successive years of crisis for maize farmers, along with the slumping prices of flaxseed and wheat, brought distress to Argentina's tenants, who were faced with land rents that had nearly doubled in eight years. The feverish expansion of pampa agriculture also bid up the price of harvest labor, with wages rising 30 percent between 1908 and 1912. And to cap things off, the price of grain bags rose in 1911, when the government began to levy an import tariff on both raw jute and jute bags.[67] Things were made worse by a rapid shriveling of agrarian credit, reflecting the hesitancy of foreign investors to continue financing Argentine development at a time when the Balkan War threatened to become a general European conflict. The credit shortage bankrupted numerous rural merchants and added to the general agrarian crisis.[68]

This cost-price squeeze struck hardest at the rich agricultural zone in southern Santa Fe and northern Buenos Aires provinces, an extremely fertile maize region that also grew much flax and some wheat. Farmland here was among the most expensive in Argentina. The Spanish and Italian immigrants who composed most of the farming population had been willing to pay the high rents as long as crop prices were high, but by 1912, when economic disaster loomed, they asked for rent reductions. Almost all landlords refused, and a wave of protest against the land-rental system suddenly erupted. On June 25, 1912, the Argentine organized farm movement was born in Alcorta, in Santa Fe province, where 2,000 farmers gathered to express their outrage and to organize a protest. Many of these farmers had taken part in European peasant movements, and this militant heritage became clear when they pledged to strike until their demands were met.[69] Striking meant not only a refusal to pay rents, but also a refusal to work, a most serious matter in a country that lived from its agricultural exports.

*See Table A.2 for a full breakdown of wheat price movements.

This was not the first time that Argentine farmers had resorted to the strike to press for reform: Santa Fe farmers had mounted a short-lived strike as early as 1893,[70] and strikes had taken place sporadically in the following years. But the 1912 strike stands out because of its size and duration, and because it marks the beginning of the modern Argentine farmers' movement.

As an emergency response to a situation that pampa farmers considered intolerable, the 1912 strike focused on the land-tenure system. The farmers also deeply resented the grain-marketing system, but there was little they could do to change it until they gained the right to longer leases. Only then could the wheat and maize producers enjoy the stability of tenure that would enable them to challenge the established grain trade.

Although there were socialists and anarchists among the strikers of 1912, no coherent ideological justification of the strike emerged. There was no social gospel of the type that was providing Canadian farmers with an ideological justification to struggle for agrarian reform. Nonetheless, churchmen—in this case Catholic priests—did play an important role in mobilizing discontent on the pampas. Indeed, José Netri, the parish priest at Alcorta, was one of the organizers of the 1912 protest. He opened his church for meetings of the distressed tenants, and as the strike proceeded he played an active mediating role. José Netri, moreover, recruited his brother Pascual, who was also a priest with a parish in a nearby town, to the movement. Several other local clerics took an active part in the agrarian movement in 1912 and 1913. During Holy Week of 1913, as *La Prensa* worriedly noted, at least three Italian priests were preaching that the farmers' cause was a "sacred crusade," that the farmers should prepare for armed resistance, and that if Argentina did not treat them justly, they should leave the country and migrate to Tripolitania. At least two priests, including Pascual Netri, were jailed for their outspoken denunciations of local authorities, who viewed the farmers' movement as seditious.[71]

But unquestionably the most important contribution that the town priest of Alcorta made to the farmers' movement was to persuade still another brother, Francisco, a Rosario lawyer, to come to the assistance of the southern Santa Fe farmers. This was a critical event in the history of the Argentine agrarian movement. Francisco Netri provided strong and decisive leadership, guiding the movement through its turbulent early years. The organization he helped to found, the Federación Agraria Argentina (FAA), became the largest and most influential farmers' organization in the country.

A graduate of the University of Naples and a specialist in rural law, Francisco Netri had followed his family to Argentina in 1897. He settled in Rosario, where he not only practiced law, but taught at a government secondary school and wrote for various newspapers. When his brother's call came in 1912, Netri, who was already familiar with the rural situation, became convinced that Argentine farmers must organize to obtain meaningful reform. He called a meeting in Rosario on August 1, while the strike was in effect, and over 700 farmer-delegates attended. In his speeches to these delegates, Netri emphasized the need for a national agrarian organization and for nonviolent tactics. On August 15 this convention formed the FAA. Antonio Noguera, a farm leader from Pergamino, Buenos Aires, who was also a member of the Socialist Party, became president. Netri's initial title was legal adviser.[72]

Netri at once turned his energies to publicity. In September he began to issue an information bulletin; in June 1913 it became the weekly *La Tierra*, the largest and most influential farmers' newspaper in the country. At first the paper grew slowly—there were only 520 subscribers in 1916—but by the late 1920's, the subscription list had reached 15,000–20,000. *La Tierra* was much smaller than its Canadian counterpart, the *Grain Growers' Guide*, but it was a fighting newspaper that reported agrarian issues thoroughly.[73]

While Netri was launching *La Tierra*, he was also locked in a struggle with Noguera over the policies of the FAA. Noguera, a colorful character who emblazoned his carriage with a sign, "Whether One Sows Ideas or Sows Potatoes, One Contributes to the Fatherland's Grandeur," wanted to align the FAA with the Socialist Party.[74] He and his fellow Socialists had been quick to sense the opportunity that the agrarian strike presented to the party, which polled few votes outside Buenos Aires and other large coastal cities. Here was a chance, they believed, to expand into rural Argentina. The party's leader, Juan B. Justo, traveled to Santa Fe soon after the strike began and spoke at various farmers' rallies. Back in the capital, Justo defended the tenants' cause from his seat in the Chamber of Deputies. But Netri considered party affiliation a serious error; he was convinced that the organization should remain a nonpolitical pressure group. By November 1912 his position had become FAA policy, and Noguera resigned from the organization. The Socialist press now turned its fire on Netri, whom it attacked as "authoritarian and personalist." Although Netri did not become the new FAA president, he held the title of director general, which in fact made him the FAA's chief executive, and he continued to edit *La Tierra*. A marvelous orator and journalist, Netri was constantly a center of controversy. An assassination attempt was made on him as early as October 1912.[75]

While Netri was occupied in organizing the FAA, the agrarian strike was spreading throughout much of the Argentine cereal belt, through central and southern Santa Fe, into neighboring Córdoba and Buenos Aires, and finally, by August, into La Pampa Territory. Tens of thousands of farmers took part, and in some areas, the strike was nearly total. Although it began among maize farmers, the strike soon spread to wheat growers as well. This was particularly true in La Pampa Territory, where farmers struck just at harvest time, in December 1912 and January 1913.[76]

Everywhere the strikers' demands were similar: rent reductions, a three-year minimum on leases (rather than the one- or two-year leases then in use), compensation to the renter for improvements he made to the land (such as planting trees or erecting buildings), and the prohibition of contract clauses that required renters to buy supplies or to hire threshing crews only from the landowners' agents. For the most part, the strikers employed nonviolent tactics. They simply refused to plow, harvest, or pay rents, although in some locations they did cut fences and burn crops.[77]

Landowners at first refused to compromise with the strikers. They appealed to the provincial police to intervene to protect non-striking farmers, and they asked the government to invoke the "Social Defense Law" of 1910 to deport striking immigrants. It is unclear whether any immigrant farmers were deported as a result of the 1912 strike, but the police certainly did use the threat of deportation as a deterrent.[78]

The farmers' strike threatened the agricultural export trade—the very foundation of the Argentine economy—and thus soon became an issue of major national concern. By early August *La Prensa*, which had been demanding since late June that the government intervene to settle the strike, found the agrarian conflict "truly alarming."[79] In the capital city the Conservative elite condemned the strike as a subversive movement. The farmers, as the agronomist Roberto Campolieti put it, were viewed "as nothing less than anarchists." And the federal government of President Roque Sáenz Peña, whose principal power base was the landed elite, did little to clarify the complex roots of the agrarian movement. Instead, the minister of agriculture told the Chamber of Deputies that outside agitators were to blame for the crisis on the pampas.[80] Nonetheless, the Sáenz Peña government took no action.* In fact, Sáenz Peña hesitated because the strike placed his government in a difficult political dilemma. The president's main goal was to bring honest elections to Argentina. For years unrest and violence had wracked Argentina's cities as the anarchist and syndicalist unions had called massive strikes. Equally serious, from the government's viewpoint, was the rapid growth of political support for the new Radical Party (Unión Cívica Radical). The Radicals' wily

---

*The government claimed that it lacked the authority to intervene, but this was an empty claim, for the 1853 constitution gave the president broad powers to take action in the provinces when their governments failed to keep order. Under this procedure the president and/or Congress could oust the existing regime and install a provisional one, which was charged with arranging new elections. The party in power in Buenos Aires had often used this procedure to advance its political interests.

leader, Hipólito Yrigoyen, refused to let his supporters vote or hold office until the federal government guaranteed honest elections. Indeed, the Radicals had threatened to overthrow the government by force. Sáenz Peña's response was to isolate the anarcho-syndicalists from the basically moderate Radicals by enacting the electoral reform that Yrigoyen demanded. Congress passed the necessary legislation early in 1912, and a few months later the first honest election held under the new laws took place. It was held in Santa Fe province, and, not suprisingly, the Radical Party was victorious.

Thus, when the agrarian strike broke out in June in Santa Fe, the Radicals controlled the provincial government. If Sáenz Peña had then intervened to stop the strike, Yrigoyen and his party doubtless would have felt that the president was using the occasion to force the Radicals out of office, and they would have begun to conspire again. This turn of events would have ruined the president's entire political strategy. So the crafty Sáena Peña let the provincial government pave the way for the settlement of the conflict—and, not least, serve as target for the local landowners' criticism.[81]

The Santa Fe Radicals knew they had to move fast, for the strike threatened to damage the province's economy severely. Governor Manuel Menchaca acted quickly, appointing a committee of three Radical leaders to study the agrarian crisis and to make policy recommendations. The committee's report, issued July 21, 1912, openly favored the strikers. "The farmers are being squeezed like lemons," read this famous report. "They can give no more." It endorsed the principal agrarian demands, including rent reductions, three-year minimum contracts, and restrictions on intermediaries. The Santa Fe landed elite was appalled at this report, and the Rosario Sociedad Rural charged the committee with Socialist tendencies. Certainly the report gave the agrarian movement powerful support, and after it was issued, landowners throughout the province, later followed by those in other parts of the pampas, began to negotiate compromise contracts with their tenants. By mid-August the Santa Fe farmers had returned to work, although the strike dragged on in other provinces. It was not settled in La Pampa until January 1913.[82]

But tension in the countryside abated only temporarily. Starting in 1912, Argentina's dizzy economic boom came to an end, and soon the onset of the First World War brought a sudden, sharp economic depression to Argentina. Although the war eventually created a high level of demand for Argentine exports, the initial impact was a period of severe economic disorganization, including a major shipping shortage and the paralysis of foreign trade for weeks after July 1914. The republic's total exports fell 23 percent between 1913 and 1914 and then revived, though not dramatically.[83] While consumer prices moved ahead rapidly, crop prices remained depressed until the 1916–17 harvest. In this bleak economic situation the tenant farmers continued to be caught in the same kind of cost-price squeeze that had hit them so severely in 1912. What was worse, in the view of the agrarian population in 1913, was that numerous landlords failed to carry out the rent reductions they had promised the previous year.[84]

As a result, strikes broke out again in 1913, and they continued to disrupt agricultural production into the following year. The worst problem in 1913 was in La Pampa, which, as a territory, was directly under the rule of the federal government. Tenants who went on strike there found landlords in no mood to bargain. Instead, the territorial police evicted dozens of farm families whose rent was in arrears.[85] The next year agrarian anger over the unfulfilled promises of 1912 boiled over again in the Santa Fe–Córdoba cereal belt. The region was the center of the FAA's support, but whereas in 1912 the FAA leaders had found the provincial governments anxious to negotiate an end to the strike so that plowing for next year's crop could begin, they now encountered hostile provincial governments that were in no mood to compromise and were bent on destroying the fledgling farmers' organization.

The FAA was a vulnerable target at the time because it continued to suffer the internal divisions that had appeared during the Noguera-Netri leadership struggle. While the landed elite carried out a campaign in the press against the FAA and its most visible leader, Netri, the Socialist Party criticized the organization's apolitical stance, and the pro-Socialist faction within the FAA remained resentful. So in 1914, when the FAA called a new strike from the original base at Alcorta, the government of Santa Fe sensed an opportunity to destroy the organization. To understand why the Radical government took this position, one must recall that the party's voting base was among the urban middle class and the native-born workers. Its leadership had strong ties to the landed elite. But the Radicals had only slight political support among the farming population.

The 1914 strike lasted for 42 days, and was marked by police harassment and the importation of strikebreakers—who were not difficult to find because of Argentina's high unemployment during the depression of the early war years. Agrarian leaders found themselves manhandled and thrown in prison. Again, the government threatened to deport strikers. But in the end the provincial governments did not succeed in destroying the FAA.[86]

This time, rather than compromise with the strikers as in 1912, the landowners used the provincial police to expel farmers who were behind with their rents. The most notorious case involved none other than Victoriano de la Plaza, Sáenz Peña's vice-president, soon to succeed to the presidency on Sáenz Peña's death that same year. A large landowner in

Córdoba, de la Plaza told his tenants to sign a no-strike clause, and when FAA members on his estates refused, the vice-president's overseers had the police throw them off his lands and confiscate their crops. Netri, who had continued to edit *La Tierra* (while also defending FAA members during the repressive wave of 1914) turned his attention to this case and publicized it widely. As a result, the federal government fired him from the teaching post he had held for fifteen years at the Colegio Nacional de Rosario, a prestigious secondary-school.[87]

Despite the militant agrarian strikes of its early years, the FAA under Netri's leadership was essentially a reformist organization. Its basic objective remained land-tenure reform, but it also mounted publicity campaigns to urge the government to provide better rural credit, transport, and education. In 1913 the FAA entered the cooperative movement and began to promote mutual hail insurance among its members.[88]

Numerous landowners throughout the cereal belt, however, viewed the FAA not as a moderate reform group, but as a threat to the prevailing socioeconomic system. They regarded Francisco Netri with hatred, and in October 1916 the controversial leader was assassinated. Netri's role in the FAA had always been a peculiar one, for he was not a farmer. But his talents as lawyer, newspaper editor, and public speaker were so valuable that he had remained the dominant voice in the organization. Although he was popular among farmers, he had made many enemies, not only among landowners, but among rival factions within the FAA. The assassin was a twenty-one-year-old orderly whom the FAA had dismissed earlier in 1916 in order to reduce expenses, but throughout Argentina farmers speculated that Netri was the victim of a plot. Although the motives behind this murder were never cleared up, Netri became the FAA's martyr.[89]

Netri's assassination threw the FAA into chaos. The organization underwent another leadership struggle and chose a new president, Esteban Piacenza, who was to head the FAA until 1947. Born in Italy, Piacenza had migrated to Argentina with his family at the age of fourteen and settled with them in Córdoba province. The family prospered, and Piacenza eventually was able to buy his own farm. He joined the FAA, strongly supported Netri's apolitical stance, and by 1916 was one of the group's most prominent leaders. A talented conciliator and energetic organizer, Piacenza ended the internal disputes that had plagued the FAA and led it into a period of spectacular growth. He believed that agricultural unionism was not enough, and that the FAA must press for meaningful land reform. If the FAA was to expand its membership base, Piacenza believed, it would have to provide tangible benefits to Argentina's farmers, particularly in practical matters such as hail and accident insurance and cooperative organization.[90] This program led to a rapid increase in membership, from 800 in 1916 to 7,000 in 1920.[91] Netri's talents had been in journalism and oratory, not in organizational matters, and in any case the controversy that had consistently surrounded him had hindered the effective organization of the FAA. Although Piacenza would endorse militant tactics when he believed them necessary, he concentrated on expanding the membership, internal organization, and concerted programs of lobbying and political pressure to induce governments to grant reforms to Argentina's farmers.

Piacenza's insistence on an apolitical stance for the FAA earned him enemies among those who believed that some form of socialism was the salvation of Argentine agriculture. One of these socialist-minded critics was Pablo Paoli, who in 1935 criticized Piacenza's tastes in clothing as excessively bourgeois; another was F. García Ledesma, who called the FAA leader a "fascist."[92] Charges like this became common among leftist Argentine writers but were wide of the mark. What these critics failed to see was that most Argentine farmers, whether tenants or small owners, were aspiring rural capitalists who saw no benefits in socialism. Although the Argentine Socialist Party avoided endorsing any kind of communal land policy and consistently supported agrarian causes in Congress, the FAA would not reciprocate with support for it—or, indeed for any party.

Agrarian Unrest and the State in Argentina and Canada

Bitter discontent swept the prairies and the pampas during the 1900–1914 period. Plagued by a serious cost-price squeeze, the Canadian and Argentine farmers responded by organizing to defend their interests and to attack the elite groups they perceived as profiting unjustly from the sweat of the farmer's brow. In the prairies this rural unrest led to the rapid rise of agrarian political pressure groups and to farmer-owned grain-marketing companies. But this kind of sophisticated cooperative organization was impossible in Argentina, where the cooperative as yet had no legal standing, and where in any case the land-tenure system robbed the rural population of the stability essential to successful cooperatives. In the pampas the farmers found that their only effective weapon was to go on strike.

The Argentine and Canadian governments responded to the prewar agrarian crisis in very different ways. In Canada the national government—both when the Liberals were in power before 1911 and then under the Conservatives—attempted to conciliate the farmers (and to attract their votes) with a reform program that created a more modern and efficient grain-marketing system. The idea behind these reforms was that they would not only benefit farmers by increasing their net returns, but benefit the entire country by raising Canadian wheat exports. The Argentine response to the agrarian crisis that began in 1912 was not reform, though the need for it was abundantly clear, but in the first instance, no response at all. Because pampa farmers for the most part lacked the vote, they were unable to induce the government to enact even a modest land-tenure and marketing reform program. When they then turned to organizing and the use of the strike, the response became repression. Particularly after 1912 the provincial governments attempted to destroy the nascent farmers' movement by persecuting its members and terrorizing its leaders.

These same patterns of political response to agrarian problems—cautious reform in Canada, at best a hands-off policy in Argentina—were to continue in the next decades, with the result that Argentine agriculture stagnated while the Canadian wheat economy rose to world leadership.

8

Economic Crisis and Agrarian Militancy, 1917–1922

The political and economic shock waves set off by the First World War reverberated strongly in Argentina and Canada, two export-dependent countries on Europe's periphery. As part of the British Empire, Canada quickly put its economy on a wartime footing and entered the conflict at Britain's side. Canada sacrificed heavily: 65,000 of its soldiers died on the battlefields of Europe, and 173,000 others were wounded. Although many people in Argentina, particularly the landed elite, sympathized with Britain and France, President Yrigoyen (who was elected in 1916) shrewdly cultivated the image of Argentina as a solitary nation, above the battle, and the country remained neutral. But Argentina could not escape the economic impact of the war. In fact, for all practical purposes, the Allies were Argentina's only export market, and they eventually mobilized the republic's agrarian resources to serve their cause. When the war ended, a sharp economic depression began in both countries that endured until 1923 and wreaked havoc with prairie and pampa agriculture.

The war caused, at most, only fleeting prosperity for Argentine and Canadian farmers. Its legacy in both cases was rural discontent and the strengthening of agrarian reform movements. Farmers began to demand that the state—which had intervened in the economy to aid the Allies during the war—respond to the plight of a nearly bankrupt postwar agrarian sector. The Canadian farmers, drawing on their cooperative tradition and their heritage of political action, were more successful in this effort than their counterparts in Argentina.

This chapter examines the great upheaval that the war and postwar years brought to the Argentine and Canadian cereal economies. We will focus on government production and price policies and their impact on the farming population, as well as on the agrarian reform movements that emerged in response to the wartime economic crisis. The underlying theme of this chapter is that the prairie farmers, despite their economic distress, were able to influence government policy more decisively in their favor than the farmers of the pampas. In a comparative perspective, the war years demonstrated the political strength of the agrarian sector in Canada and its weakness in Argentina.

Allied Economic Power and the Wartime Wheat Trade

As the war began the Allied powers' reliance on imported wheat came into sharp relief. Because cheap and plentiful bread was essential to a successful war effort, wheat became the most important element of Allied food policy. The British government achieved one of its basic wartime goals—the avoidance of bread rationing. Overseas wheat was particularly important to Britain, which imported 78 percent of its total consumption in the immediate prewar years, but France and Italy also needed vast amounts of foreign wheat to survive. Prewar imports of wheat by the Allies had averaged 364,000,000 bushels yearly, but by 1917 the voracious requirements of the war had pushed their demand up to 600,000,000 bushels. The overseas suppliers were able to provide these huge amounts only once, in 1915–16, but they did succeed in sending enough wheat to prevent Allied starvation.[1] Although Lord Curzon later stated that the Allies "floated to victory on a sea of oil," he might well have included the critical role of imported wheat as a factor in victory.[2]

Despite these vast import requirements, the Allies at first were confident that they could continue to procure enough wheat through the private grain trade, just as they had done before the war. But the cataclysm that the war unloosed soon made reliance on the private sector impossible. Russia, which prior to the war had been the world's largest exporter, was cut off from Western Europe in 1914, and by 1917 Russian exports ceased entirely. North America then became the prime source of Allied wheat. Bumper crops in North America sufficed to supply the Allies in 1915, but by 1916 lower crops in the United States and Canada, along with an ever-increasing shipping shortage, brought a wheat crisis to Allied Europe. With unrestricted submarine warfare threatening to destroy more and more shipping, wheat prices soared. Britain and France faced starvation unless their governments could devise an effective wheat policy.[3]

Recognizing the dimensions of this crisis, the British government decisively abandoned laissez-faire in the food sector. On October 10, 1916, it closed the Liverpool Wheat Market and created the Royal Commission on Wheat Supplies, with sweeping powers to purchase, sell, and control wheat and flour supplies. A month later Britain, France, and Italy signed the Wheat Executive Agreement, which in effect ended private imports and made the British royal commission the sole purchasing agent of all three countries. In 1917 and 1918 the commission took over buying for the smaller Allied powers (Greece, Belgium, and Portugal), as well as for the neutrals. The royal commission, which opened purchasing offices in Winnipeg, Buenos Aires, and New York City, thus concentrated unprecedented powers of food procurement and supply in its hands, and it would have as great an impact on Argentina and Canada as it had in Europe.[4]

While the royal commission was still organizing, the Canadian grain trade was descending into chaos. By early 1917 the wheat futures market was in such disorder that the solvency of several of the Winnipeg Grain Exchange's member firms was in jeopardy. Unless the government acted quickly, the interests of Canadian farmers and consumers, to say nothing of the Allies, were threatened in a wave of speculation. Thus, in June 1917 the Borden government suspended the private grain trade and established a government grain-marketing monopoly, the Board of Grain Supervisors. The new board had extensive powers, including procuring and allocating grain, fixing prices, and making transportation arrangements. It is important to note that the government invited prominent prairie farm leaders, including Henry Wise Wood, president of the United Farmers of Alberta, and Thomas A. Crerar, president of the United Grain Growers, to sit on the board along with figures from the grain trade, the railway workers' unions, and the government. The farmers, of course, wanted to set the 1917–18 crop price as high as possible, suggesting at least $2.40 a bushel, but that figure proved unacceptably high to the United States, which believed it essential to set the same price on both sides of the border. The board finally settled on a price of $2.21 a bushel for No. 1 Northern wheat. Although not as much as the farmers wanted, that price was far more than the $1.30 the government had suggested (and farmers had indignantly rejected) in March 1917, when discussions about a set price began. During the second year of price controls, 1918–19, the board raised the price a little, to $2.24½.[5] The Canadian wartime grain-marketing system worked reasonably well. The grain board effectively delivered the export crop to the Allies while protecting the interests of Canadian producers and consumers.

But this kind of conciliation of interests did not occur in Argentina. The Allies and the Argentine government did not bargain from positions of mutual trust. Argentina's position as a neutral power and its refusal to break diplomatic relations with Germany kept relations between London and Buenos Aires cool. In the early years of the war, Britain tried to induce the recalcitrant Argentines to end their neutrality by procuring most of its wheat from the United States and Canada. The British carried out this policy of economic pressure in the name of saving on shipping costs, but that argument fell apart in 1916–17, when Britain began to import wheat from Australia and India while continuing to avoid Argentine wheat. Another factor that caused deep friction between London and Buenos Aires was the predominance of German firms or firms with strong German connections in the Argentine cereal trade. London introduced a Black List policy that severely restricted the participation of German companies in British grain imports. And once Britain did begin to import Argentine wheat heavily, London allowed only its Royal Commission on Wheat Supplies to procure grain in Argentina.[6]

The situation was different in the beef trade, for Britain had no alternative sources of supply for chilled beef other than Australia, whose production was far smaller than Argentina's. Argentine beef exports to the United Kingdom boomed. But while the pampa cattle industry prospered, Argentine wheat agriculture suffered from the British market restrictions of the early wartime years. Farmers could not shift profitably from wheat to maize, for Argentine maize was not in demand in wartime Europe. Maize exports fell in both quantity and value, and by 1916 maize overflowed everywhere in Argentina. When coal and petroleum shortages appeared, power plants and railways burned the cobs for fuel.[7] With maize dead and wheat sluggish, 1915 and 1916 were years of agrarian depression in Argentina. Everywhere landowners were shifting from cereals to cattle, a trend that made economic sense for the elite but forced thousands of farmers off the land and caused widespread discontent.

Finally, the Argentines capitulated to the British to get the wheat crop moving. In June 1916 Buenos Aires offered to let the British government supervise the grain trade and pay for the grain after the war. But the negotiations that began over this initiative broke down after October 1916, when Hipólito Yrigoyen, the newly elected president, insisted that Britain should pay cash. Because world wheat supplies were still plentiful in 1916, the British responded to Yrigoyen by refusing to import pampa wheat and by intensifying purchases in Canada and Australia.[8]

By 1917, however, world wheat supplies began to tighten, and the Allies grew anxious about the availability of Argentine cereal supplies as well as about Argentine exports to neutral countries like Spain—exports that might reach Germany. Late in the year, when the North American crop proved much smaller than expected and starvation loomed for Europe, Argentine wheat supplies suddenly assumed critical importance. President Yrigoyen, however, still refused to cooperate with the Allies and, in fact, temporarily banned wheat exports in the name of protecting domestic consumers. The Allies now unveiled their muscle and forced Argentina to do their bidding. Herbert Hoover, President Wilson's food commissioner, declared that it was "absolutely imperative" to purchase Argentine grain "at the earliest possible moment" and "at reasonable prices," and approved the tactic of banning U.S. coal and farm machinery exports to Argentina. The British already had suspended coal exports, so U.S. coal was vital to Argentina; without it the economy would have ground nearly to a halt. Another lever the Allies held was hundreds of millions of dollars in Argentine loans whose repayment fell due in 1917, loans that the financially pressed Yrigoyen government wanted to renew. Hoover made it clear that to avoid drastic economic consequences, Argentina would have to reserve its entire grain surplus for the Allies.[9]

With this object in view, the Allies opened discussions with the Yrigoyen government in late 1917. Argentina's bargaining position was poor. Given Allied intransigence, Argentina had no alternative market for its cereals, and no other source of the coal it needed to avoid economic collapse. In the end the Yrigoyen government agreed to loan 200,000,000 gold pesos (£40,000,000) at 5 percent annual interest to Britain and France in return for their agreement to purchase at least 2,500,000 tons of cereals and to provide Argentina with an unspecified amount of coal. Yrigoyen reluctantly agreed to a minimum price, set by the royal commission, of 12.45 paper pesos per 100 kg of wheat.* He refused, however, to suspend sales to neutral countries. Pampa farmers were appalled at these agreements and charged that the government fixed wheat prices much too low, but the Argentine Congress approved the agreement on January 18, 1918.[10]

The 1918 agreement effectively meant that the royal commission's minimum price "commanded the market," as one of the commission's agents later stated. Argentina had no alternative but to accept the price because the Allies effectively forced neutral countries like Spain and Mexico to deal with the commission if they wanted Argentine wheat. In June 1918 the Allies even went so far as to intercept cable traffic between Argentina and Spain at the Rio de Janeiro relay station. The royal commission was determined to prevent Argentine wheat from moving to Germany via neutral Spain.[11]

Through the 1918 agreement the Allies were able to import, at fixed prices, about 2,500,000 tons of Argentine wheat, 1,000,000 tons of maize, and smaller quantities of oats, linseed oil, and flour. This food filled a crucial gap in the food requirements of Europe, which in 1918 verged on starvation.[12] But the agreement left a great deal of bitterness in Argentina, particularly among an agrarian population that believed it had not received an adequate price. The extent of this bitterness became clear in 1919, when Yrigoyen asked Congress to approve another 200,000,000-peso loan, this time without a minimum price, which the British refused to accept. The Argentine government proposed to finance cereal purchases by printing new paper money. Yrigoyen's request scan-

---

*This price was equivalent to U.S. $1.47 a bushel, or about 80 cents less than North American farmers received.

dalized agrarian opinion. *La Tierra*, the FAA's newspaper, pointed out that Argentine cereals would have no problem selling in the tight 1919 market. Numerous critics charged that the farmers would gain nothing through the second agreement, and that its inflationary effects would damage the entire public. Although Yrigoyen attempted to defend the project on the moral grounds of assisting the Allied powers, Argentine public opinion was outraged and the plan never passed Congress.[13]

As *La Tierra* had predicted, demand and prices for Argentine wheat grew rapidly in 1919 and on into 1920. The royal commission continued to buy heavily in Argentina, and in July 1919 wheat futures rose sharply. The cereal export trade boomed to the extent that by early 1920 a wave of price speculation enveloped Argentine domestic food supplies. To protect consumers the Yrigoyen government in June 1920 placed a heavy export tax on wheat and flour, and then in August banned exports altogether. Pampa farmers, however, received few financial rewards from the speculative bubble of 1920. It was the grain merchants and speculators who benefited.[14]

Because the Canadian government acted decisively in 1919, prairie agriculture escaped the wave of speculation that plagued Argentina at the end of the war. When the legal authority of the Board of Grain supervisors was ending in the summer of 1919, private grain traders, claiming heavy pecuniary losses, urged the government to lift all marketing controls. The Canadian Council of Agriculture sharply objected, and pointed out that because the Allies still controlled buying, an open market would create rampant speculation. Nonetheless, the Borden government reopened the private markets on July 21, 1919. The result was disastrous, for a speculative frenzy surrounded the market, and wheat futures rose to such dizzy heights that experts predicted the crop would not move that fall. This took place exactly at the same time that the speculative frenzy began in Argentine wheat. Speculation, in other words, might well have prevented prairie farmers from selling the 1919–20 crop, which in turn would have thrown the Canadian banking and railway systems into crisis. After only one week, the government closed the markets again and established a one-year compulsory wheat board, which unlike the wartime board was organized as a pool.

The government adopted this plan after consulting closely with the CCA. All farmers had to sell to the board, which made an initial payment of $2.15 a bushel to the farmers, sold the wheat to the best advantage, and then returned the additional proceeds to the farmers at the end of the crop year. The eventual total payment was $2.63. This system was not without its hitches. Many farmers mistrusted the board's "participation certificates" and sold them to speculators; others felt that they could have done better in the open market. But on balance the board protected the interests of not only the farmers, but the prairie economy in general. Nonetheless, the Borden government was philosophically opposed to marketing controls in peacetime, and in August 1920 it discontinued the board. The private grain trade began to function again, a move that led to some sharp criticism from the agrarian sector. Wheat prices fell sharply, and only two months later the CCA urged the government to re-establish the Wheat Board. The Borden government refused, but the pooling system of 1919–20 became the marketing model that prairie farmers would follow during the 1920's.[15]

The First World War was a unique period in the history of the modern grain trade. For the first time the free market in cereals disappeared. The Allied powers used their economic power and their command of the seas to create a monopsony—monopoly buying power—in the world wheat trade. Farmers in the exporting countries had to accept whatever price the Allies would pay, for exporters had nowhere else to sell. But within this monopsonistic market the farmers of Canada did considerably better than their counterparts in Argentina. Prairie farmers used their political strength to lobby for a high minimum wheat price, and the Canadian government, which was making huge contributions of men and money to the war, was in a relatively favorable position to bargain with the Allied powers. First the Allies refused to buy Argentine wheat, and when they did begin to buy it heavily, they offered a price much lower than Canadian farmers received. In other words, to a much greater extent than in Canada, Argentina's farmers were prevented from enjoying the high side of the price cycle.

The Plight of Pampa Agriculture During the War

"The War has not established once and for all the prosperity of our farmer; it has only saved him from going further back. It has brought no permanent result; it has increased our prices but not our profits."[16] As this comment by an experienced observer of the Argentine agrarian scene suggests, Argentine farmers indeed faced continuing economic uncertainty during the war. Although wheat prices were higher than usual, the cost of such vital factors of production as rents, jute bags, harvest labor, and freight transport made farming more economically risky than ever. For the most part the Yrigoyen government maintained a hands-off attitude on agricultural problems, and as a result a new wave of agrarian strikes broke out in 1919. Unlike the movement of 1912–13, this one saw the eruption of widespread violence on the pampas.

Land rents in the pampa provinces had fallen rapidly between 1912, when Argentine agriculture entered a period of depression, and 1916. But under the impact of higher wartime prices and of a major boom in cattle raising, rents then began to move up rapidly, and by 1919 cash rents in Buenos Aires and Santa Fe provinces had reached the highest levels in history. In Buenos Aires the average cash rent jumped from 12 pesos a hectare to 25 pesos between 1916 and 1919; in Santa Fe the increase was even sharper, from 10 pesos to 25 pesos.[17] The labor market also tightened during the war, forcing up rural wages. Immigration into Argentina had ceased in 1914, and the steady flow of golondrinas—Italian seasonal harvest laborers—could no longer be counted on. In fact, there was a net emigration from Argentina during the war. Although heavy unemployment plagued the cities, the government made no systematic effort to recruit farmworkers. Rural wages thus began to rise, especially in 1918 and 1919, and as we shall see, when farmers attempted to resist this additional cost of production, a bitter strike resulted.[18]

The primitive agricultural marketing system and especially that eternal bugaboo of Argentine grain farmers, the jute bag, added to the pampa farmers' financial distress. The cost of bags was becoming exorbitant, with a price jump from 35 centavos each during the 1916–17 harvest to 1.2 pesos the following year. Farmers were irate, and the government attempted to placate them by turning the problem over to the Royal Commission on Wheat Supplies, which imported the bags and sold them to farmers (at 75 centavos each).[19] But the royal commission was not as adept at managing the bag business as it was at grain buying. The bag-distribution system was unwieldy, and farmers had to pay cash in advance, a provision that forced poorer farmers to deal with speculators. By late 1918 the shortage of bags became so serious that some crops were rotting on the ground. Farmers were furious, and the lowly jute bag became a symbol of the government's organizational weakness in the agricultural sector.[20]

In 1919 the Yrigoyen government only made things worse when it severed the agreement with the royal commission and went into the bag business itself. The government spent 25,000,000 pesos to buy bags, which it planned to sell to farmers at cost plus 10 percent, but the bags arrived too late for the 1919–20 harvest, and they turned out to be 25 percent smaller than ordinary bags. The upshot of all this was that the government had 29,000,000 unsold grain bags, and the whole expensive affair had done nothing to ease the farmers' problem of getting the crop to market.[21]

A third economic burden that the agrarian population of the pampas had to bear during the war was a new tax on exports. When this type of tax is placed on commodities whose price is set in the world market, it is always the producer who is hardest hit, for he is unable to raise his price to meet the added tax burden. Export taxes, consequently, have aroused nothing but contempt among Argentine cattlemen and farmers, who have heard their government, since the time of the First World War, argue that the nation's perennial fiscal crisis makes such taxes necessary. In reality, the government has been appropriating a large part of the agricultural surplus that results from the difference between Argentina's low cost of production and the world market price. The export tax has been one means that the state has used to slow the rate of capitalization in the countryside.

The Yrigoyen government proposed an export tax as a temporary, one-year emergency measure in 1917, when the government faced financial crisis. Import duties, traditionally the government's largest source of revenue, fell by half between 1913 and 1915, and the general economic depression of the early war years cut deeply into other revenues. But in the face of these falling revenues, Yrigoyen did not curb government spending like his predecessor, de la Plaza, who had cut the budget 20 percent. On the contrary, he raised the total budget 10 percent in 1916—even though revenues that year were sufficient to finance only 65 percent of the government's outlay. The government desperately needed more revenue, and Yrigoyen proposed a program of tax increases and new taxes, including an income tax, but Congress would agree to only one part of this program, the export tax, which it approved in January 1918.[22]

Because the FAA and especially the powerful Sociedad Rural vehemently objected to this tax, the government presented it as merely a levy on "excess profits" and inserted clauses in the legislation that would trigger the tax only when world market prices rose above certain levels. But farmers and cattlemen, who knew all too well that the Argentine cost of production also rose along with world market prices, rejected the government's explanation as sophistry. Although the tax was designed to fall more heavily on wheat and maize than on meat exports, Pedro Pages, a deputy who also was a leader of the Sociedad Rural, led the attack against it in the Chamber. He predicted that it would intensify Argentina's agrarian crisis, and charged that Yrigoyen was allowing Argentina's urban nouveaux riches to escape their fair share of the tax burden.[23] But the government put together enough votes from the cities and the interior provinces to pass the export tax. This allegedly temporary tax, which remained on the books until 1933, left Argentina in the unique position of being the only major agricultural exporting country to levy an export tax.* (The U.S. constitution specifically prohibits such a tax.)[24]

The republic's wheat farmers had to bear a new and heavy burden because of the export tax. It cost them 28,000,000 gold pesos between 1918 and 1921 alone, and it extracted an additional 20,000,000 from their pock-

---

*The government repealed the tax in 1933 but reimposed it during the Perón dictatorship. It has remained in force ever since. In 1984 the export tax amounted to about 25 percent of foreign grain sales (*New York Times*, May 2, 1984, p. 34).

ets between 1922 and 1931. To add insult to injury, the government levied other charges on wheat exports, including a "statistical charge" of 16 gold centavos per ton, plus 5 centavos for stamped paper and 2 centavos per ton for the Society for the Protection of Free Labor, an employer-backed union-busting organization in Argentina's seaports. The government, in other words, financed part of the cost of strikebreaking by taxing the rural producers.[25]

But though Yrigoyen taxed agriculture heavily, he did not reciprocate by spending on the agricultural sector. By 1920 the government's total budget was 50 percent larger than in 1916, but Yrigoyen *cut* the Ministry of Agriculture's budget 25 percent during the same period. This reduction hit the government's agricultural research and education programs particularly hard.[26] In effect, the Yrigoyen government was presaging the policy that Perón was to follow 30 years later—increased taxes on agriculture to finance the urban bureaucracy as well as public education and welfare programs that were also concentrated in the cities, and that built up the government's power base.

The Revival of Pampa Agrarian Unrest and the Government's Response

By 1919 the burden of all these increases in the cost of production—land rents, harvest wages, bags, and taxes—was becoming unbearable for Argentina's farmers. Their prices were not keeping pace with the cost of production. Indeed, the 1918–19 wheat and maize prices, which the British Royal Commission was effectively able to set as the only major export buyer, were much below the 1916–17 levels (see Table A.2). To make things worse, heavy rains damaged the 1918–19 wheat crop, and yields were substantially below average. A prolonged port strike in the opening months of 1919 disrupted the export trade and intensified the distress of the agrarian sector.[27] By March 1919 farmers were beginning to default on rent payments, and landowners started ejecting tenants in arrears. The FAA appealed to President Yrigoyen for a moratorium on rent payments and for emergency rural credit. When the president made no response, the FAA issued a strike call.[28] In contrast to earlier Argentine agrarian conflicts, this time the FAA did not limit the objectives of the strike to short-term improvements like rent reductions. In 1919, in addition to contract renegotiations, the farmers' organization demanded a more fundamental and long-range land reform that would enable tenants to become owners.[29]

The 1919 tenant strike was a militant and sometimes violent affair that clearly illustrated the extent of Argentina's rural malaise and the lack of government empathy for the plight of agriculture. The Yrigoyen government, in fact, portrayed the farmers' movement as part of a larger revolutionary threat. The government was nervous, for Buenos Aires was the scene, in January 1919, of a huge anarcho-syndicalist general strike that Yrigoyen broke with repression, violence, and considerable bloodshed. Although the Semana Trágica, as this weeklong strike has been known ever since, was not the revolutionary movement that the government claimed, the events of that January led to an anti-Bolshevik paranoia among Argentina's upper classes.[30]

There was no connection between the Semana Trágica and the farmers' strike that began three months later except that a severe cost-price squeeze plagued both the rural and the urban population. The Yrigoyen government at first tried a conciliatory policy by sending Minister of Agriculture Alfredo Demarchi (who was an industrialist by background) on an inspection tour of the cereal belt. Demarchi assured the farmers that the president was looking after their interests, offered short-term credit for seed and sowing expenses, and urged them to return to work. The tenants ignored these pleas and on several occasions shouted Demarchi off the platform.[31] When they failed to go back to work, the government turned to police repression. The Ministry of Agriculture issued circulars announcing that outside agitators were the cause of the strike. The provincial police of Buenos Aires received authority to arrest anyone making "seditious" remarks. By May and June of 1919 the repressive campaign was in full swing as police broke up FAA meetings, jailed rural leaders, and sent strike organizers to Buenos Aires to await deportation.[32] *La Vanguardia* reported that the territorial police of La Pampa employed barbaric measures. "The prisoners are shackled and chained with their arms over their heads. They are kept isolated for 10 or 12 days and receive no food for three or four days at a time."[33] At a number of locations, striking farmers burned crops and destroyed machinery, but the level of repression was unusually high for a generally peaceful rural strike.[34]

While fear of a radical revolution may explain the government's harsh policy, some suspected that Yrigoyen was playing a double game. This at least was the interpretation of Sumner Welles, then attached to the U.S. embassy at Buenos Aires. Welles reported that Yrigoyen had hired a Catalan labor agitator to foment rural unrest with the aim of discrediting Yrigoyen's enemy, the governor of Buenos Aires province, José Camilo Crotto, and preparing the ground for a presidential intervention. Whatever the government was up to, it did quell the strike by June 1919, but Yrigoyen's policy had raised agrarian resentment to bitter levels.[35]

The landed elite for the most part responded to the 1919 strike with sharp criticism of the tenant farmers. If Argentine agriculture was in distress, wrote one landowner, the roots of the problem lay not in the land system, but in the farmers, who "lack skill, order, economy, and industriousness." In a similar fashion, the Sociedad Rural's official publication portrayed tenants as indolent drones who wasted time and money on cards, billiards, and alcohol.[36]

Despite these efforts to portray the strike as the result of the agrarian population's sloth, the events of 1919 made the pampa land-tenure question a major public issue. It was an issue that had been developing for several years. Since 1912 the small but active Socialist Party's congressional delegation had been introducing legislation to regulate rental contracts and to force landowners to compensate tenants for physical improvements to rental property. But Congress pigeonholed these proposals, and the executive branch ignored them.[37] When Yrigoyen came to office in 1916, he attempted to deflect attention away from the tenancy reform issue by proposing a *ley del hogar*, or homestead law. In September 1917 Congress sanctioned this legislation, which provided for grants of public land to farmers. But Yrigoyen's Ley del Hogar was weak and essentially meaningless. Constitutionally, the federal government had jurisdiction over public lands only in the national territories. But almost all areas suitable for cereal farming were in the provinces and not the territories, which, with the exception of part of La Pampa, were located either in the far north or in Patagonia. The 1917 law, moreover, provided for a maximum land grant of 200 hectares, an area that, in most of the territories, was far too small to be economically viable. And in La Pampa, where a 200-hectare farm might be enough for profitable cereal growing, almost no good public land remained. As early as 1903 Minister of Agriculture Wenceslao Escalante had reported that "the state does not possess lands fit for agriculture." This was even more the case in 1917.[38]

In any event, the Ley del Hogar never went into effect, for after Congress passed it, Yrigoyen refused to sign it until the congressmen made certain changes. Specifically, Yrigoyen, who was always apprehensive about the impact of immigration in Argentina, believed a clause in the law that allowed unnaturalized immigrants to apply for land grants invited a virtual foreign takeover of the territories. The Homestead Law fell victim to the poor working relations between the president and the Congress that plagued Yrigoyen's first term (and especially the period before 1920, when his party controlled neither chamber). Congress refused to make any changes, and the Ley del Hogar remained a dead letter.[39]

In 1919 a reform coalition composed of Socialists, Progressive Democrats, and dissidents in Yrigoyen's own party emerged to demand not only tenancy reform, but a land tax that would force the division and sale of the more inefficient large estates.[40] The inspiration for this plan came from Lisandro de la Torre, an intellectual and political reformer who headed the Progressive Democrats, a recently founded and vigorous party based in Santa Fe province. In a major policy speech in 1919, de la

Torre, who was a landowner himself but was convinced that land reform was essential to pampa agriculture, argued that no real differences separated the Radicals from the Conservatives. He condemned Yrigoyen's rural policy as "ultra-conservative." Referring to the chaotic countryside of 1919, de la Torre argued that the government's programs were in fact "factors promoting anarchy and threats to the social peace."[41] In May 1919 Governor Crotto of Buenos Aires, a reformist Radical who was on poor terms with Yrigoyen, asked the provincial legislature to sanction a land-reform law along the lines de la Torre suggested. The Buenos Aires proposal called for a large issue of provincial bonds to pay for the expropriation of estates near railway stations that could be divided and sold to farmers. If Argentina did not soon act, Crotto warned, its agriculture faced "decadence."[42]

The growing political pressure for land-tenure reform induced President Yrigoyen to introduce legislation in 1919. He asked Congress to require owners to make tenancy contracts for at least three years and to compensate tenants for improvements. He also proposed an agrarian bank to extend rural credit, a reform of the National Mortgage Bank charter to permit it to make loans to small farmers who wished to buy land, and a system of arbitration boards to settle rural contract disputes.[43] Congress acted only on the Mortgage Bank proposal. In practice, this reform's effects were minimal. Between 1921, when the law went into effect, and 1930, the bank made a total of 4,097 mortgage loans to small farmers, and in many of these cases, farmers were sold poor lands at high prices.[44] The rest of Yrigoyen's 1919 program died in congressional committees and received no further support from the president. And the Buenos Aires legislature, controlled by Radicals loyal to Yrigoyen, shelved the governor's land-reform proposal.

The government's suppression of the 1919 tenant strike and its failure to enact even minimal tenancy reform might have led to a new outburst of protest among the republic's farmers, but two events intervened to ease tensions. One was a rapid rise in prices for the 1919–20 harvest to record or near-record levels. The other was a strike of farmworkers that threatened to disrupt the harvest and destroy this rare opportunity for farmers to recoup their losses of the last few years. The context of this strike was the end of golondrina migration—which had traditionally supplied abundant harvest labor—and the new reliance of farmers on unemployed urban workers to harvest the crops. Workers who migrated from the cities for the 1919–20 harvest often brought with them the anarcho-syndicalist traditions of proletarian militancy that were widespread and popular at that time in urban Argentina. When the harvest began, labor organizers from the Federación Obrera Regional Argentina (FORA), the anarcho-syndicalist federation, appeared in the countryside, where their appeals for higher wages, a twelve-hour shift, and a two-hour noon-day rest quickly attracted a large following.[45] Farmers, many of whom had suffered severely in 1919, rejected these demands, and the result was a farmworkers' strike that spread throughout the entire cereal zone in December 1919 and January 1920.

This became one of the most bitter and violently repressed strikes in Argentine agrarian history. Not only the farmers but also grain exporters, landowners, and the urban economic establishment demanded a quick end to a strike that threatened to disrupt or even prevent one of the best harvests in Argentine history. Yrigoyen, who vowed to do everything possible to end the strike, asked provincial governors to use their police forces to ensure "the right to work." The government also allowed paramilitary forces, called "patriotic brigades," to repress the strike.[46] The result was a wave of violence and bloodshed on the pampas as the police and the brigades broke up meetings, arrested strike activists, and burned literature. The press reported ugly incidents, such as the arrest and deportation of 153 workers after a crowd stormed a rural police station to free an arrested striker, and the indiscriminate shooting by brigade members of men riding on rural freight trains. This White Terror, as the anarcho-syndicalist *La Protesta* called it, quickly crushed the strike, and by the end of January the harvest was under way again.[47]

Argentina's farmers vigorously applauded the government's use of force to destroy the incipient labor movement among pampa farmworkers. President Piacenza of the FAA made his position clear from the start: the workers' demands were "impossible," the strike was caused by anarchist agitators, and the government must suppress it. Farmers, said *La Tierra* editorially, were "true martyrs," who had risen from poverty. "We have frequented neither houses of prostitution, nor gaming establishments, nor horse races," while the workers were a shiftless lot with a high proportion of "degenerates and vice addicts."[48]

This farmworkers' strike had a number of far-reaching consequences. For one thing, the volatile labor situation on the pampas strengthened the determination of Argentine cereal farmers to introduce labor-saving harvest machinery as much as possible. Second, the strike virtually ended any cooperation between the urban-based labor movement, which condemned the rural repression of 1919 and 1920, and the organized farmers' movement. And third, the events of that season demonstrated that on the rural labor issue the interests of pampa farmers were identical with those of the landed elite and the cereal exporters. Thus farmers showed no sympathy for legislation that reformers occasionally proposed to improve the lot of rural labor. The federal government enacted no such legislation until 1944, when Perón began to organize rural labor as one of his bases of power. Like Canadian farmers, the farmers in the pampas could be militant and radical when advancing their own interests but joined the forces of conservatism when a labor movement appeared that they perceived as a threat.

When Argentina's Congress convened for its 1920 sessions, pampa farmers had high hopes for the tenancy reforms that they long had been advocating. For the first time since Yrigoyen had come to the presidency, his party controlled the Chamber of Deputies and, because the President himself had proposed tenancy reform in 1919, the prospect for agrarian reform legislation looked bright. But the farmers' own political impotence, along with the indifference of the urban population to their plight and the hostility of the elite to any meaningful change in rural property relations, kept the agrarian reform legislation of 1920 weak and ineffectual. The Chamber of Deputies passed a tenancy reform bill, which called for four-year-minimum contracts, but in general it was weaker than Yrigoyen's 1919 proposals, and it contained major loopholes, such as limiting the law's jurisdiction to farms under 300 hectares rented directly from owners. Landowners could evade the law's intent by renting plots larger than 300 hectares or by renting through intermediaries. Also notable was the legislation's failure to require written contracts.[49]

The Senate, which Yrigoyen's Radical Party did not control, refused even to consider the tenancy reform legislation passed in the lower house. To protest this procrastination, the FAA took the step, unprecedented among Argentine farmers, of organizing a demonstration in the capital. About 1,400 farmers converged on Buenos Aires on August 26, 1921. When Yrigoyen refused to meet them or to receive a delegation, the farmers held open-air protest meetings by night and publicized their cause by day with a march down the Avenida de Mayo through the heart of the business district. Their rude clothing standing out quaintly in a capital that worshiped the latest trends in London and Paris fashions, the demonstrators gathered in front of Congress, where several senators appeared and assured them that the upper house would act soon. The crowd then dispersed peacefully.[50]

The far different reception accorded the farmers' march on Ottawa that the Canadian Council of Agriculture had sponsored eleven years earlier (see Chap. 7) highlights the political strength of the farmers there and the weakness of their pampa counterparts: in Ottawa Prime Minister Laurier had met with the marchers, the governor-general had entertained them, and the House of Commons had invited them to its sessions; in Buenos Aires Yrigoyen would not see the marchers, most congressmen ignored them, and they were not allowed into the congressional buildings.

The pampa farmers' political weakness became clearer the month after their march on Buenos Aires. The Senate passed the tenancy reform legislation, complete with loopholes, that the Chamber had approved the year before. This legislation, known as Law 11,170, stipulated that contracts should be for four-year periods; guaranteed farmers the right to retain personal property in case of debt foreclosures; prohibited landowners from requiring tenants to sell to, buy from, or insure with any specific parties; and required landowners to compensate tenants for physical improvements.[51] Although this legislation was less than the FAA had wanted, the farmers' organization considered it at least a step in the right direction and was therefore perplexed when Yrigoyen obstructed it and delayed issuing the decree necessary to give it the full force of law. Thus, although Congress approved the tenancy reform in September 1921, it did not actually go into effect until February 1922. *La Tierra* condemned Yrigoyen's delay as an insult to the republic's farmers.[52]

In theory the 1921 tenancy law was highly significant, because "for the first time in Argentine legislation in matters regarding land tenure, a limitation on the right of property appeared."[53] But in fact the law's loopholes were so gaping that it had little practical impact until it was reformed in 1932. Argentina's farmers gained no reform legislation of any consequence during the decade of agrarian struggle following the 1912 Grito de Alcorta. They remained a politically marginal and impotent group without important political allies on any issue other than the repression of rural labor. The Conservative and Radical parties were indifferent to their plight, and Argentina's first democratically elected president largely ignored them. When Yrigoyen left office at the end of his six-year term in October 1922, *La Tierra* called him an "Indian" and "a strong enemy of farmers, as are all Indians."[54] Meant as a slap, this reference to Yrigoyen's mixed racial background (his mother was a mestiza) clearly demonstrated the deep alienation and mistrust that existed between the FAA and the government. The government's failure to work together with the organized farmers to improve and reform Argentine agriculture during the economic crisis of the First World War and the succeeding years suggests that the decline of the country's export agriculture, which became so noticeable under Perón, began during the Yrigoyen government.

The Prairie Wheat Boom and Wartime Canadian Politics

The First World War brought a period of prosperity, accompanied by a rapid increase in wheat-growing and production, to the Canadian prairies, but the armistice left the region no better off than before and in some respects worse. At the end of the war a severe agricultural depression descended on the prairies and ignited a regional protest movement so powerful that it changed the face of Western Canada. The economic and political results of this six-year period of boom and bust were enormous.

TABLE 8.1
*Area Under Wheat Cultivation
in Argentina and Canada, 1914–1923*
(acres)

Year	Argentina	Canada
1914–1915	15,500,000	10,300,000
1919–1920	17,400,000	19,100,000
1922–1923	16,300,000	22,400,000

SOURCE: De Hevesy, pp. 344, 394–95.

To protect themselves against the violent wheat price fluctuations that characterized these years, farmers urged the federal government to continue the wartime Wheat Board. When Ottawa refused, the prairie provinces began to search for other means of enforcing "orderly marketing," a search that resulted in the formation of a huge voluntary wheat-pool system, which promised to increase the farmers' returns by avoiding the middlemen of the grain trade. At the same time the reluctance of the traditional Liberal and Conservative parties to take decisive action to meet the grievances of the West sparked a massive political revolt against the East, the creation of a third party, and the accession to power of farmers' movements in two of the three prairie provinces.

The roots of this prairie revolt lay in the prewar period, when grain growers had banded together to form their own elevator companies and farmers had taken an increasingly militant and critical stance toward the agrarian policies of the federal government. When the war ended, prairie farmers revived this tradition of organization and protest, and in so doing, they altered the course of Canadian history. In the long run the prairie reform movement of the early 1920's increased the political and economic strength of the West and laid institutional foundations for the long-range viability of Canadian wheat agriculture.

High prices and pressing Allied needs for food stimulated Canadian farmers to expand wheat production rapidly.[55] The area under cultivation in the prairies almost doubled during the war, an increase that far exceeded the expansion of wheat cultivation in the pampas. Low prices, rising costs, uncertain government policy, and the rapid wartime growth of the cattle industry all tended to restrict the growth of Argentine wheat production. But in Canada a great speculative spurt in wheat planting took place (Table 8.1).

It was during the First World War that Canada's wheat production moved decisively ahead of Argentina's for the first time. And Argentina never again caught up. But although Canada became the world's premier wheat exporter, the impact of this greatly increased production on the prairie economy was mixed at best. Everywhere farmers brought new land into cultivation, either by tilling previously unused land or, more commonly, by buying or leasing additional land. The number of new farms rose rapidly—92,000 homesteads were added to the rolls between 1914 and 1918, and the total number of farms rose 28 percent.[56] As the fever to produce more and more wheat grew, yields dropped steadily, to fall by half between 1915 and 1918. The prairie farmer was becoming a "wheat miner." Farmers neglected proper summer fallowing techniques and seeded hundreds of thousands of acres that should have been left fallow. Some farmers avoided fall plowing "by sprinkling kerosene on their fields and simply burning off the stubble," a technique that would hardly help soil productivity.[57]

The prairies, which had long concentrated on wheat, now became even more a monocultural region. By 1916 wheat and its companion crop, oats, accounted for 90 percent of the field-crop acreage. Nor did the war do anything to diversify the prairie economy. While wheat growing boomed, little if any new industry took root in the West. As John Herd Thompson put it, "The West did poorly at the pork barrel" in Ottawa, where the federal government was distributing hundreds of millions of dollars in munitions and supplies contracts. Indeed, in the prairies industry lost ground to agriculture as an employer of labor between 1911 and 1921.[58]

Viewing the wheat boom from the East, the intellectual Stephen Leacock called the prairie farmer a "war drone," who used his profits for "pianos, victrolas, trotting buggies, moving pictures, pleasure cars, and so on." This was an unjust and indeed a jaundiced viewpoint. While consumer consumption (often long deferred) did increase in the war years, this period, as the distinguished Alberta historian Lewis Thomas noted, was one of "precarious prosperity."[59] Most of the profits from high wartime prices went to finance further expansion. Indeed, prairie farmers went further and further into debt during the war. Land prices rose rapidly: an acre of Canadian Pacific land that sold for $13.55 in 1915 brought $21.53 in 1917. Because agricultural labor was not only scarce but increasingly expensive (wages doubled or more during the war), farmers invested heavily in implements and machinery, much of which also was financed through loans. (In his vivid novel *Fruits of the Earth*, Frederick P. Grove portrayed the precarious wartime financial situation of prairie farmers in the figure of the main character, a farmer named Abe Spalding: "Economically the war had been hard on him. The price of wheat had been fixed for the farmer; for nobody else had the price of anything been fixed; by legislation, the farmer had been the prey of all preying interests."[60])

Throughout this period rural credit remained tight and expensive, which added to the prairies' long-term debt and made diversification into other potentially profitable rural industries very costly. Outside of the banks, which seldom loaned for more than a few months at a time, Canada had no organized rural-credit program, and farmers had to make their own arrangements with land salesmen, implement dealers, and loan companies. In 1917 the credit squeeze became so great that all three provincial governments passed legislation to establish long-term credit programs to provide loans at cost to farmers. But the Alberta government, in a deep financial crisis, never implemented the 1917 legislation, and the programs in Manitoba and Saskatchewan were able to make only a few million dollars in credit available.[61]

Then, in 1921, the wheat boom suddenly collapsed. As we have seen, the government refused to extend the Wheat Board and returned the grain trade to private hands in 1920. Shortly afterward wheat prices began to fall sharply. They dropped from $1.99 per bushel (No. 1 Northern) in 1920 to $1.29 in 1921, and they fell still further in the next two years. But the wholesale prices of the goods the farmers consumed declined much more slowly—wholesale prices in 1923 were 50 percent over their level of 1914, whereas the average wheat price was only 8 percent higher. To make things worse, railway rates remained high, for the Cabinet in 1916 had rescinded the Crow's Nest rates. And another severe blow was the 1921 U.S. Fordney-McCumber tariff, which struck hard at Canadian exports and virtually ended meat shipments from the prairies to the United States.[62]

Many prairie farmers found themselves with huge debts and mortgages taken out to produce wheat that now sold at prices below the cost of production. As unpaid bills piled up, "the sheriff appears to have free range among us garnishing our wheat at the elevators," complained a letter writer in the *Grain Growers' Guide*. "Economic conditions . . . were worse than they had been in 30 years," in the words of the historian Paul Sharp.[63] This economic devastation created fertile ground for agrarian political revolt.

Agrarian unrest did not stem from any particular prairie opposition to Canada's war effort. Indeed, in 1914 the prairie provinces, like the rest of the country, supported Canada's entry into the war, and in many instances supported it enthusiastically.[64] Although the carnage—the very high Canadian casualty rates—that took place on the battlefields of France in 1916 and 1917 shocked and dismayed the nation, the Borden government and the bulk of the Canadian people gritted their teeth and vowed to see the war through to successful conclusion.

Nonetheless, during the early years of the war, a strong undercurrent of opposition to the economic policies of the Borden government continued to run through the prairie provinces. Prior to the war and following the Liberal defeat of 1911, the idea of a Western third party had begun to emerge, and one of the first proponents of this idea was that pioneer of prairie progressivism, E. A. Partridge. In a 1913 Non Partisan Manifesto, Partridge had condemned the "rich grabbers of fat, public-service franchises, and the choicest parts of the public domain" who were prospering while the farmers' position became "continually more insecure."[65] The idea of a third party gathered strength in 1916, following the victory that year of the Nonpartisan League in North Dakota, an event that attracted great attention in the prairies. *The Grain Growers' Guide* called the North Dakota victory "an outstanding example of the power possessed by farmers when thoroughly and efficiently organized." The next year the NPL appeared in Saskatchewan and Alberta with a militant reform program, and some of the candidates it backed were elected to the Alberta and Saskatchewan legislatures.[66]

This grass-roots political movement, although it professed support for the war effort, thoroughly alarmed not only the Borden government, but also the opposition Liberals. The great conscription debate of 1917, however, temporarily dampened third-party talk, or as W. L. Morton put it, "war hysteria" disrupted agrarian organization.[67] But the disruption proved to be only temporary. By early 1917 not nearly enough men were volunteering to make up the army's heavy battlefield losses, and the Borden government decided that the draft, or conscription as it is called in Canada, was essential if the Dominion was to help prosecute the war as it should. This was a momentous step, for the war was not popular in Quebec. The French Canadian enlistment rate was already lower than the national rate, and bitter opposition to the draft mounted in French Canada. On the other hand, prairie farmers, along with most of the rest of English Canada, supported conscription on the premise that the West had contributed men very heavily, and that the manpower burden should be shared nationally. Nonetheless, many voices in the West declared that a "Union" or Liberal-Conservative coalition government must accompany conscription to ensure that it was free from party politics.[68]

A Union government, with Liberals as well as Conservatives in the Cabinet, would have at once conciliated Westerners to conscription and reduced the chances of the emerging Non Partisan League or some other prairie third party being elected to power during the war. It would also have enabled the federal government to present a unified front to the Canadian people at what promised to be a moment of political crisis. Borden suggested a Union government to the Liberals, headed by former Prime Minister Laurier, in early 1917. But Laurier, who was unalterably opposed to conscription and believed that it might precipitate civil war, rejected Borden's overtures. Nonetheless, many English-speaking Liberals did not agree with their leader, and when the conscription bill passed

Parliament after a series of bitterly debated sessions in the summer of 1917, numerous Liberals voted with the Conservative majority.

As a result, by late summer 1917 conscription had become the principal Canadian public issue. In the prairies it overshadowed the region's economic grievances and eclipsed the third-party movement. And nationally it marked the beginning of a deep split in the Liberal Party between French Canadians and the many English Canadians who were now ready to support a Union government. Those Western Liberals who remained unconvinced that Union was necessary dropped their recalcitrance in September 1917, when the Borden Parliament passed the draconian Wartime Elections Act, which disenfranchised all former citizens of enemy countries who had become naturalized after 1902. A parallel law enfranchised overseas soldiers and their female relatives; it also gave soldiers the right to vote in any constituency they chose or to become a "voter at large" and let the government apply the ballot to any constituency it wished. As R. MacGregor Dawson commented, the purpose of these acts was "to give the vote to those who would support the government" and "to take it away from those who would oppose it." Because many Western Liberals relied heavily on the foreign-born vote, the Wartime Elections Act left them no alternative but to support a Union government if they wished to stay in office.[69]

The 1917 disenfranchisement of "enemy" immigrants did not cause great objection among Western Canadians. Although the *Grain Growers' Guide* did protest the Wartime Elections Act, letters to that newspaper's editor revealed a strong sense of resentment among English-speaking farmers. One letter urged readers to look at "those aliens in our midst who live snug and secure on their well-stocked farms" while their neighbors sacrificed for the war.[70] Indeed, a spirit of ethnic antagonism against Central and Eastern European immigrant farmers had been building up for many years. "We do not want the scum of Continental Europe, we do not want men and women who have behind them a thousand years of ignorance, superstition, anarchy, filth, and immorality," editorialized the *Guide* in 1909. One reason why this ethnic hostility was growing so strong was that the grain growers' associations were finding these immigrants very difficult to organize. Another reason was the conviction that they took little interest in politics, "except when they are paid for their vote," a reference to a not-uncommon practice.[71]

By the end of September 1917 most English Liberal leaders had settled with Borden, and on October 6 he announced his new Union cabinet, which contained ten Liberals (five from the prairies) and twelve Conservatives. The prime minister then called a general election for December to confirm the new government. Unionist candidates carried out a vigorous campaign in the prairies and proposed a number of social and economic policies that were popular in the region, including female suffrage, reform of the Western railways, support for the cooperative movement, and increased taxation of income (Borden had already passed an income tax). Shortly before the election the government announced that farmers and their sons, as well as farmworkers, would be exempt from the draft. In the December 16 election the Unionists swept the prairies with 68 percent of the civilian vote, winning 41 of the region's seats compared with two for the Liberals. The once-mighty Liberal Party was reduced to little more than a Quebec rump.[72]

This political integration of the wheat-producing prairies into the Conservative party with its high-tariff protectionist policies would not outlast the special circumstances of the war years. Borden did indeed give the prairies representation in the Union government, but of the five cabinet ministers from the region, only one, Thomas A. Crerar, was a farmer. Crerar, who was "the new government's most important Western prize," was appointed minister of agriculture, a choice that was initially received favorably in the prairies.[73] But the Union government soon began to lose support in the West, for only a few months after the election, the Borden cabinet canceled the draft exemptions of farmworkers aged twenty to twenty-two. And Crerar's support dwindled as a result, though he privately opposed the government's conscription policy. Later, in 1919, when the Union cabinet ignored the position of its Western members on the tariff issue and presented a budget that kept the tariff—including the duties on agricultural machinery and implements—at high levels, the prairies protested vehemently. As we have seen, Borden did extend the Wheat Board for one year in 1919, thus meeting one of the demands of the West, but the region viewed tariff policy as the acid test of the Union government's intentions. Crerar resigned as minister of agriculture over this issue on June 14, 1919, and led a dissident group of nine prairie Unionists into opposition in the House. This group became the parliamentary nucleus of a new third-party movement, the National Progressive Party.[74]

The prairies were not reassured by Borden's successor as prime minister. The exhausted Borden retired in 1920, and to replace him, the Tories chose Arthur Meighen of Ontario, a leader of the Conservative Party's ultra-high-tariff wing. When it became clear that the prairies had little influence in the new Meighen cabinet, the West abandoned the Unionists, and prairie political protest revived stronger than ever.

Prairie Political Revolt and the Third-Party Movement

Between 1919 and 1921 a full-fledged regional political revolt against Eastern economic and political domination swept the prairies, with the result that the traditional parties lost control of two of the three provincial governments and the new Progressive Party became, at least temporarily, a major force in federal politics. As a distinct economic region in Canada, the prairies were thus able to adopt a distinct political position, something that was impossible in Argentina, where the pampas were the center of national life rather than on the periphery, where the farmers were overshadowed in power by the ranching elite and outnumbered by the urban population, and where most farmers were not naturalized citizens and could not participate in politics in any case. Ultimately, the prairies accumulated enough political power to influence federal economic policy decisively on several key issues important to the long-term viability of the prairie wheat economy. An examination of prairie politics thus helps clarify Canada's relative success as a wheat-growing nation.

The postwar agrarian revolt was particularly powerful in Alberta, where the United Farmers of Alberta (UFA), which by 1919 had 19,000 members, decided to enter politics and developed a sophisticated ideology of economic group politics. The Alberta case is particularly important because this was in many respects the most radical of the movements in the three provinces and had a special impact on the National Progressive Party. The UFA had excellent leadership. Its president between 1916 and 1931 was Henry Wise Wood, a man who strongly shaped prairie politics. Born in Missouri in 1860 and an emigrant to Alberta in 1905, Wood became a Carstairs wheat farmer and an early activist in the UFA, where his integrity, intelligence, hard work, and political astuteness soon made him a leader. The prestige of "the uncrowned king of Alberta," as his biographer calls him, was enormous, especially among farmers—although he never held political office.[75] Wood had consistently opposed a farmers' political party, a position he justified by pointing to the weak showing of third parties in the United States. But when rank-and-file desire for political participation emerged in the UFA locals, along with pressure from sectors of the organization influenced by the Non Partisan League (which had elected two members to the Alberta legislature in 1917), Wood gave way.

In 1920 he outflanked the NPL, which had advocated a farmer-labor party separate from UFA control, by deciding that the UFA itself would enter politics. He took over the NPL idea of constituency control instead of cabinet domination and then formulated a political theory, which became UFA doctrine, that stressed that if farmers were to enter politics, they must do so not as a new political party, but as an organized economic group. Parties, Wood believed, were inevitably controlled by plutocracies and caucuses, which compromised the interests of the voters. Wood instead visualized a society in which political organization would follow the lines of economic class interest and political groups would represent occupational or industrial groups. Moreover, Wood believed that direct democracy—the accountability of representatives to the electorate—was essential to the success of this kind of group-based politics. Wood was a staunch believer in private property and a very religious man, but as W. L. Morton points out, his political theory was "revolutionary in concept." He envisioned the destruction of the competitive social order and its replacement by a producers' cooperative economy.[76]

In sum, Wood developed a rather elegant theory to express a fairly simple idea that Albertans widely accepted: the Eastern economic elite controlled the Liberal and Conservative parties, and neither of them were responsive to the needs of the province. As one farmer put it, the "pussyfooted, hifalutin' bunch o' political windbags in the East . . . don't care a damn about us hayseeds out west!"[77] The solution was to organize a new political movement directly responsible to the citizens of Alberta and to take power. The UFA, the organization of the largest economic group in the province, would exclude the professional politicians and be the vehicle for this movement.

One of the thorniest problems for Wood was finding a way to accommodate the province's labor movement in his system of interest-group politics. Although small in number, miners and urban workers were concentrated in a few ridings in Calgary, Edmonton, Medicine Hat, and other large towns, and their votes might be critical to the UFA's electoral success. Although Wood opposed the idea of opening the UFA to labor, he eventually accepted the idea of cooperating with labor groups in ridings where they were strong in order to overthrow the traditional parties. The leading proponent of this idea of farmer-labor cooperation was William Irvine, a former Social Gospel minister who had become an NPL leader and newspaper editor, as well as a shopman for the CPR. By 1920 Irvine had won his point with Wood, and in 1921 he was elected to the federal Parliament with UFA support as a Labour member from East Calgary. His impact on the new Progressive Party (and later on the Canadian Socialist political movement) would be profound.[78]

While the UFA was occupying itself with these doctrinal matters, it was also organizing to enter and win elections. The party's appeal was stronger in the province's southern and central agricultural heartland than in the mountains, the cities, or the northern regions, which contained a large pro-Liberal continental European population. But the UFA's electoral task was made much easier by political circumstance, for in many respects Alberta was ready for political change. For one thing, the agricultural depression of the early 1920's greatly increased rural discontent with the established political system. In addition the Liberals, who had been in power since Alberta gained provincial status in 1905, had engaged in grandiose and scandal-ridden programs of railway development in the north that had virtually bankrupted the province by 1914 and left it with a huge debt. And finally, many Albertans perceived the opposition Conservatives as too close to the hated CPR and to Eastern financial interests.[79]

The UFA elected its first member of the legislature in a by-election in 1919, but the turning point for the organization came in 1921, during a by-election for the federal riding of Medicine Hat. There the UFA candidate, Robert Gardiner, faced a Unionist candidate whom the Meighen government was determined to elect. Meighen threw his full resources into the race for this traditional government seat, but Gardiner, who invited the local labor movement to cooperate with him, won a stunning 79 percent of the vote. Then, in the 1921 Alberta provincial election, the UFA won 39 of 61 seats in the legislature. These shocking defeats shattered the Unionist government, whose Liberal and Conservative components split up, and forced Meighen to call a general election in late 1921. The UFA carried all twelve Alberta ridings in that election.[80] Thus, in only two years, the UFA had captured the government of Alberta, which it was to hold until 1935, and had changed the direction of Canadian politics.

The year after the UFA's victories, the United Farmers of Manitoba took control of their province's government. As with the Liberals in Alberta, financial scandals had devastated the Manitoba Conservatives, who long had held power, and the Liberals who replaced them in 1915 never gained the farmers' confidence. In 1920 the Manitoba Grain Growers' Association changed its name to the United Farmers of Manitoba (UFM) and began to enter local political contests.[81] Also in Alberta, the idea grew in rural Manitoba that "farmers must . . . elect farmers to represent farmers," but unlike the UFA, the UFM did not find a willing political ally in the provincial labor movement. Indeed, deep distrust between the large labor movement centered in the metropolis of Winnipeg and the farmers resulted from the Winnipeg general strike of 1919. Although the organized farm movement was initially sympathetic to urban labor, which was plagued by inflation and unemployment, rural Manitoba viewed the strike with "shocked indignation." The *Grain Growers' Guide* became convinced that the Winnipeg strike "was wrong" because many of its leaders "preached the worst doctrines of Bolshevism," and some of them preached the "establishment of the Russian Soviet system of government in Canada." After the strike was over, "farmers' juries" convicted several strike leaders of seditious libel.[82]

This assessment of the famous general strike was neither adequate nor fair, and much bad blood existed afterward between the UFM and the new Independent Labour Party (organized in 1920), which elected members to both the provincial assembly and the federal Parliament. The cooperative electoral arrangements with labor that the UFA created were much weaker in the more polarized rural–urban society of Manitoba.[83]

Although the UFM on the whole lacked labor support, its emphasis on nonpartisan politics and occupational representation gained widespread rural support among a farming population that distrusted labor radicalism on the one hand and Eastern domination on the other. The UFM elected 12 members to the legislature in 1920 and 24 in 1922, when it took control of the government. Immediately after the 1922 elections the UFM recruited John Bracken, a farmer who headed the Manitoba Agricultural College, to become premier. Bracken, who had no political experience, who disliked partisan politics, and who "admitted that he had never cast a vote," brought nonpartisan government to Manitoba.[84] As in Alberta, the farmers' government introduced a number of reforms favorable to agriculture, including an income tax (which fell primarily on urbanites), lower land taxes, and a strengthened rural-credit program. And also as in Alberta, the UFM in power insisted that the primary responsibility of legislators was to their constituents and not to the cabinet, and thus grassroots democracy prevailed in Manitoba. But there the resemblance ended, for the new government could not stress a policy of group or class representation, since the UFM did not have a majority in the legislature before 1927 and had to rely on other political groups for support. In 1928 the UFM withdrew from politics altogether, and from that point on, the Bracken "nonpartisan" government in fact constructed an informal entente with the Liberals.[85] The situation in Manitoba was thus quite different from that in Alberta, where Wood and the UFA consistently opposed collaboration with the traditional parties. And as we shall see, the Manitoba-Alberta cleavage had serious consequences at the level of federal politics.

Only in Saskatchewan did the traditional party system withstand the assault of the nonpartisan farmers' political movement. Indeed, in that province the Liberal Party remained in control of the government continuously between 1905 and 1929. Unlike the Alberta Liberal Party, which was corrupt and incompetently led, the Saskatchewan party enjoyed able leadership—and leadership, moreover, that identified closely with agrarian interests and kept closely in touch with farmer opinion.[86] In fact, many Saskatchewan Liberal leaders were also prominent members of the SGGA. When the Non Partisan League emerged as a third-party threat in 1917, the Liberals sought to broaden the participation of farmers in the party by seeking agrarian leaders as candidates and by bringing some of them into the cabinet. To maintain its legitimacy among the farmers of Canada's leading wheat province, the Saskatchewan Liberal Party in 1920 divorced itself from the national party. Thenceforth the Saskatchewan Grits, as they styled themselves, would cooperate with the federal party only when it was clearly in the province's interest to do so.[87]

The political rebellion against the traditional parties that swept the prairies during the postwar years did not stop with control of the provincial governments. The agrarian radicals had another goal as well—the creation of a third party in federal politics that would become a decisive force for social and economic reform and for protecting and promoting agriculture. As we have seen, the new National Progressive Party emerged in 1919, when Thomas Crerar and a group of other prairie Unionist MPs rebelled against the Borden government's budget and formed a new opposition political bloc. The Progressives became the federal voice of the agrarian and labor political movements that seized control of the prairie governments (and of Ontario) by 1922. But though they were initially successful in elections at the federal level, as the creature of provincial movements whose ideology and goals varied greatly the party was weak and divided. Its Alberta wing retained the commitment to "direct democracy" and continued to emphasize the UFA's principle of representing the special interests of the farmers and their allies. The Alberta group at Ottawa was thus frequently in conflict with the Manitoba Progressives and with Crerar, the Manitoban who helped found the party and led it until 1922. Crerar envisioned the Progressives as a party with national objectives, and he thus hoped to impose some centralized policy direction over the Progressive MPs. But the party's divisions made this unity impossible and laid it open to the strategy of Prime Minister Mackenzie King, whose goal was to attract as many Progressives as possible, particularly from the Manitoba and Saskatchewan groups, back into the Liberal ranks.[88]

Although the Progressives did not become an enduring third federal party, their success at the polls in the early 1920's temporarily gave them important political leverage, which they used to enact major reforms for prairie agriculture. After the government's stunning defeat at the 1921 Medicine Hat by-election, Meighen called a federal election for December 1921. This was to be a particularly interesting contest, for the Tories and the Liberals returned to their traditional rivalry, and the National Progressive Party fielded candidates throughout the prairies, in rural Ontario, and also at some ridings in the Maritimes and British Columbia.

The Liberals had a new leader. William Lyon Mackenzie King was one of the few English-speaking Liberals who had remained loyal to Laurier when the party split over the issue of conscription and participation in the 1917 Union government. Although he had served as Laurier's minister of labour between 1908 and 1911, he was not well known nationally, was still relatively young, and had little political experience. But Laurier, who correctly judged Mackenzie King's potential as a man who was both politically astute and very intelligent, made him his heir apparent, and after Laurier's death in 1919 the Liberal convention (composed primarily of the party's Quebec wing) chose King as leader.[89]

King faced several awesome challenges. In order to rebuild his shattered party, he had to overcome the animosity between Anglophone and Francophone Liberals that had developed over the conscription issue. At the same time he had to face a strong and determined Conservative Party, whose leader, Arthur Meighen, was not only an able politician and brilliant orator, but also a political and personal rival of King's since their student days together at the University of Toronto. Finally, he had to confront the Progressive political rebellion that threatened to destroy the Liberal Party's traditional power base in rural Canada. Dealing with the Progressives was a particularly serious challenge, for the Liberals themselves were divided between a protectionist and a low-tariff wing. King thus had to waffle on the tariff issue, which Canada's farmers considered of critical importance.[90]

Mackenzie King's ultimate success in meeting these challenges testifies to his supreme ability as an artful political strategist and crafty government leader. King realized that the success of his party and his government depended on attracting the West back into the Liberal fold and in particular on co-opting the Progressive leadership. He succeeded because he combined careful attention to party organization with concessions to the West on critical economic and political issues.[91]

At first, however, the prairies had Mackenzie King on the defensive. When he visited the region in 1920, he was treated like "an unknown visitor from some distant land."[92] By then the Progressives were already the ascendant prairie political force. In 1920 the Canadian Council of Agriculture called a conference at Winnipeg to discuss the political future of the Progressive movement. There Crerar, who wanted a national party organization, clashed with Wood, who insisted that the party organization be left in the hands of the existing provincial associations. Wood triumphed on this point and also on the crucial issue of constituency control over Progressive MPs. Crerar then met with the independent farmer group he led in Commons, who now formally took the name National Progressive Party. Because of the decisions adopted at Winnipeg, however, Crerar became leader of a weak and decentralized party, something that he adamantly opposed but accepted for the time being. In December 1920 the CCA confirmed Crerar's parliamentary Progressives as the representatives of the farmers' political movement and recognized him as the party leader.[93]

But weak and disorganized as the party was, it adopted a platform of great appeal to Canada's agrarian population. This platform was the CCA's "New National Policy" of 1918, which called for a program of sweeping economic and political reforms, including, among other things, free trade with Britain within five years, ratification of the 1911 reciprocity treaty with the United States, public ownership of the railways and utilities, graduated income and inheritance taxes, and the introduction of the initiative, referendum, and recall.[94] As a result, the 1921 federal elections saw a bitterly contested three-way battle for the prairie vote. Meighen went on the attack against the farmers' political movement. Not only did the farmers' groups promote class rule, he declared; they were also addicted to "Socialist, Bolshevistic and Soviet nonsense," and to "freak government." On top of that, said Meighen, they were even allied with "the Winnipeg and Vancouver seditionists."[95] This political line hardly endeared Meighen to the farmers. The *Grain Growers' Guide*, which called him "reactionary," published blistering editorials against the Tory leader and his high-tariff platform.[96]

Mackenzie King took an altogether different tack. He refrained from directly criticizing the Progressives on the assumption that ultimately many of them would rejoin the Liberals. He therefore confined himself to attacks on Meighen, to vague statements about the tariff, and to an appeal for Canadian unity.[97] But King's moderation fell on deaf ears in the West, which voted solidly Progressive in 1921. Of the 43 prairie seats, the new party carried all but five. All told, it won 65 seats, picking up 21 seats in rural Ontario. (It won only one contest east of the Ottawa River.) The Conservatives won 50 seats, and the Liberals 117. There were also three independents elected, so the Liberals were one short of a majority in the House of Commons.[98]

This was a stunning electoral victory for the new Progressive Party, which now technically held the balance of power in the House and would continue to do so until 1926. The survival of the new Liberal government depended on support from the Progressives. This support, however, did not prove difficult for King to obtain. For one thing, outside of the Albertans, few Progressives were willing to bring the King government down out of fear that a Meighen government would replace it. In addition, because of the Progressives' failure to create a strong, unified organization, they could not as a party exercise the balance of power. This organizational weakness enabled King to bid for the support of individual Progressive members on key issues. Indeed, he looked forward to the eventual reunification of the Progressives with the Liberals, and he invited Crerar to join his cabinet. To stay in power and to attract Progressives back to the Liberal fold, King adopted a strategy of conciliating the West without fundamentally weakening his political strength in the East. The result was a politics of compromise through which the prairies gained a measure of reform.[99] We will examine these reforms more fully in later chapters, and here focus only on King's conciliatory policy toward the West during 1922, his first year in office.

Three issues that particularly concerned Progressive MPs as the 1922 sessions opened were the tariff, railway policy, and the reestablishment of the wartime Wheat Board. Although King eventually made tariff concessions to the West, he did not do so in 1922 or 1923. During these years his financial policy, formulated by Minister of Finance William S. Fielding, reflected the strong Quebec protectionist bloc in the Liberal Party.[100] Rebuffed on the tariff, the West did gain satisfaction from King's railway policy. The prime minister refused to consider returning the newly nationalized railways to private interests, something the West strongly opposed, and he strongly supported 1922 legislation restoring the Crow's Nest Pass rates on grain and flour.[101] King, then, could not yet maneuver on the tariff issue, but he could and did deliver railway-rate reform to the West. And on the third issue, the restoration of the Wheat Board, King also acceded to a widespread Western demand.

But on this issue the Progressives were not united. Support for a Wheat Board was strong in the UFA and among smaller farmers in Saskatchewan. This group argued that although the board could not prevent a slump in the world market, it could prevent extreme price fluctuations, and it might be able to deal directly with European governments. Their principal opponents were the farmer-owned grain companies and the farmers of Manitoba, many of whom believed that a uniform price would cut into the profits they enjoyed because of their closer proximity to markets. Crerar, who was president of the United Grain Growers, opposed the board in the name of free trade, a position that cost him a great deal of support in the Progressive ranks.[102]

Although King was philosophically opposed to interfering with the private grain trade, he realized the strength of the sentiment for a wheat board in the West. King began by appointing a House committee to study the constitutionality of a compulsory board. When this committee reported negatively, Agriculture Minister Motherwell urged King to at least support a board for moving the 1922 crop. The prime minister agreed to act, and his government prepared legislation, passed in June 1922, to create a one-year compulsory board, provided that at least two of the three prairie provinces enacted complementary legislation. The board envisioned in this legislation, however, did not have all the powers of the wartime board. For one thing, the law required the provincial governments, not Ottawa, to meet any deficits incurred. And for another, it did not grant the board control over railway transportation and sales to flour mills.[103]

Saskatchewan and Alberta quickly passed the necessary enabling legislation, but in fact the 1922 Wheat Board was stillborn. The premiers of these two provinces were unable to find any qualified executives to head the board. Most of the top men in the grain trade adamantly opposed the whole idea and refused to take part, and others objected that the board's powers were insufficient to ensure success. In Manitoba the legislature later voted against the idea. Premier Bracken, who gave it only lukewarm support, allowed the principle of constituent responsibility to guide members of his party, and the measure lost. Without Manitoba the plan was impractical, and it died. But new plans for a voluntary wheat "pool" immediately sprang up to replace the board idea.[104]

Thus, as the first year of King's government ended, he had mended some of his fences in the West. The Crow's Nest rates were popular, and the Wheat Board bill had been at least a gesture to the agrarian radicals. But it was not so much the divided Progressive Party that was responsible for this legislative success as it was King's own desire to restore the federal Liberals as a legitimate party in the prairies. Late in 1922 Crerar, who had found his position as Progressive leader a profound exercise in frustration, resigned the post amid a continuing heated exchange of views with Wood, who charged that Crerar favored "political autocracy."[105] The new leader, Robert Forke of Manitoba, would be no more successful at organizing the party than Crerar had been, and this leadership vacuum would enable King to continue his policy of wooing back the Progressives.

Summary

Despite the disunity of the Progressive Party, the postwar political rebellion had some notable achievements to its credit. Progressives had taken control of two provincial governments and strongly influenced policy in the third. And at the federal level they had achieved the restoration of the Crow's Nest rates, which were important to the competitive position of prairie agriculture, especially when prices were low. Prairie farmers, moreover, could look back at their success during the First World War, when the Union government had taken their interests into account in formulating policy for the wartime wheat boards. The experience that prairie agrarians gained in this compulsory marketing scheme would be of great value as the West began creating its own voluntary wheat pools.

The achievements the prairie farm movement gained during the wartime and postwar years come into sharper focus when compared to the experience of Argentine farmers in the same period. In the South American nation the government had not protected the farmers' interests during the war, and as a result pampa farmers received a much lower wheat price than was the case in Canada. The Yrigoyen government, much more beholden to the British than the Canadian government, raised rail rates after the war rather than lowering them (Chap. 6), and Argentina also enacted an onerous export tax. And Yrigoyen did nothing to redress the grievances of Argentina's farmers other than to pass a mostly ineffective tenancy reform.

Thus, while the prairie farmers of Canada gained in political power during the war and immediate postwar years, the pampa farmers of Argentina remained politically ineffectual. In the democratic political atmosphere of Canada, the rural producers were able to strengthen their position, while in the more autocratic system of Argentina, reform could only come from above, and it did not come in the 1917–22 period. As the postwar decade got into full swing, Canada's farmers would draw on their experiences and their successes to build new institutions to improve their economic situation. And on the production side Canadian agriculture continued to move ahead while Argentine agriculture entered what was to be a long-term period of stagnation, a trend that the government did nothing to arrest. It is to this period, which set the scene for Canadian agriculture during the Depression, the Second World War years, and the postwar period, that we will turn in the following chapter.

9

Tradition and Innovation
in the 1920's

The 1920's were a period of gathering crisis in the farming regions of the big agricultural export nations. By 1925 world wheat prices had rebounded strongly from the sharp drop that followed the First World War, but then they began to fall steadily, a trend that reflected ever-increasing world production. Prairie farmers responded with a massive campaign to "pool" their wheat through a cooperative, farmer-controlled market program designed to increase the producer's returns by eliminating the middlemen in the grain trade. The Canadian wheat pool was initially successful, and the enthusiasm it sparked among prairie farmers led to attempts to organize an international pool among farmers in the other major exporting countries. Although Canadian agrarian leaders emphasized that the international pool idea aimed at the "orderly marketing" of the world's wheat, in fact this plan envisioned an international cartel of wheat producers in the Big Four exporting nations. Pool enthusiasts, in other words, hoped that cooperative marketing among the world's wheat farmers would lead to more stable and possibly higher world wheat prices. To succeed, this plan required the participation of Argentina's farmers, but the pooling idea did not take root in the pampas. In fact, the Canadian scheme for an international wheat pool was impossible, given the structure of pampa agriculture.

This chapter examines and compares Argentine and Canadian responses to the deteriorating world wheat market in the mid- and late 1920's. In Canada the federal government made significant concessions to prairie agriculture, but the pooling experiment that the farmers devised was the central feature of the Canadian wheat economy during the 1920's, and we will focus on it closely. We will examine the failure of Canadian initiatives to bring Argentine farmers into an international pooling scheme against the background of traditional land tenure and marketing schemes in the pampas. Still another important concern of this chapter is the Argentine government's failure to launch any new initiatives to modernize or reform the agrarian sector. The chapter thus revolves around the twin themes of innovation and tradition, for the 1920's were a decade of intense—and largely fruitful—experimentation in Canada and a period of lost opportunity in Argentina.

The World Wheat Market

By the mid-1920's wheat was beginning to swamp the world market. Total world production, excluding the USSR, was on the average 12 percent higher in 1925–29 than in the preceding five-year period. Total net exports of wheat rose by exactly the same 12 percent between 1923 and 1929. Wheat production in the United States remained stable, and Russia, for all practical purposes, stopped exporting wheat, but Australia, Canada, and Argentina all substantially increased production and exports, as did the wheat-exporting countries of Eastern Europe, especially Hungary, Rumania, and Yugoslavia.[1] At the same time, several of the Western European countries that traditionally imported wheat embarked on programs to protect and develop their agrarian sectors and to reduce their dependence on food imports. Particularly notable in this regard was Italy, where Mussolini in 1925 proclaimed "the battle for grain," and sharply increased the tariff on imported wheat in order to stimulate domestic production. As a result, Italian wheat production increased 17 percent between 1919–24 and 1924–29. France followed suit by doubling its wheat tariff in 1927, and French wheat output grew steadily. By 1929, as one Canadian wheat pool official noted, France "practically disappeared as a buyer of imported wheat." Germany acted on similar lines, imposing a wheat tariff in 1925. Europe achieved its production gains not only by tariff protection, but also by programs to increase agricultural productivity, and as a result, the average yield per hectare in Europe as a whole rose 13 percent between 1923 and 1929.[2]

European countries that had traditionally imported wheat together produced nearly 100,000,000 bushels more wheat during the second half of the 1920's than in the first half of the decade. To demonstrate how much wheat was coming on the market, this extra 100,000,000 bushels represented about two-thirds of Argentina's entire wheat exports in 1926–27. But at a time when nearly everyone was producing more wheat, the average consumer was eating less of it, for as the industrial cities prospered during the 1920's, their inhabitants tended to shift away from bread and to substitute more expensive proteins. Although per capita wheat consumption increased in China and Japan, the trend in Europe and North America was in the other direction, so that average world per

TABLE 9.1

Percentage Change in Average Acreage, Production, and Yields of Wheat in Argentina and Canada, 1909–1929

Category	1909/10–1918/19	1919/20–1928/29	Percent change
Acreage			
Argentina	16,300,000	18,000,000	+10.0%
Canada	11,900,000	21,800,000	+83.2
Production (bushels)			
Argentina	153,900,000	225,000,000	+31.6
Canada	222,600,000	374,200,000	+68.1
Yields (bushels/acre)			
Argentina	9.4	12.5	+33.0
Canada	18.7	17.2	−8.0

SOURCE: De Hevesy, pp. 344, 394–95.

capita wheat consumption fell from 2.81 bushels in 1909–14 to 2.76 in 1924–29. Particularly significant was the fall in consumption in Great Britian, the world's largest importer, where the per capita figure declined from 5.63 bushels in 1909–14 to 5.16 in 1924–29. Also noteworthy was that per capita consumption was falling steadily in all the Big Four wheat exporters except Argentina. Because of the growing world population, total world consumption of wheat increased in the 1920's, but at a slower rate than the rise of production.[3]

These trends produced a wheat glut on the world market. World wheat stocks on hand rose from 557,000,000 bushels on August 1, 1922, to 919,000,000 bushels on the same date in 1929.[4] Wheat prices dropped steadily. They fell by about half between 1920 and 1923 (see Table A.2), recovered considerably in 1924 and 1925, and then dropped steadily again in the late 1920's. This price decline would have been much more serious except for the fact that each of the Big Four suffered a bad crop between 1924 and 1927.

To save themselves in this deteriorating market situation, Argentine and Canadian farmers made one rational response (other than leaving farming entirely, as many would do in the 1930's): increase production. And since additional land was available in both countries, farmers typically expanded production not by more intensive farming of the land they already occupied, but by bringing more land into cultivation. Table 9.1 shows that although total cultivated wheat area rose in both countries during the 1920's, it expanded much more rapidly in Canada. The demand for more land in the prairies pushed wheat agriculture into dry marginal areas on the frontier of settlement, such as the Palliser Triangle (near the U.S. border in southwestern Saskatchewan and southeastern Alberta). Wheat yields in these marginal areas typically were lower than in the richer, wetter parts of the prairies, and as a result average yields per acre of Canadian wheat fell substantially during the 1920's (Table 9.1).

But yields in Argentina, far from dropping during the 1920's, rose a healthy 33 percent over the previous decade. This rise was due partly to generally favorable weather and partly to improved technology, particularly scientific seed selection, that Argentina adopted during the 1920's. But another important factor contributed to Argentina's increasing yields. During the late 1920's the market for Argentine beef suffered when numerous European countries adopted import regulations and quotas on foreign meat, and in response Argentine landowners shifted some of their estates from cattle to cereals. This shift meant that pampa wheat agriculture did not have to expand to distant marginal lands to expand production, as was the case in Canada. Instead, farmers were able to move on to rich, fresh land where the soil had been naturally fertilized by the alfalfa previously planted for cattle feed.[5]

The net result of bringing so much new pampa and prairie land into cultivation during the 1920's was that total wheat production in both countries spiraled rapidly upward, a trend that only tended to depress world wheat prices still further. As the decade drew toward its end, both Canadian and Argentine wheat farmers found themselves in a vicious cycle, for to stay afloat financially, they produced more wheat, but as they did so, prices fell still lower. Although Argentine farmers were enjoying higher yields than previously, they faced the same problem of overexpansion as their Canadian competitors.

The Political Context for the Canadian Wheat Economy

Despite the fact that all three prairie provincial governments were closely allied with the organized farmers' movements throughout the 1920's, prairie agrarians could expect little effective government support to solve the problem of falling prices. The provincial governments' financial resources were limited, and their powers too narrow to confront the wheat-marketing problem, which was national and international in scope. The federal government, which was under the control of Mackenzie King and his Liberals throughout the decade, maintained the cautious stance toward the West that King had followed since 1921. The prime minister's major goal remained to destroy the new Progressive Party by co-opting its leaders and luring its less radical members back into the Liberal fold through timely concessions to prairie interests. By 1927 King had succeeded brilliantly in this endeavor, and while he continued to make political concessions to the West, his economic policy remained basically laissez-faire. A brief analysis of King's policies toward the West will demonstrate his strategy and will highlight the message the prairies received from Ottawa after the failure to revive the Wheat Board in 1922: if the farmers wished to improve their position in the marketplace, they would have to take action themselves.

In his desire to reintegrate the prairie Progressives into the Liberal Party, King faced a serious dilemma, for the West insisted on tariff reform and the Liberal Party contained a strong protectionist wing centered in Ontario and Quebec. This wing of the party had two elderly and powerful voices in the Cabinet—Sir W. H. Fielding, the minister of finance, and Sir Lomer Gouin, the minister of justice. King could not afford to alienate these party stalwarts, a fact that his standpat budget in 1923, which contained no significant tariff reductions, made clear. This budget, which the *Grain Growers' Guide* called "a ruthless and unjustifiable betrayal of public trust," earned King no plaudits in the prairies.⁶ The next year, however, King saw an opening to achieve the "political breakthrough" he desired. Fielding and Gouin both retired, and King was able to bring in a man of more flexible views, James Robb, as minister of finance and to call on Ernest Lapointe, an extremely able Quebec politician and a close ally, as minister of justice. King's 1924 budget then proposed a moderate tariff reduction, and as he had hoped, this reduction won him the support of the moderate Progressives. The more militant Progressives, mostly from Alberta, formed the "Ginger Group," which refused to have anything to do with the government, but King had found the political support he needed, and meanwhile the Progressives had suffered a deep and lasting schism.⁷

The prime minister also moved to satisfy the West on other matters important to agriculture. One issue was railway rates. Although Parliament had restored the Crow's Nest rates in 1922, King had asked the Board of Railway Commissioners to determine the actual scope of the rates, and in 1924 the board revoked them on all commodities except grain and flour moving east from the prairies. This was not enough for the West, which demanded the restoration of the entire structure of rate reductions. In December 1924 King convinced the cabinet to restore the rates the board had set aside, but in 1925, after a Supreme Court decision, the government restored only the rates on grain and flour, although it also extended the grain rates to westbound cargos to Pacific Coast ports. The important political fact here is that although the West had not gained all it wanted, the region had "won the major point," and King had appeared to take its side in the dispute.⁸ To deal with another issue, alleged improprieties in the private grain trade, King appointed a Royal Grain Inquiry Commission in 1923. It recommended several amendments to the Canada Grain Act that gave farmers a greater voice in designating how private grain companies would sell their wheat. King then guided these amendments through the House.⁹

TABLE 9.2

Canadian 1925 and 1926 Federal Election Results: Seats in the House of Commons

Party	National totals			Prairie totals		
	Pre-1925[a]	1925	1926	Pre-1925[a]	1925	1926
Liberal	116	101	116	3	20	23
Conservative	49	116	91	0	10	1
Progressive	63	24	12	37	22	8
United Farmers of Alberta	—[b]	—[b]	11	—[b]	—[b]	11
Independent Liberal	0	1	1	0	0	0
Liberal Progressive	0	0	10	0	0	8
Independent	1	0	1	0	0	0
Labour	2	3	3	0	0	0
TOTAL	231	245	245	40	52	51

SOURCE: Neatby, pp. 75, 169.

[a]Column shows the distribution of seats on the eve of the 1925 elections. At that time, there was a total of 235 seats, but four were vacant. In 1925, as a result of the 1921 census, the total rose to 245.

[b]Included in Progessive.

King, however, had miscalculated the political impact of these reforms in the West. He called an election for October 1925, and the Progressives, although weaker than in 1921, fielded candidates in 32 prairie ridings. Despite King's attempt to define the issue as national unity, the main issue in the 1925 election, which pitted him against both the Western Progressives and the Conservative protectionist Arthur Meighen, was the tariff. This presented King with a cruel dilemma, for while the West agreed that the prime minister had not gone far enough in the direction of reform, industrial Ontario responded enthusiastically to Meighen's rousing protectionist campaign.[10] The result of the election was a near disaster for King (see Table 9.2). The Liberals greatly improved their standing in the West, particularly in Saskatchewan, but the Progressives still held 22 of the region's seats, and the Conservatives picked up 10. King had not done well enough in the West to balance his defeat in Ontario and the Maritimes, with the result that the Liberals carried only 101 seats nationally.

But the crafty prime minister managed to snatch victory from the ashes of defeat. Rather than resign, he decided to let Parliament decide the fate of his government, and he used the interval between the October election and the opening of Parliament in January 1926 to maneuver for the support of the 24 Progressive members who held the balance of power. King promised to restore control over their natural resources to the prairie provinces, to support an old age pensions act, to approve several important changes in immigration legislation, and to complete the long-desired railway line to Hudson's Bay. He further wooed the moderate Progressives with his 1926 budget, which sharply reduced the tariff on automo-

biles. These promises kept his government alive in 1926, and he strengthened his appeal in the West in June by bringing the most powerful Liberal
in the prairies, the Saskatchewan premier, Charles Dunning, who was
strongly committed to completing the railway to the bay, into the Cabinet as minister of railways. King himself, who had lost his Ontario seat
in 1925, sought out a safe riding in Prince Albert, Saskatchewan, a seat
he occupied until the end of his career. As W. L. Morton comments,
"The West having so long refused to come to King, King was at last
going to the West."[11]

King's victory soon came unglued, however, for during the 1926 sessions, the Tories exposed a major customs scandal that reeked so strongly
of liquor smuggling to the United States that numerous Progressives deserted the prime minister. This proved to be only a temporary setback. A
confused and short-lived Meighen government followed, and the Progressives defeated it as well, amid cries from King that the governor-general was unconstitutionally meddling in Canadian politics. The result was
another general election, held in September 1926.[12] King now had found
his political formula—the cause of Canadian independence from royal
political interference. Avoiding any discussion of the messy customs
scandal, he took the offensive and campaigned lustily on the theme of
"Independence vs. Colonialism." And this time he won a decisive victory. Table 9.2 shows the Liberals winning only 116 seats, well short of a
majority, but the Liberal Progressives were ready to support the government, a fact that became clear when their leader, the Manitoban Robert
Forke, entered the cabinet in October as minister of immigration. The
remaining Progressives were divided into the hard-line Ginger Group,
centered in the UFA, and a small leaderless group that drifted back to the
Liberals because it had nowhere else to go. The 1926 election, then,
marked the end of the Progressive Party, which was already weakened by
ideological conflict, poor leadership, and the return of prosperity, as a political force in Canada. It was also a "shattering blow" to the Conservatives, whose leader, Arthur Meighen, resigned soon afterward. It reestab-
lished the Liberals on the prairies, and it signaled the beginning of the
long dominance that King would hold over his party and the nation—a
dominance interrupted only by the Depression.[13]

Although William Irvine of the Ginger Group argued that King's
"bowels of mercy" closed after he obtained his majority, in fact the prime
minister kept the promises he had made to the prairies. The government
completed the Hudson's Bay Railway, returned control over natural resources to the provinces, and passed not only an old age pensions act, but
also a farm loan act. It is true that initially the government supported the
farm-loan program with parsimonious budget appropriations, that it
made no provision for short-term credit, and that few farmers benefited

from the program as late as the Second World War.[14] But the Canadian Farm Loan Act was at least a beginning and a precedent for the future. In the late 1920's King continued to strengthen his political position in the West. In 1929 he made his former minister of railways, Charles Dunning of Saskatchewan, minister of finance, and brought in Thomas Crerar, whom he long had hoped to attract to his cabinet, as minister of railways.[15]

The agrarian political revolt ended with King's 1926 election victory, but King probably would not have won without the reforms he offered to the prairies. The Progressive movement had gained policy concessions—particularly on tariff and railway issues—that strengthened the infrastructure for wheat production in the prairies. Except for the amendments to the Canada Grain Act, Mackenzie King did not take the federal government further into the area of wheat-marketing policy—he was still far too much the laissez-faire liberal to entertain such an idea—but he did redress some of the West's outstanding economic grievances and reintegrate the region into the national political system. Thus, the 1920's began a long period, interrupted only briefly during the early 1930's, of political balance between the agrarian West and the industrial East. When King returned to power in 1935, after five years out of office, he would again strive to maintain this balance—a policy that the Liberals maintained until 1957. This balance was still unsatisfactory to the prairie West, for King did not give the region all it wanted, but he did deliver half a loaf. The success of this performance will come into sharper relief later in this chapter, when we examine the failure of Argentine governments during the same period to act on behalf of pampa agriculture.

The Canadian Wheat Pools

To attempt to escape from the dilemma of falling prices, prairie farmers did more than merely increase production. They formulated a strategy to eliminate the grain-trade middlemen, certain that in this way they could increase returns from the sale of their wheat. Canada's wartime experience with the government Wheat Board, which had successfully marketed Canadian wheat without recourse to the private grain trade, provoked, as we have seen, a widespread demand in the prairies to continue the board in the postwar years. When the 1922 Wheat Board legislation proved a dead letter (see Chap. 8), agrarian leaders began to search for some kind of substitute marketing plan. What they settled on was a co-operative voluntary marketing plan in which the members' wheat would be "pooled" and sold directly to European importers. The first of these marketing groups appeared in 1923, and thereafter the Canadian wheat pools swept the prairies and attracted massive support among the region's farmers. Indeed, the pools became the central institution of prairie agriculture during the 1920's.*

Cooperative marketing was well known in Western Canada by 1920, for it was already in use in the United States, where pool marketing was particularly successful among California orange and raisin producers. The apostle of this "bold new vision of collective marketing" was the San Francisco lawyer Aaron Sapiro. This "spellbinding" orator, who "employed a fine sense of drama" in his public appearances, had served on the staff of the California State Market Commission, which was instrumental in organizing the first producer pools in that state.[16] Sapiro then threw his enormous energy and formidable public speaking ability behind a nationwide campaign to spread his message that cooperative marketing was agriculture's hope of salvation. Under his "Sapiro Plan," producers would join a farmer-controlled cooperative and sign an enforceable long-term contract with it. This pool would then proceed to market the product vigorously by seeking new buyers and guaranteeing a high-quality, standardized product. Sapiro promoted this plan zealously. In the words of Grace Larson and Henry Erdman, "He made the marketing of a barrel of apples more exciting than a tale from Boccaccio, and the signing of a co-operative agreement as vital to social justice and progress as the Magna Charta." Sapiro appeared frequently before farmers' groups (for a high fee) to spread his message that, through pooling, farmers could not only eliminate middlemen, but also control a major part of the supply and thus manipulate prices and markets in their favor. By 1920, when he convinced wheat growers in Washington State and Idaho to form a pool, Sapiro was becoming a nationally known figure. In the early 1920's he earned between $40,000 and $60,000 a year from the fees he charged co-ops.[17]

The message that this super salesman spread among North American farmers was easy to understand. Under the Sapiro system, each farmer signed a contract (usually for five years) to deliver all his production of a certain crop to the pool. The contract gave the pool the authority to sell the crop as it wished, presumably to the best possible advantage. The pool's selling agency would follow the principle of "orderly marketing," meaning that it would attempt to avoid seasonal price declines by spread-

*There were four organizations in the wheat pool, a term referring to the movement as a whole. By 1924, farmers in Alberta, Saskatchewan, and Manitoba had established three provincial pools. The term wheat pools refers to these provincial organizations, which owned grain elevators and kept memberships. The marketing association for all three provincial pools was called the Central Selling Agency, often referred to simply as the Central. The pool was a producer-controlled alternative to the open-market system, which did not survive the 1929–31 crisis in the world economy; the three provincial pools continued on as grain elevator companies but their international marketing system disappeared with the Depression. See Vernon Fowke, *National Policy*, Chaps. 11–13, pp. 196–255, and J. F. Booth, "Cooperative Marketing," pp. 44–98.

ing the sale of the crop evenly over the whole year. Upon delivery of the crop, the pool paid the producer an advance (based on the expected price), and after the sale the farmer received his share of the remaining surplus, minus deductions for sales expenses and reserves. Here was the old device of the patronage dividend, long used among retail cooperatives, adapted to marketing.[18]

The organization that popularized the wheat pool idea on the prairies was the Farmers' Union, a new association founded at Ituna, Saskatchewan, in 1921, as a reaction against the "conservative" policies of the established grain growers' associations. Stimulated by the emergence of the One Big Union idea among industrial workers in the West, the radicals who formed the Farmers' Union employed the vocabulary of the class struggle. As its preamble stated, "In the struggle over the purchase and sale of farm produce, the buyers are always masters—the producers always workers. From this fact arises the inevitable class struggle."[19] According to the Farmers' Union, the established SGGA not only had mistakenly trusted the government to solve agrarian problems; it had also fallen under the dominance of the now wealthy Saskatchewan Co-operative Elevator Company. More bluntly, it charged that the SGGA had become an elite organization of well-off farmers, and portrayed itself as the representative of the mass of poor "dirt farmers." Despite its radical rhetoric—or perhaps because of it—the Farmers' Union made little headway until 1923, when it seized on the idea of the contract wheat pool. The campaign to organize such pools propelled the new group "from obscurity to a position of influence in the rural communities."[20]

Once the Farmers' Union started a grass-roots wheat pool movement, the three established grain growers' organizations (the UFA, the SGGA, and the UFM) were quick to rush aboard lest they be left behind. At a July 1923 meeting, the three groups agreed to support the formation of provincial wheat pools.[21] But in fact the Farmers' Union had forced this decision because it already had invited the famous Aaron Sapiro to a mass meeting in Saskatoon. Sapiro, who also appeared that year in Calgary as guest of the *Herald*, arrived in Saskatoon on August 6, 1923, and in a hard-hitting speech the next night, he took the Canadian wheat belt by storm. The fiery evangelist of the pool movement told his overflow audience to "stop dumping and start merchandising" and to "organize by the commodity and not by the locality." After assuring the prairie farmers that a Canadian pool would have "a distinct and definite effect in stabilizing the wheat market of the world—upwards," he proclaimed that New Zealand and Australia were ready to join them. Then, he continued, "we would have to find some way to organize Argentina, and tie up the surplus there with the Canadian and Australian surplus." Next would come the turn of the U.S. wheat belt, especially Kansas. Within five years, Sapiro asserted, "a world wheat pool would be organized—and the consumer would not be hurt!" Sapiro's optimism regarding international organization was irresponsible and, as we shall see, led to some serious mistaken assumptions in Western Canada. But he was a salesman for his cause, and as he concluded, he thundered an emotional appeal to his audience to "think of your children! . . . If you only will to do it, no power can stop you."[22]

The historian Frank Underhill, who then lived in Saskatoon, thought Sapiro offered "the best revival meetings" he had ever attended. And when Sapiro returned the following year and told a Regina audience that an international wheat pool would be functioning within three years and influencing world prices, he stirred "pool fever' to new heights. The *Grain Growers' Guide* warned that it was "unwise to build extravagant hopes upon the advantages of a world monopoly of wheat," but in the euphoria of the moment, few heeded the *Guide*'s advice. Nor did they listen to academic economists, who were already warning that Sapiro's proposals might be economically unsound.[23] When J. B. Craig, the president of the Winnipeg Grain Exchange, referred to Sapiro as an "alien demagogue," and when the *Regina Morning Leader* and the *Saskatoon Star* attacked his integrity, many farmers concluded that a man who so incurred the establishment's wrath truly did point the way to a better future for Canadian agriculture.*

The campaign to organize a wheat pool began in earnest after Sapiro's visit of August 1923. Organizers from the Farmers' Union fanned out across Saskatchewan, but they found the season too late and the harvest too near for success that year. But in Alberta, where Henry Wise Wood and the powerful UFA embraced the idea, a pool did indeed begin to function in 1923. The provincial UFA government contributed $5,000 to begin the membership campaign and agreed to guarantee the pool's obligations to the banks, whose loans would be necessary to finance the advance payments to farmers. Wood himself led the canvassing effort in his province. The success of the membership campaign—in Alberta as well as in the other provinces—resulted, at least in part, from the image of righteousness and justice evoked by the organizers. They emphasized the democratic organization of the pool (members elected the board of directors) as well as the pool's goals of increasing the farmer's income by freeing him from "speculators" and "middlemen." The Alberta Pool signed up 45 percent of the province's acreage after a whirlwind campaign and sold 34,000,000 bushels of wheat for 29,000 members in the 1923–24 crop year.[24]

*S. W. Yates, *The Saskatchewan Wheat Pool*, pp. 92–100. See also "Unwarranted and Undignified," *Grain Growers' Guide*, Sept. 19, 1923, p. 3. Sapiro was later to sue no less than Henry Ford for $1,000,000 after Ford's *Dearborn Independent* made anti-Semitic attacks on him and charged that co-op marketing was a Jewish conspiracy. Ford settled out of court. (Grace H. Larsen and Henry E. Erdman, "Aaron Sapiro," p. 267.)

Having achieved this remarkable success in Alberta, pool organizers, accompanied by Wood, fanned out across the other two prairie provinces in early 1924. Saskatchewan was critical, for it produced over half of all prairie wheat. As in Alberta, the two Saskatchewan farm organizations loaned organizational funds, and the provincial government advanced $45,000. The pool encountered some bitter opposition from the farmer-owned Saskatchewan Co-operative Elevator Company, which regarded the new marketing experiment as a threat, but the pool organizers worked with the fervor of a religious revival—and indeed, Wood commented that "the Wheat Pool was as much a religious institution as the church."[25] In hundreds of rural schools and town meeting halls in the farm communities that dotted the prairies, organizers met the farmers, explained the advantages of the pool arrangement, and signed up members. By June 6, 1924, the Saskatchewan Pool "went over the top" with 50 percent of the province's acreage and 46,500 contracts.[26] Success in Manitoba was slower—cooperative marketing schemes always enjoyed less support in that province—but the UFM backed the pool, and by June 1924, with nearly 40 percent of the province's wheat farmers signed up, the Manitoba Pool went into operation.[27]

Thus, in the short space of a year, a huge grain-marketing organization emerged and was well on the way to becoming the world's largest wheat exporting agency. In 1924 each of the provincial pools incorporated as the Alberta (or Saskatchewan or Manitoba) Co-operative Wheat Producers, and the three pools then formed a central selling agency, the Canadian Co-operative Wheat Producers, Ltd., to market the crop. A. J. McPhail, a leader in the Saskatchewan Pool, was elected president of the Central and immediately threw his remarkable energy and organizing ability behind the new agency. (Indeed, he devoted the rest of his life to the wheat pools, and his early death in 1931 was probably hastened by utter exhaustion.) Henry Wise Wood, who became vice president, placed his enormous prestige firmly behind the Central.[28]

The leadership now worked vigorously on a number of fronts to establish the wheat pools securely. For one thing, they continued to promote massive membership campaigns. In this effort they were aided by rising world wheat prices in 1925, for pool members, organizers were careful to point out, received an average of several cents more per bushel than non-members. In 1925 the Alberta Pool launched an ambitious campaign to secure 10,000 new members, and ultimately even luminaries like Edward, the Prince of Wales, who owned a ranch near High River, signed up. In Saskatchewan the pool controlled 71 percent of the province's acreage by October 1925.[29]

Despite this success, many farmers refused to join the pools. Some wanted the immediate cash payments that the private grain trade provided, others (particularly in Manitoba), thought they could "beat the pools to market," and still others distrusted the pooling idea as an encroachment on personal liberty.[30] A particularly weak spot in the organizational effort was the large and mostly unassimilated population of continental European immigrant farmers. Like the other cooperative prairie movements, the pools were primarily an Anglo-Saxon movement. Ukrainians, particularly in Manitoba, were hard to organize, partly because of an ingrained skepticism about cooperative ventures, partly because of a generalized distrust of Anglo-Saxons, and partly because of the language barrier. To get around the last problem, the Alberta Pool printed placards in Ukrainian, but its contracts were only in English. Although some Central European groups, like the Doukhobors, contributed many pool members, others—particularly the German Catholics of Saskatchewan—maintained their group isolation and were very hard to organize.[31] By 1927 the rapid growth of pool membership ceased, and a few radicals in the organization began to suggest that all farmers should be compelled to join.

A second major problem the pool managers faced was the physical organization of a storage and marketing system. Initially, they made arrangements to use the line elevators of the United Grain Growers and the Saskatchewan Co-operative Elevators, but in 1925 the pool leased its own terminal elevators at Thunder Bay, and it later acquired terminals at Vancouver and Prince Rupert as well as a transfer elevator at Buffalo. By 1930 the pool controlled 11 major terminal elevators. Meanwhile, the indefatigable McPhail supervised the opening of a head office at Winnipeg, formed branch offices in other major Canadian cities, New York, and Paris, and established a network of agencies in 15 wheat-importing countries throughout Europe and also in Mexico, Brazil, and China. This vast organization enabled the pool not only to sell on the Winnipeg Grain Exchange, but also to sell between 40 percent and 60 percent of its wheat each year directly to European millers. In 1926 the pool supplied 60–80 percent of the wheat imported by France, Italy, Norway, and Greece.[32]

The capstone of the pool's physical organization was the purchase (by the Saskatchewan Pool) of the Saskatchewan Co-operative Elevator Company, which made it the owner of the largest elevator system in the world. From the beginning, numerous influential figures in the pool movement, particularly among the Farmers' Union leaders, argued that the pools must acquire their own elevator network to be successful. The farmer-owned elevator companies, explained one pool official, "sound good on paper" but "were in reality huge joint stock companies. . . . Dividends were distributed not on a basis of grain delivered but on a basis of the stock held." The pool members, in other words, were frustrated at enriching the wealthy farmers who were shareholders in the elevator companies. Moreover, there were frequent complaints that the elevator companies gave pool members poor service. In the final analysis, the provincial pools needed to control the movement of their grain and to give their members good service at cost.[33]

To this end, in 1925 the pools began to construct their own elevators, but this was a slow process and one that would not give the organization access to many smaller towns. Stimulated by the Farmers' Union, pressure built up among the membership to buy the Co-op Elevators.[34] Although McPhail hesitated, fearing the burden of debt involved, he could not resist the wave of sentiment in favor of a purchase. Again, support from the provincial government was crucial to the Saskatchewan Pool's success. McPhail discussed the elevator issue with Premier Dunning, who agreed to back the purchase politically. The provincial pool would finance the purchase by deducting two cents per bushel from each farmer's returns. Each pool member thus acquired equity in the elevator system and was entitled to a patronage refund out of elevator profits. Although the Co-op Elevator Company originally did not want to sell, agrarian opinion—which in Saskatchewan meant political pressure—forced it to do so in April 1926. The sale price was fixed at $11,000,000, an amount that allowed the Saskatchewan Pool to acquire not only the Co-op's 451 line elevators, but also its terminal facilities and the transfer elevator at Buffalo.[35]

The Saskatchewan Pool, after acquiring a huge elevator network that gave it access to most of the province's grain shipping towns, continued to build new elevators of its own to reach still more shipping points, and the other two provincial pools also expanded their elevator networks. By 1930 the wheat pool elevators reached 76 percent of all prairie shipping points—and fully 91 percent in Saskatchewan.[36] As Table 9.3 shows, this vast elevator system accounted for nearly one-third the total elevator capacity in the prairie provinces.

Reflecting on the acquisition of the Co-op Elevators, McPhail wrote to a friend: "We are living under entirely changed conditions and the man or men who try to set themselves up against the majority wish of the people are very apt to get steam rolled out of existence. The people today . . . are doing more thinking for themselves than ever before."[37] As McPhail suggested, the rise of the wheat-pool organization reflected the deep democratic impulse that prevailed in the prairies. The majority of the agrarian population was demanding to participate in the shaping of the West's economic destiny, and this mass participation had resulted in the formation of a huge cooperative enterprise. This kind of mass democratic participation in political and economic decision-making, which was totally absent in the Argentine pampas, marked one of the most fundamental differences between the wheat economies of the two countries.

TABLE 9.3
The Prairie Wheat-Pool Elevator System, 1936

Category	Manitoba	Saskatchewan	Alberta	Three-province total
Total grain elevators				
Number	718	3,243	1,792	5,753
Capacity (bushels)	22,741,650	101,439,150	66,417,610	190,598,410
Pool grain elevators				
Number	153	1,085	429	1,667
Capacity (bushels)	6,303,000	36,500,000	16,591,000	59,394,000
Pool elevators as percent of total				
Number	21.31%	33.46%	23.94%	28.98%
Capacity	27.72%	35.98%	24.98%	31.16%

SOURCE: Canada [7], 1: 17.
NOTE: These figures had not changed appreciably since 1930 (see MacGibbon, *Canadian Grain Trade*, 1932 ed., p. 344).

Indeed, one of the most significant results of the campaign to acquire the Co-op Elevator Company was the revitalization—along increasingly militant lines—of the Saskatchewan organized farmers' movement. Shortly after the elevator sale was completed, the rapidly growing and vigorous Farmers' Union—which had been the dynamic force behind the elevator movement—amalgamated with its older and more sedate rival, the SGGA, to form a new organization, the United Farmers of Canada (Saskatchewan Section). In reality, the new organization absorbed the older one, for it set nearly all the terms for the merger, agreeing only to tone down the Marxist class-conflict language in the preamble to its constitution.[38] Once organized, the UFC (SS) moved energetically to enroll new members. Among its leaders were radicals who were dedicated to promoting social and economic change to favor the mass of wheat producers. Its zealous fervor (one leader told a convention, "I send you men and women forth as missionaries to begin building the world of tomorrow today"), its democratic ethic (men and women members had equal rights), and its vigorous campaign among hitherto neglected ethnic groups (especially the Ukrainians, among whom it used native-speaking organizers) led to success.[39] The most militant radicals in the movement, however, were convinced that the UFC (SS) was too conservative, and formed a separate though allied organization, the Farmers Educational League. It openly espoused class struggle and promoted its vision of a cooperative socialist future. The FEL remained a small group, but it attracted a great deal of attention.[40] Meanwhile, UFC (SS) membership reached 30,000 by the end of the decade, and the organization became a major force in Saskatchewan politics.

TABLE 9.4
Growth of the Canadian Wheat Pools, 1924–1930

Crop year	Wheat deliveries (bushels)		Pool deliveries as percent of total	Pool membership[a]
	Total	Pool		
1924–25	211,926,000	81,394,000	38.41%	91,195
1925–26	351,683,000	186,938,000	53.15	122,385
1926–27	338,541,000	179,950,000	53.15	136,212
1927–28	399,654,000	209,871,000	52.51	149,590
1928–29	468,392,000	244,248,000	52.15	130,283
1929–30	238,048,000	121,963,000	51.23	140,161

SOURCE: Canada [7], 1: 20.
NOTE: These figures follow the data supplied by the Board of Grain Commissioners in 1937 and revise slightly the data reported in most secondary works.
[a]Calendar year, beginning in 1924.

By the late 1920's the Canadian wheat pool had become the dominant grain-marketing organization in the prairie provinces. As Table 9.4 shows, the number of members grew steadily until 1927, when a high of nearly 150,000 was achieved; the figure declined slightly in the next year and then revived at the close of the decade. Beginning in 1925, these members delivered over 50 percent of all the wheat marketed in Western Canada. (The wheat pool also organized a parallel pool network to market the coarse grains—barley and oats—produced in the prairies.) The pools achieved this success because they brought tangible benefits to Western Canadian farmers. They brought about economies of scale that cut the cost of selling, as well as the cost of elevator handling. They improved the position of the weakest sellers, those who had less than a carload of grain to sell and who had been compelled to accept lower prices from the elevator companies. And their careful treatment of grain, along with their vigorous merchandising policy, improved the prices farmers received for the lower grades of grain. Profits made from the mixing and raising of grades went to the farmers, not to the private grain trade.

Despite this rapid rise to success, the Canadian wheat pool faced a number of serious problems. About half the prairie farmers refused to join. The three provincial pools were financially vulnerable, as prairie farmers would learn during the 1930's, when the pools went bankrupt and were saved only by government financial aid. The pool also had numerous powerful enemies—not only among the private grain traders in Canada, but also among European buyers, some of whom hoped to "bust the pools." Perhaps most fundamentally, the pool could not achieve its long-term end of raising the market price of wheat without the cooperation of farmers in the other major exporting countries. It was to this task that the leaders of the Canadian pool turned in 1926. The Canadians would soon learn that international producer cooperation was extraordinarily difficult to organize.

Attempts to Build an International Wheat Pool

By the time of the 1925 Canadian wheat harvest, a certain euphoria prevailed among the organizers and backers of the prairie wheat pools. They were not only successful; they were rapidly growing, and their management was building a solid physical plant. Moreover, there was a widespread expectation among members and leaders alike that their success was only the first step toward the organizing of an international wheat pool that would include farmers in the other Big Four exporting countries. This multinational wheat exporters' co-op, it was assumed, would enable Australian, U.S., Argentine, and Canadian farmers to avoid price competition and to raise the world price of wheat to a more profitable level. In other words, what the Canadians began to envision was a worldwide cartel. Indeed, as one grain trade expert pointed out, some kind of international cooperation was essential for the Canadian pool to succeed over the long run. Otherwise, the Canadian organization would be "but one big private trader . . . with the potentialities and all the perils of bigness."[41]

Aaron Sapiro, as we saw earlier, had emphasized that "when we talk cooperative marketing we say this: we are interested in raising the basic level of the price of wheat."[42] But Sapiro was irresponsible in spreading this message. He implied that organization in other Big Four countries would be as easy and as rapid as it had been in Canada, and in 1923 he had predicted that a world wheat pool would be operating within five years. It was not without reason that McPhail noted in his diary that "some of his [Sapiro's] remarks were very pretty and others very misleading, but he seemed to put them all over. I am convinced that he is one of the most dangerous men, perhaps the most dangerous, to the cooperative movement." By the time McPhail wrote this, many of the U.S. pools Sapiro had instigated were already out of business.[43]

But Sapiro's remarks made a deep impression on many Canadian pool members. As early as 1923 a group of Farmers' Union members stated in the *Grain Growers' Guide*: "We firmly believe that as soon as we have the farmers of Canada, the United States, and Australia affiliated in one organization we will be able to regulate the selling price of farm produce according to the cost of production. . . . The law of supply and demand is a farce." Some voices of caution were raised, among them the *Grain Growers' Guide* itself, which warned that "it is unwise to build extravagant hopes upon the advantages of a world monopoly of wheat," and pointed out that the world wheat price was set by the lowest-cost producers, who were not necessarily Canadian.[44] But at Regina in 1924 Sapiro had specifically mentioned including Argentina—a country neither he nor most Canadians knew anything about—in an international pool.[45]

The Canadian pool leaders moved quickly to implement the international organization they had in mind. In February 1926 they sponsored an International Wheat Pool Conference at St. Paul, Minnesota. The *Pioneer Press*, that city's leading newspaper, reflected the optimistic tone of the meeting in a front-page headline: "Americans, Canadians, Australians to Plan Control of 75 Percent of Globe's Output."[46] Delegates from the United States and Australia attended, but it was clearly a Canadian show. Henry Wise Wood set the tone of the discussion when he urged an international cooperative movement to raise the world price of wheat 50 percent so that farmers could keep up with the rising cost of production. Wood also emphasized that the "orderly marketing" system of pooling would bring farmers a better "average price" because the pools sold wheat all year long instead of disposing of it quickly after the harvest. The conference was designed to publicize the advantages of wheat pooling, and it did not go beyond that charge, to formally constitute an international organization. But it was successful in fostering interest among Australian and U.S. delegates; the chairman of the South Australian delegation asserted that when pools were properly organized in his country and the United States, the international co-op would seek the admission of Argentina and other wheat exporters. Before the delegates adjourned, they agreed to meet again the following year.[47]

The Canadians soon learned, however, that organizing workable wheat pools in Australia and the United States would not be a simple matter. As in Canada, the pool movement in Australia dated from the First World War, when the government had set prices. Voluntary pools had begun with the 1921–22 crop, but the movement declined after handling the majority of the wheat crop between 1921 and 1923. The strongest support for the pool idea centered in the states of Western Australia and South Australia, where, as in the prairies, resentment against "exploitation" by the populous Eastern states prevailed. The Canadian pool sent several delegates to Australia, and in September 1926 Henry Wise Wood himself sailed Down Under to discuss pool organization. He was received with particular warmth in the Western states, but Wood also found that there was strong opposition to a contract pool, fanned by powerful merchants and shippers, throughout the country. And indeed merchants were able to pay farmers higher prices for their wheat than the infant pools. The Australian pool movement briefly revived (apparently because of the Canadian example) in 1926–27, only to decline again. Nonetheless, many farm leaders were impressed with the Canadian experiment, and Australians continued to participate in international pool conferences.[48]

Nor was the situation in the United States particularly encouraging. Although wheat pools appeared in at least one state (Washington) as early as 1920, they grew very slowly. In 1923, the "record high" year for U.S. wheat pools, they handled only 3.4 percent of the total crop.[49] The enormous geographic scope and diversity of wheat farming in the United States, along with the divided nature of the U.S. farmers' movement, an extremely well-financed anti-pool campaign that the private grain trade mounted, the huge domestic market for wheat that gave farmers numerous sales outlets, and a relatively weak commitment to large organizations among U.S. farmers, who preferred local cooperative action, all combined to keep the pool movement small and scattered. Pools were fairly successful in some states, particularly North Dakota, but there was no effective national organization or national central selling agency.[50]

The Canadian Pool and the Argentine Wheat Economy

The Canadian pool's initial contacts with Argentina revealed the enormous obstacles it faced in any attempt to integrate that country's farmers into an international wheat-marketing scheme. But Canadian promoters of the international pool idea realized that Argentine participation would be essential. As the *Grain Growers' Guide* pointed out, "Argentina will be Canada's greatest competitor in the world's markets."[51] This prediction would not prove correct, but Canadians in general knew very little about the agricultural situation in Argentina. Although the Royal Bank of Canada had a branch at Buenos Aires, there was no Canadian embassy in the capital, and until 1927, when Ottawa began sending trade commissioners to Argentina, Canada's knowledge of its South American competitor was gleaned from grain trade reports and travelers' comments. Because of this lack of knowledge, mistaken assumptions about the Argentine wheat economy were commonplace. Leaders of the Canadian wheat pools realized that they needed an agent on the scene in Argentina to study their competitor and, hopefully, to interest Argentines in the international pool concept.

The man chosen for this task, W. J. Jackman, was a UFA leader who had migrated from England to homestead in Alberta in 1902. A fluent linguist who spoke Spanish well, Jackman was the logical choice when the Central's leaders decided to send an agent to Argentina.[52] Jackman left New York on August 21, 1926, and arrived in Buenos Aires three weeks later. He immediately got to work. He traveled widely throughout the pampas, consulting grain trade officials, leaders of farmers' associations (including Esteban Piacenza of the FAA), prominent figures in the Sociedad

Rural, and members of the Anglo-Argentine community. Everywhere he went he discussed the success of the Canadian wheat pool, as well as the possibility of Argentine participation in an international arrangement.[53] Jackman soon concluded that Argentina's wheat economy was in a primitive state of development, certainly in comparison with Canada's. His pessimistic views about Argentine wheat growing became clear when he returned home at the end of 1926. He wrote two long reports to pool officials; both were subsequently published.[54] Jackman was sent back to Buenos Aires in 1927, where he served as the Central's Argentine representative (at a $6,700 yearly salary) until 1931. He sent back another series of reports on the Argentine wheat economy, at least one of which also was published.[55]

These publications represented virtually the only accessible information that Canadians had about the Argentine wheat economy in the late 1920's. But Jackman's reports were in some respects both misleading and incomplete. They were typical of the one-sided observations that countless North Americans long had made about Latin America, with their stress on poverty and backwardness, and they tended to emphasize the weaknesses of Argentine wheat agriculture without a careful examination of its strengths. Jackman dismissed the marketing system as extremely primitive. The land-tenure system, with its mass of renters and concentration of ownership in big estates, was "possibly the worst feature of rural life in the Argentine."[56] Jackman, however, neglected to note the flexibility of use that the Argentine land system enabled. And he also colored his analysis of Argentine grain growing with a generous dash of racial prejudice. Although he conceded that the Argentines were "very fine people with many admirable qualities," he believed that the "less vigorous" Southern Europeans of Argentina were not as capable as the farmers of "Northern European blood" in Canada.[57]

Jackman studied the question of Argentine agrarian cooperation closely, and he noted "the greatest interest in our organization, and a keen desire to emulate what we have done." Although he found "the obstacles in the way of the formation of a Pool . . . formidable," he concluded that "they can be overcome . . . if a little assistance is given from our end." Apparently with himself in mind, he noted in his first report that if the pool would send "a man of the right personality" to organize pampa farmers, "I believe an Argentine Wheat Pool is possible in the near future."[58] How this pool would function, given the marketing and land-tenure patterns of Argentina, Jackman did not venture to say. It was evidently this appeal that led to his appointment as the pool's Argentine representative. After he returned to Argentina in 1927, he apparently came to like the country, for he remained there until his death in 1945.

Jackman's unrealistic analysis of the future of wheat pooling in Argentina contributed to the general air of optimism that prevailed among the Canadian advocates of cooperative marketing in 1927. The apparent success of the Canadian pools, along with favorable reports such as Jackman's, seemed to convince prairie farmers that an international pool was sure to come sooner or later, and that creating it would be mostly a question of educating farmers to the benefits of cooperative marketing. G. W. Robertson, the secretary of the Saskatchewan Pool, expressed this view with near-religious fervor: "The co-operative marketing movement is on the march. . . . It will continue to go ahead and no power under Heaven can stop it."[59]

This optimism was particularly noticeable at the Second International Wheat Pool Conference, held at Kansas City, Missouri, in May 1927. It was a much larger and better planned affair than the previous year's St. Paul conference. A total of 210 delegates attended, and they came not only from Canada, but also from most of the Great Plains states, as well as from Italy, Australia, and the USSR. Jackman, who was back in Argentina at this time, reported that no pampa farm leader was able to attend, but he sent along a note from the FAA leaders expressing their desire "to join the movement of joint action."[60]

Buoyed by reports that the pool movement was reviving in the United States and growing in Australia, the conference proceeded on an air of optimism that sometimes bordered on naïveté. Emphasizing the advantages of pooling, C. H. Burnell of Manitoba told the conference that "the market price is a price made by the pool," and that consumers need not fear the pool movement because, as the Canadian experience showed, only middlemen and speculators had anything to lose.[61] The aim of a worldwide wheat co-op was not the exploitation of consumers, McPhail explained, but the stabilization of wheat prices at levels comparable with the prices of other essential commodities. Nonetheless, as J. E. Brownlee, the premier of Alberta, noted, "the ultimate result" could not be achieved in Canada until the other exporting countries adopted cooperative marketing.[62] Although Jackman sent a report from Argentina that was more realistic than his earlier ones, observing that "the prospect of pooling in the immediate future is not very good," few delegates seemed to regard Argentine recalcitrance as important, and the idea of organizing an international cooperative pool movement received wide support.[63] The conference concluded by adopting a resolution to hasten "international co-ordination" to "permit the world surplus to be placed on the market in an orderly manner." Another resolution recommended the establishment of a permanent international bureau to coordinate statistical information and to carry on the worldwide pool movement between congresses.[64]

But no one took the initiative to establish this international wheat growers' secretariat, and when the Third International Wheat Pool Conference met at Regina in 1928, it was apparent that the optimism of the preceding two years was beginning to dwindle. The McNary-Haugen high tariff bill then before the U.S. Congress was creating much ill will in Canada, and it cast a pall over the meeting. No dramatic breakthrough in organizing farmers was reported from either Australia or the United States, and again Argentina failed to send delegates to the conference.

Nonetheless, McPhail assured the conference that Canada had "a practical monopoly of the world's exportable surplus of hard spring wheat," simply blinking the fact that (as had been mentioned at the 1927 conference) Argentina's Barusso wheat closely resembled Canadian spring wheat and sold for less.[65] Another ominous note for the future of the pool movement came from the consumers' representatives who attended. Although the growers emphasized that the goal of the pool was not to raise prices "unduly," and that the producer and the consumer should "walk hand in hand towards the realization of the co-operative commonwealth," delegates from the cooperative wholesale societies of England and Scotland, which purchased wheat from the Canadian pool, were not so sanguine. One British delegate believed that pooling might in fact raise consumer prices, and that cooperative wheat marketing would succeed in the long run only if the Canadian pool would support a cooperative agency for the wholesale purchase of imports in Britain. This suggestion fell on deaf ears, for the Canadian wheat pool was interested only in co-operative selling.[66] In any case, the 1928 conference adjourned without resolving to strike out in any new directions.

The idea of international cooperative wheat marketing was in fact already doomed. The wheat-pool movement enjoyed mass support only in Canada, and even there half the prairie farmers had not joined. In the United States and Australia, the pools remained small and grew slowly. And in Argentina, the pool movement did not exist except in the imagination of a few promoters.

What was notable about the Argentine wheat economy during this period was that it performed so well, given the failure of the government to correct the glaring abuses in the marketing and land-tenure systems that had produced strong unrest among the republic's farmers since 1912. As noted earlier in this chapter, many landowners in the 1920's confronted the stagnating world beef market by taking their property out of cattle production and using it instead for cereal growing. This flexible use of the land enabled Argentine farmers to raise acreage, production, yields, and exports substantially in the late 1920's. Argentine agriculture was capitalizing on its strengths, but its weaknesses remained unchanged during that decade. Because the traditional land-tenure and marketing systems remained intact, the cooperative movement developed slowly;

and without a viable structure of rural co-ops, the development of wheat pools on a national scale, to say nothing of the international scale, was highly improbable.

Argentina's failure to reform its agrarian economy during the 1920's was certainly not a result of ignorance. Some of the republic's most prominent intellectuals and statesmen realized that the vast expansion of international trade that had enabled the Argentine wheat economy to develop so spectacularly prior to the First World War had ended, at least temporarily, and that the world wheat economy was becoming increasingly competitive. Appreciating that Argentine export agriculture enjoyed certain advantages, these voices warned that the republic must not become complacent, that it must instead work to strengthen the wheat-export sector so it could maintain its position in the world grain market. The Canadian pool's agent in Argentina, W. J. Jackman, did much to foster this reevaluation of Argentina's agrarian economy. The biting reports he published (in one, for instance, he wrote that the Argentine storage and marketing systems resembled those in use "in the days of the Pharaohs") touched off extensive discussion among the proud Argentines.[67] As the *Review of the River Plate* put it, Jackman's Argentine-Canadian comparisons were "most galling to men possessed of sincere feelings of national pride and patriotism." *La Prensa*, in a leading editorial, agreed with this pessimistic assessment and noted that "the government must bear most of the blame" for the inefficient marketing system.[68]

Argentina's best-known economist of the era, Alejandro E. Bunge, used the occasion of Jackman's visit in 1926 to publish a major article in which he compared the Argentine economy to that of Canada, "perhaps the most apt country for this kind of comparison, for it offers the most similarities with the Argentine Republic." Bunge was most unhappy with his findings. "Economic decadence" was beginning to prevail in Argentina, which had enjoyed more rapid development than Canada prior to 1908. This was particularly true of the agricultural sector, said Bunge. Canada's wheat production, which had been lower than Argentina's in 1908, was now almost double their own. What was worse, he stated, Argentine production and exports of cereals in recent years had been falling per capita. Although Bunge, who was both the chairman of the Economics Faculty at the University of Buenos Aires and the editor of the prestigious *Revista de Economía Argentina*, was a champion of industrial diversification and greater economic self-sufficiency, he did not believe that Argentina needed to sacrifice its agriculture to industrialize. What the country required, he argued, was a decisive change in government policy, particularly toward land tenure, to promote agricultural production and exports. The result, he predicted, would be that "within a few years, perhaps no more than a decade, our production per capita would double and Argentina would be one of the most prosperous and progressive countries in the world."[69]

Jackman's propaganda in favor of an international wheat pool aroused interest in Argentina but did not touch off any real attempt to develop a cooperative marketing system. Some agricultural economists believed that the Canadian pool was an example for Argentina to imitate. On the other hand, at least one prominent agronomist, Emilio Coni, pointed out that the Canadian pool sent Jackman to Argentina to defend its own interests and cautioned that Argentina might have nothing to gain from an international marketing arrangement.[70] All these writers admitted that the realities of pampa agriculture made an Argentine pool out of the question for the time being. The real problem, as Bunge argued, was the government's failure to adopt any consistent policy to develop the republic's agriculture and to correct the serious land-tenure and marketing abuses that impoverished so many of the republic's farmers. As Manuel Ortiz Pereyra, a leader in the rural co-op movement, put it, the farmer had become "that poor old ox" at the bottom of the Argentine economic structure. Few people entered farming anymore, he said, because "any other work is better." The fault, he emphasized, was that of the government, which ignored the plight of agriculture.[71]

Agrarian Reform Politics During the Alvear Presidency

The proposals of Bunge and the others for government action to promote the modernization and development of Argentine agriculture had little chance of enactment given the harsh realities of Argentine politics in the 1920's. Hipólito Yrigoyen's first presidency ended in 1922, but the Radical Party chieftain continued to cast a long shadow over Argentine political life during the six-year term of his successor, Marcelo T. de Alvear. Although he was a wealthy aristocrat (he once served as president of the new exclusive Jockey Club), Alvear had been a member of the Radical Party since his student days, but he had not become one of the party's leading figures. In fact, during the Yrigoyen government he had spent all his time outside the country, serving as ambassador to France, a country he deeply loved. In this capacity, he remained largely out of touch with domestic Argentine politics, which probably explains Yrigoyen's decision to support his candidacy in 1922. The outgoing president passed over more prominent Radical aspirants and, apparently hoping to continue to dominate Argentine politics while out of office, backed the relatively unknown Alvear.[72]

But once elected, Alvear proved to be very much his own man, and he soon clashed with Yrigoyen over key appointments, as well as over Yrigoyen's free-spending ways. This dispute deepened, and by 1924 the party suffered a devastating schism when the Alvear faction broke away entirely and formed a separate party, the Unión Cívica Radical Antipersonalista, a reference to Yrigoyen's "personalist" leadership. This schism cost the Radical Party its majority in the Chamber of Deputies and led to several years when open partisan warfare, rather than substantive discussion of Argentina's pressing problems, dominated the halls of Congress. The Antipersonalists aligned with the Conservatives, Socialists, and Progressive Democrats against Yrigoyen's Radicals, but this coalition had little in common except hostility to the former president and his followers, and as a result Congress deteriorated into a forum for the venting of political spleen for the rest of Alvear's term. It transacted very little business and enacted little legislation after the 1924 split. Instead, as the London *Times* noted, the legislators spent their time in "futile floods of oratory."[73]

Although political vendetta stalemated Congress, the Alvear government moved ahead quickly on the administrative front to rationalize and reorganize the government's departments and to reduce the favoritism and corruption that had characterized them during the Yrigoyen presidency. Alvear clearly appreciated the need for the government to act to promote agrarian development, and to head the Agriculture Ministry, he named the capable and farsighted Tomás A. Le Bretón, who had been Yrigoyen's ambassador to the United States. Working energetically, Le Bretón eliminated unnecessary positions, fired corrupt bureaucrats, and rooted field employees out of their urban swivel chairs and sent them to the countryside where they belonged. What was particularly important about the Agriculture Ministry under Le Bretón's direction was that it began to emphasize the technical and scientific aspects of agrarian development that previous Argentine governments had neglected. Le Bretón contracted Leon M. Estabrook, a Canadian expert, to head the statistical bureau, and under his direction soon began to publish reliable agricultural data for the first time.[74] Although the ministry continued to devote inadequate attention to agricultural education, it did begin to promote scientific seed experimentation and the distribution of the products of this research, something that had received less support in Argentina than in Uruguay or even in backward Paraguay. European experts arrived to carry out research in plant genetics, while the government began an informational campaign to urge farmers to use better seeds. As a result of these innovations, Le Bretón's ministry acquired a favorable reputation among farmers.[75] But as we shall see, political circumstances forced this capable minister out of office in 1925.

The Alvear government's achievements in agrarian legislation were far less impressive than its administrative accomplishments. This record did not reflect the president's lack of interest, for Le Bretón's progressive views influenced Alvear strongly. The president introduced major legislation to reform a number of major agrarian problems, and he certainly took a much more active interest in agriculture than Yrigoyen had done. Nonetheless, most of his program failed to become law. As already noted, the factionalism and rancor that pervaded Congress in the 1920's made legislation difficult to pass, but an analysis of the fate of agrarian reform in this period must also take into account the fundamental fact that the landowning elite still wielded great political power. And as in earlier periods of Argentine history, the landowners subordinated agriculture to cattle raising—despite the fact that agricultural products were a more important export item.[76]

While the cattlemen were well organized and were experienced veterans in the use of the political system to gain their economic ends, pampa farmers remained politically weak, particularly in comparison to the skilled political leverage that Canadian prairie farmers employed at both the provincial and the federal level. The FAA, the principal farmers' association in Argentina, grew rapidly between 1920 and 1930, increasing its membership from about 7,000 to nearly 32,000, but it did not become a powerful political pressure group.[77] During the 1920's the FAA actively carried out educational and propaganda activities on behalf of farmers, and by mid-decade it was beginning to experiment with a few small colonization and land-distribution plans. The circulation of its newspaper, *La Tierra*, rose from 4,000 in 1919 to 17,000 in 1924. The FAA grew prosperous, to the point where, in the late 1920's, it spent 2,500,000 pesos (over a million dollars) to erect a new six-story headquarters building in Rosario. This "palatial" structure (to use Jackman's term) symbolized more than anything the FAA's turn away from its youthful radicalism.[78] It had become an organization of larger, more prosperous farmers, and under the leadership of President Esteban Piacenza it resolutely avoided alignment with any political party, a fact that prompted sharp criticism from leftist agrarians. Criticism of the FAA's apolitical stance was nothing new: as we have seen, it dated back to the organization's earliest days. But now the FAA's critics denounced it as a conservative organization that had become part of the established rural power structure.[79] Nonetheless, the FAA's apolitical stance was hardly surprising, for the large bulk of its members—and the large bulk of all pampa farmers—remained unnaturalized and were thus unable to vote.

Criticism of its political stance was not the FAA's major problem, however. More important was new competition in the form of two cooperative associations that appeared in the 1920's. One of these groups, the Asociación de Co-operativos, was primarily a grain-marketing organization. The other, the Co-operativo Nacional de Productos Argentinos, whose 5,000 members were mostly farmers, founded the only co-op bank in the country.[80] But despite this rivalry, the FAA remained by far the largest farmers' organization and the only one with a national structure.

The continuing disenfranchisement of pampa farmers was politically costly to the agrarian movement. They did not constitute an important segment of any political party, with the possible exception of Lisandro de la Torre's small Santa Fe–based Progressive Democrats. Lacking any real power base of their own, pampa farmers were politically at the mercy of the landed elite and of the other major interest group in Argentine politics—the increasingly numerous populace of Argentina's large coastal cities. But as in most countries, urbanites—whether middle class or working class—were seldom concerned with agrarian questions beyond a desire to keep prices down. So the pampa farmers found few allies in Yrigoyen's Radical Party—whose power base was primarily urban. The other faction of the Radicals (the Antipersonalists) agreed on little more than opposition to Yrigoyen and generally did not share Alvear's interest in agrarian problems. Although the Socialists, the other major urban party, did take an active interest in agrarian affairs, they held too few seats in Congress to wield much influence on most issues. That left, besides the small Progressive Democratic Party, which was responsible to agrarian concerns, the large Conservative Party—an alliance of provincial political machines that could still mobilize a large vote. The Conservatives zealously defended the traditional agrarian order and backed agrarian legislation only when it would benefit the interests of the landowning class.

During Alvear's first two years in office, he urged Congress to approve some major agrarian reform proposals dealing with land tenure, the cooperative movement, and the marketing system. If enacted, these proposals would have established the legal structure for fundamental changes in pampa agriculture. But for the most part Congress emasculated Alvear's program, and as a result, the traditional land-tenure and marketing systems remained intact in the 1920's. An analysis of the fate of Alvear's program will reveal the extent of the difficulties and obstacles that confronted any attempt to alter the distribution of power and wealth in rural Argentina.

The pampa land-tenure system cried out for attention. As we have seen, in 1921 Congress passed a tenancy reform act (Law 11,170), but it soon became apparent that this law contained such large loopholes that landowners were able to evade its purpose entirely. As one farmer bitterly observed, *"hecha la ley, hecha la trampa"* (when the law was passed, so were the loopholes).[81] Perhaps the greatest loophole was that the law applied only to plots of 300 hectares or less. All a landowner had do to escape it was to rent in parcels of 301 hectares or more, which in fact frequently occurred. As in the years before 1921, landowners continued to refuse to indemnify tenants for property improvements and continued to force renters to purchase bags, machinery, and insurance at excessive prices.[82] On several occasions during the Alvear presidency, congressmen introduced legislation to revise the law, but all of these proposals became hostages to political infighting and died in committee.[83] As a result, the tenancy problem continued to fester until 1932, when the plight of the farmers became so serious that Congress tightened the enforcement of Law 11,170.

Although in practical terms the tenancy-reform question was extremely important to the mass of Argentina's farmers, Alvear took little interest in it, and he did not attempt to force Congress to take the loopholes out of the 1921 law. The president's attitude on this question is difficult to explain, for he clearly believed that a reform of pampa land tenure was essential. Perhaps he feared that a campaign to end tenancy abuses would deflect attention from one of the cornerstones of his agrarian program, his land reform. But by neglecting the issue, Alvear missed the opportunity to do something practical that would quickly help the mass of Argentine farmers while retaining the advantages of the renting system. Instead, he pressed for a major land reform, something that was bound to provoke powerful opposition. The result of Alvear's policy was that the pampa land-tenure system remained unchanged.

In 1924 the president introduced legislation providing for the sale of 100,000,000 pesos' worth of bonds a year to fund the government's purchase of large estates suitable for agriculture, which would be subdivided and sold to individual farmers. This sweeping proposal, which would have radically altered the pampa economy and society, was the brainchild of Le Bretón, who was convinced that Argentine agricultural production and exports were stagnating dangerously under the traditional land system. He believed that the insecurity of tenure not only made agriculture an unattractive way of life, but also encouraged speculative planting and low yields.[84] Leon Estabrook, Le Bretón's Canadian adviser, summarized the advantages of this plan (which Estabrook probably helped formulate) in a letter to the *Review of the River Plate*:

The benefits that flow from subdivision and colonization of agricultural lands are fully demonstrated in Argentina itself. One simply has to compare the population, production, wealth, prosperity, good roads, schools, automobiles, high standard of living, cooperation, public spirit, and optimism of the Department of Castellanos in the Province of Santa Fe [the location of successful nineteenth-century colonies] with those of any other department or *partido* in the republic to realise the tremendous differences that proprietorship of small holdings of agricultural lands makes in the economic life of the people.[85]

Farmers throughout the pampas wholeheartedly agreed and endorsed Alvear's plan. Within ten months over 14,000 telegrams arrived in Congress to urge passage.[86]

And the farmers had a good deal of support from other influential sectors of Argentine society. Several prominent economists and agrarian experts, including Alejandro Bunge, applauded Alvear's scheme and in-

voked the example of Canadian land policy, which the agronomist
Francisco Bórea called "farsighted, energetic, and well organized" as a
model for Argentina. The agronomist Pedro Marotta saw land distribution as "true economic nationalism," and argued that "by distributing
land, we will prevent Lenins from arising in Argentina."[87] Support for
the president's plan also came from the British-owned railways, some of
which had been advocating land redistribution since 1921. Particularly
vocal in this regard was Viscount St. Davids, chairman of the Buenos
Aires and Pacific Railway, who wanted to increase freight traffic by developing agriculture along his company's routes. As he told his company's annual meeting in 1921: "Cattle are no good to a railway. Where you
get a great cattle ranch you get a few special trains with cattle in the
course of the year but you get nothing else. On a great cattle ranch there
are perhaps half a dozen peons. . . . It is nothing. What does help you is
close cultivation and particularly grain." St. Davids went on to tell the
meeting that the government should settle farmers by buying up and sub-
dividing land if necessary.[88] Indeed, the railways were already making
clear their commitment to raising pampa agricultural productivity: several of the British companies instituted their own agricultural extension
and educational services during the 1920's, and these services functioned
effectively.[89]

Like Bunge, Bórea, and Marotta, British economic interests in Argentina that backed land reform were keenly conscious of the Canadian
homestead model—and of Canada's competition with Argentina for potential agrarian immigrants. Commenting on a scheme that the Canadian
Pacific Railway was formulating to attract Central European immigrant
farmers to the prairies, the Buenos Aires *Herald* in 1923 argued that "Argentina has yet to learn that what she needs more than anything else is a
subdivision of opportunities" and asked, "If the CPR can settle farm hands
in Canada is there any reason why a similar plan would not prove equally
feasible in Argentina?"[90]

Such strong statements from the British economic elite no doubt influenced Alvear, who was a pronounced Anglophile, to introduce his 1924
land legislation, but the president's plan encountered a frigid reception
from the Sociedad Rural, the association of large landowners that was the
most powerful interest group in Argentina. The Sociedad's official journal flatly opposed Alvear's proposal. Not only was it too vague, said the
journal; it also was unnecessary, for many farmers were becoming
wealthy enough to buy at market rates.[91] And indeed, as we saw in Chapter 4, the number of farmer-owners did rise substantially in the 1920's.
The landowners' opposition guaranteed tough political sledding for Alvear's plan. The U.S. embassy, which kept close watch on the agrarian

situation in this South American competitor, reported (in the words of Vice Consul E. K. Ferrand): "The fate of the project is doubtful. The present administration does not possess a working majority in Congress and . . . it must not be forgotten that the large landowners are the real rulers of the country today. . . . It is generally felt that this class will not subject willingly to measures of this nature."[92]

As Ferrand predicted, the project soon died. Although Le Bretón lobbied vigorously for it in the Chamber's Agrarian Legislation Committee, that group pigeonholed the legislation in late 1924, and it never came to the floor for a vote.[93] And the following year, in an attempt to repair the disastrous schism in the Radical Party, Alvear abandoned Le Bretón. The most militant Antipersonalists in the president's cabinet were urging him to strengthen his faction's political position by sponsoring a federal intervention in Buenos Aires province, an Yrigoyen stronghold. Their goal was to prevent the election of an Yrigoyen Radical in the 1925 gubernatorial elections, but Alvear was appalled at the results of the party schism. He began to plan for the eventual reunification of the Radical Party, and he therefore backed a fusion candidate for the governorship whom he hoped both Personalists and Antipersonalists would support. This policy of conciliation with Yrigoyen caused a cabinet crisis: Interior Minister Vicente Gallo, the architect of the party schism, resigned in protest in July 1925, followed by Le Bretón in September.[94] One may speculate whether Yrigoyen, who never had backed land reform and who owned a number of ranches himself, might have demanded Le Bretón's departure as part of the price for an easing up in the Personalists' political warfare against the president. The sources are unclear on this point, but in any case the 1924 land-reform proposal remained dead and buried for the rest of the Alvear presidency. Le Bretón's successor, Emilio Mihura, was a professional engineer who gave the Agriculture Ministry competent technical leadership, but he was not interested in rural structural change.

Yet President Alvear remained personally convinced that the tenant farmers must have greater access to landownership, and to achieve this end he attempted to induce the British railways to begin their own settlement scheme. In March 1927 Alvear invited officials of the British railways to meet with him at the Casa Rosada to discuss reviving his stalled land reform. He asked the railways to consider forming a consortium with a fund of 40,000,000 pesos for the purchase of land. (Because of the high cost of pampa land, this sum would have been enough to buy land and provide improvements for about 1,250 families). This colonization company would then sell 50- to 100-hectare plots to farmers at cost plus 10 percent. These "colonists" would have the option of long-term repayments. The companies, which apparently considered the plan primarily as good public relations, signed a preliminary draft agreement, but they balked at providing such a large land-purchase fund. Instead they agreed to provide only a tenth of what Alvear proposed—about 4,000,000 pesos.[95] But even this reduced program soon bogged down, for the companies became involved in a rate dispute with the government (Chap. 6), and by the time the dispute was settled, the 1928 presidential elections were approaching. The incoming Yrigoyen government adopted conciliatory policies toward the railways, but Yrigoyen was uninterested in land reform and did not press the railways to revive the rural-settlement plan. Alvear's ambitious land-reform proposals thus came to naught. The Buenos Aires and Pacific Railway managed to settle 40 Italian immigrant families in 1927, but that was the only result.[96]

The Alvear government was more successful with its cooperative policies. Argentine rural cooperatives (and in fact all cooperative societies) had developed slowly, particularly in comparison with the flourishing cooperative movement that was the hallmark of the Canadian prairies. As late as 1925 there were only 92 rural co-ops, whose total membership was about 23,000, in Argentina. Most of these restricted their functions to selling basic consumption goods and insurance policies; only a few experimented with marketing. All but a few farmers were thus still dependent on traditional marketing systems, which tied them to the rural merchants who extended them credit and later purchased their crops at low prices. The exploitative relationships between rural businessmen and pampa farmers that we have examined during the prewar period continued unabated through the 1920's. In 1926 the co-op movement leader Manuel Ortiz Pereyra charged that farmers were being impoverished by the "parasitic voracity" of these rural merchants. He provided an example: "There are merchants . . . who plot with the insurance companies to force farmers in regions where hail has never been recorded to buy policies against this risk, under the threat of shutting off credit to them, and these merchants retain half of the premiums gained in this fashion."[97]

Argentine agrarian experts considered the stunted growth of the cooperative movement one of the most lamentable aspects of the republic's agrarian development, but most of them agreed that the movement would never flourish as long as short-term tenancy was the central feature of Argentine farming.[98] Although the cooperative had established strong roots among the Jewish farmers of Entre Ríos, most of whom were owners or were buying land from the Jewish Colonization Association, Entre Ríos was an exception. Even in the districts of Santa Fe and Córdoba provinces where nineteenth-century land-distribution schemes had created a rural small-holding class, and one might therefore expect to find the cooperative flourishing, agrarian cooperation remained weak during

TABLE 9.5

Growth of the Argentine Agrarian Cooperative Movement, 1925–1939

Year	Number of co-ops	Membership	Year	Number of co-ops	Membership
1925–26	92	22,775	1932–33	251	38,219
1928–29	143	25,098	1935–36	163	29,229
1930–31	161	23,408	1938–39	163	37,048

SOURCES: Argentina [4], p. 282; Bórea, *Tratado*, p. 268.

the early 1920's. The reason for this was that Argentina had no basic legislation defining the legal identity, rights, and responsibilities of cooperatives. and that without such legislation, they remained a risky venture.

As part of his agrarian program, Alvear proposed two bills to Congress in 1924 to give co-ops legal status and to make financing available to them. These proposals authorized cooperatives to construct grain elevators at rural railway stations and empowered the National Mortgage Bank to loan funds for this purpose, as well as for the purchase of land to subdivide among co-op members. The Alvear program also gave the president authority to set up a new bureau of the Agriculture Ministry to register cooperatives, inspect their finances, and promote the cooperative movement.[99]

The cooperative legislation was the only agrarian reform that the Alvear government succeeded in enacting. If it had been accompanied by the president's land-reform plan, as he had apparently intended, pampa agriculture would have acquired the legal structure essential for the growth of a class of well-established farmer-owners. But as we have seen, the land-redistribution plan died in committee, and without it the cooperative legislation was of limited utility. It helped farmers who were already owners (and who lived in areas where owning rather than renting prevailed), but it did little or nothing for the great mass of tenants. Apparently because the cooperative legislation did not directly threaten the existing distribution of wealth and power in the countryside, Congress approved the two new laws in 1926. Although opposition from rural commercial interests surfaced during the debates on the two laws, none of the major political parties actively opposed the legislation, and two of them—the Socialists and Alvear's Antipersonalists—backed it strongly. The Sociedad Rural did not actively oppose the legislation.[100]

The 1926 laws, as Table 9.5 shows, did not spur a rapid expansion of the Argentine rural cooperative movement. By the end of the 1920's the number of rural co-ops had increased substantially, but most of these new associations were small, for total cooperative membership grew only about 10 percent. The failure of cooperatives to grow more rapidly reflected the fundamental fact that short-term land tenancy and the cooperative movement were basically incompatible.

What the 1926 laws did accomplish was to strengthen existing cooperatives, which now had access to government financing. Several co-ops in Santa Fe, Córdoba, and Entre Ríos secured grain-elevator loans from the National Mortgage Bank, and four years later Argentina's first rural grain elevator began business at Leones, Córdoba. By 1937 cooperatives marketed 197,000 tons of wheat and 519,000 tons of all types of grain— still only a small fraction of total production.[101] The provision that permitted cooperatives to buy land was little used in the late 1920's, partly because of the high cost of land and partly because of the disorganized and apparently corrupt condition of the National Mortgage Bank. As we saw in Chapter 8, a 1919 law enabled the bank to make loans to individual farmers; the 1926 legislation extended this authority to cooperatives. But during the whole of the 1921–30 period, the bank made only 4,097 loans for land purchase, and a mere 565 of these were made after 1926. The problem was that the bank did not have the power to acquire and sell lands; it was only allowed to make loans on lands whose owners had previously defaulted. Much of this land was marginal at best, and very little of it was in areas where co-ops functioned.[102] Like every other government institution on the pampas, the National Mortgage Bank existed primarily to serve cattlemen, not farmers.

The third major agrarian-reform issue that faced the Alvear government was Argentina's grain-storage and marketing system. As we have seen, the pampas lacked grain elevators, wheat was transported in bags, and the result was an inefficient and costly system that not only imposed a severe financial burden on the farmers, but also prevented any effective method of standardizing and grading grain, as its Canadian and U.S. competitors had been doing for some time. For many years Argentine economists and agricultural experts had been urging the government to promote the construction of a network of grain elevators, and the congressional archives bulged with moribund legislative proposals that dated back at least to 1912.[103] In 1924, as part of their agrarian-reform package, Alvear and Le Bretón introduced legislation to build such a network and to control the cereal trade "in the collective interest," but like the land-reform legislation, this proposal died in committee.[104] The government took no further action until the Sociedad Rural became interested in the grain-elevator question.

The powerful landowners' association opposed land redistribution as a cure for the pampas' agrarian ills, but the Sociedad was well aware that something had to be done to increase agricultural efficiency—and also the farmers' returns so as to avoid a decline in the land-rental market. After

W. J. Jackman published his scathing condemnation of the pampas' grain-marketing system, grain elevators became a major public issue, and for the first time the Sociedad seriously investigated the question. As Luis Duhau, its president, later explained, "Mr. Jackson [sic] has been unjust with the Egyptian kings. They knew how to listen to the advice of Joseph and they built enormous graneries. . . . A simple dream was enough to show them the gravity of the problem of grain storage." But in Argentina, he continued, "we scarcely have more than a few miserable sheds where the rats eat and infest our grain."[105]

Duhau was so concerned about the storage problem that in 1927 he made a trip to the prairies to study Canada's grain elevators, and when he returned to Argentina, he published a pamphlet to express his conviction that "we must change our system."[106] "I arrived in Canada," he related, "with the idea that the main advantage of grain elevators was that they made bags needless." But, Duhau continued, "reality soon taught me" that the "foundation stone" of the Canadian system was "free competition, thanks to the provisions of the Grain Act." If Argentina was to advance its agricultural economy, it must undertake a fundamental reform of its storage and marketing systems, and this should include elevator construction and a national grain law on the standardization and grading of wheat. For too long, he emphasized, "we have stumbled from one bonanza to another." Now the state would have to act.[107]

Faced with this demand for action, in early 1928 the Alvear government named a special commission to survey the various groups involved in the grain trade about possible courses of action. The FAA, the Sociedad Rural, and the railway companies all agreed that grain elevators were "essential." The government's survey concluded that rural elevators would save 30 centavos for each 100 kilos shipped, primarily because bags would no longer be needed and because labor costs would be reduced. Railway rolling-stock productivity would rise 11 percent, and less damage to the grain would occur. Despite these economic advantages, the government's survey found that the powerful Grain Exporters' Exchange opposed any national elevator plan. Bunge y Born was willing to support a system of elevators, but only on the condition that the private grain firms operate the elevators, a condition that the farmers' organizations resolutely opposed. Indeed, the survey revealed strong disagreement among the supporters of elevators about whether the farmers, the state, or the grain trade should own and operate them.[108]

With only a month to go before his presidential term expired, Alvear asked Congress for approval to sell bonds in the amount of 80,000,000 pesos to finance a state-owned elevator system that would be operated by an autonomous government commission to be named by the president. Apparently Alvear took this action only to register his views on the elevator question, since Congress refused to consider such important legislation until the new president was inaugurated, and his views on the matter were clear.[109]

Policy Paralysis During the Second Yrigoyen Presidency

Agrarian reform and development continued to receive low priority during the second Yrigoyen presidency, which lasted from 1928 to 1930. Hipólito Yrigoyen, who had maintained his popularity among the native-born masses despite his seventy-six years, swept to power with a large majority of the popular vote, and the magnitude of his victory stirred hopes that at long last the government would enact much-needed reforms. The Radical press fed these hopes: *La Epoca* assured its readers that the new administration intended to "aid the noble producers" by promoting the "development and protection of agriculture."[110] But the aged Radical leader was unable to fulfill these great expectations, for he spent most of his second term engaged in violent political vendettas. Yrigoyen's primary goal was to gain a Radical majority in the Senate, and to this end he engaged in controversial provincial interventions that soon had Argentine politics in an uproar. Practical matters, such as agrarian legislation, had to await the resolution of this political crisis.

Indeed, the second Yrigoyen presidency was a period of virtual policy paralysis in the agrarian sector. The president proposed no land reform or colonization policies, nor did he suggest controls or reductions of railroad rates, which farmers considered exorbitant. Despite widespread support for the construction of a grain-elevator system, Yrigoyen put forward no legislation on the subject.

The only two agricultural reforms that the administration did propose were programs that had long been bogged down in the political muck of the era: revision of the 1921 land-rental act and creation of a system of rural credit. In September 1928, despite the opposition of the Sociedad Rural, the Radical-controlled Chamber approved the administration's proposal to strengthen and reform Law 11,170. A few months later the Senate approved the reform, but the upper house introduced some modifications in the Chamber's version.[111] But from that point on, tenancy reform fell victim to the intense political hatreds that polarized Congress into warring factions during 1930. The Chamber, which spent all its sessions between May and August debating the credentials of three anti-Yrigoyen Radicals legally elected from Mendoza and San Juan provinces, was unable to constitute itself and could transact no business. Eventually, the president had his way, and the Chamber refused to seat the three deputies-elect, but vital legislation, including consideration of the Senate's modifications of the land-rental law, was left waiting.[112] A similar fate befell the administration's proposal to establish a national agrarian bank. Passed by the Chamber in 1929, the proposal was not considered by the Senate that year, and in 1930 the upper house held no sessions, for it could not legally convene until the Chamber had constituted itself.[113]

Mired in inaction on the legislative front, the Yrigoyen government also failed to administer existing agricultural legislation effectively. The Ministry of Agriculture, which had shown some signs of life under Alvear, slid back into incompetence and corruption. Although the ministry's share of the government budget rose slightly in 1929 and 1930, the quality of its services declined. Agricultural statistics, which the *Review of the River Plate* described as in a state of "collapse," exemplified the ministry's sad state. The publication of many important statistics was abandoned, and other data appeared so tardily as to be useless.[114]

The one occasion during Yrigoyen's second term when he vigorously used his executive powers in the agricultural policy area occurred in late 1928, when bitter strikes of harvest workers broke out in Santa Fe for the first time since 1920. As in his first term, Yrigoyen made it clear that he would not tolerate rural labor unrest that might threaten production and exports. The strike, called to press for higher wages and the recognition of rural unions, led to violence and threatened to stop agricultural production in southern Santa Fe.[115] After the police failed to enforce the "right to work," the national government took the highly unusual step of sending two army regiments to Santa Fe without any request for assistance from the provincial government. Justifying this tough approach by blaming the trouble on "professional agitators," Yrigoyen swiftly broke the harvest workers' strike. The FAA, which had opposed occasional congressional proposals during the 1920's to improve the rural workers' living and working conditions, applauded Yrigoyen's use of the army.[116]

But the FAA had little else good to say about Yrigoyen's second term. Indeed, by mid-1930 farmers were echoing other segments of Argentine society concerning the aged president's failure of leadership in the midst of a national economic crisis. When a military coup unseated Yrigoyen on September 6, 1930, the FAA shed nary a tear for the deposed president. Yrigoyen, editorialized *La Tierra*, was a man incapable of ruling, who had "vegetated, . . . surrounded by deceiving and ignorant advisers." Argentine politics, the FAA newspaper concluded, had done nothing fundamental to aid or develop agriculture.[117]

The 1920's: A Summation

La Tierra's conclusion was accurate. Neither the Alvear regime nor the second Yrigoyen government had acted decisively to confront the major problems that beset pampa agriculture. During 14 years of Radical Party rule, the Argentine government had passed little legislation of importance to farmers and had enforced little of what was passed. The Alvear government did make important initiatives in land-tenure and marketing policy, but the political weakness of that government prevented action. And the second Yrigoyen presidency was in almost constant political crisis. Again, agrarian legislation was put on hold.

As in earlier years, Canadian agrarian affairs during the 1920's contrasted with the situation in Argentina. While Argentine farmers remained too politically weak to bring about even minimum reforms, the farmers of the Canadian prairies gained significant policy concessions from the Mackenzie King government. While the co-op movement lagged on the pampas, it flourished on the prairies, and the biggest cooperative of them all, the wheat pool, appeared to be very successful. While traditional—and increasingly antiquated—institutions prevailed in Argentine agriculture, the decade of the 1920's was a period of fruitful innovation in Canada. This contrast reflected the distinct differences in the political systems of the two countries: the agrarian producers in Canada had political power and did not hesitate to use it, whereas their powerless counterparts in Argentina were at the mercy of governments that did not take much interest in agrarian affairs.

10

Conclusion

By 1930 the main patterns of agrarian development in the two New World grassland regions had become readily apparent. Some of these patterns were similar. Over the past half century the prairies and the pampas had developed at an astonishing rate. As late as 1880 the two regions had been vast, mostly empty oceans of waving natural grasses, populated only by the Indians and the métis (or their gaucho counterparts in Argentina). During the decades that followed, the demographic and economic structure of the prairies and the pampas was transformed. Millions of immigrants rushed in to make Canada and Argentina world leaders in the production and export of cereals, especially wheat. The open grasslands were gone forever, and in both countries a rural society of small farmers emerged.

The prairies and the pampas became great agricultural regions in response to the rapidly rising world demand for cereals. But in fact two very different kinds of agrarian societies emerged in Canada and Argentina. The prairies were a society primarily of small owner-operators. Although most were immigrants, they quickly became citizens (indeed, they had to become citizens in order to receive title to their homesteads). They formed vigorous and large cooperative movements. Through these voluntary associations as well as through politics, the prairie farmers lobbied hard to protect and advance their interests. And by the 1920's they could claim some major victories. On the cooperative front, they had formed a gigantic farmer-controlled marketing plan, the Canadian wheat pool. On the political front, they had scored notable victories, including the Crow's Nest Pass Rates (which kept railway freight charges down), the Canada Grain Act (which set up a government-administered system of grading and standards that was the best in the world), and a series of outstanding government rural research institutions and experimental stations (whose work greatly improved the feasibility and profitability of prairie agriculture).

The pampas, in contrast, were a society of tenants who rented land on short-term contracts and who moved about frequently. As in Canada, these farmers were overwhelmingly immigrants, but few became citizens. Because only a minority of pampa agrarians owned their land and formed stable rural communities, the cooperative movement long remained small and weak. Unable to vote (and faced with governments that erected numerous barriers to citizenship), pampa farmers remained a marginal group in Argentine society. Cooperative marketing associations were unknown, and farmers had to rely on a primitive, exploitative marketing system. In the absence of proper storage facilities and standardized grading, Argentine wheat suffered from an inferior reputation in the world market—at least in comparison with the Canadian product. And it was not only in marketing where Argentina was behind Canada. During the 1880–1930 period the Argentine government enacted no significant legislation to develop agriculture or to reform the rural institutional structure. Railway rates were high and constantly increased. Agricultural research and technology were little developed.

Why did such vast differences occur in two societies that emerged in new countries at about the same time, and in response to similar market forces? Reliance on dependency theory, which would have us believe that the raw-material-exporting countries are exploited economically by the world economic metropolis, does not explain much in this case. For Canada and Argentina not only exported wheat to the same markets and at the same world market prices; both also prospered greatly during the period of the wheat export boom. By the end of the 1920's per capita incomes in both countries were higher than in many of the countries in Europe.

Dependency theory also argues that the nature of a country's export trade and the nature of the foreign investments made in the country strongly affect its domestic political system. Put simply, the argument is that the political elites in raw-material-exporting countries do the bidding of the industrialized powers that are the main sources of investment as well as the principal markets. In the case of Canada, this aspect of dependency theory finds little confirmation. The government forced the powerful Canadian Pacific Railway (owned primarily by foreign investors) to lower grain rates in 1897 and to restore those low rates in the 1920's. And the British loathed Prime Minister Mackenzie King, who campaigned successfully for reelection in 1926 on a platform of "Independence vs. Colonialism." Perhaps one can argue more effectively that the imperial connection decisively influenced Argentine politics. During the First World War the British forced the Yrigoyen government to sell wheat at terms that were widely regarded in Argentina as scandalous. And various Argentine governments, in decisions that were highly unpopular, allowed the British-owned railways to raise their rates. The British could wield a powerful threat over Argentina—to close or restrict their markets to food imports from outside the Empire.

But there were definite limits to how far the British could push this strategy. At home political opposition to taxing the workers' food was a potent argument, while in Argentina the hundreds of millions of pounds of British investments could be taken hostage by an Argentine government should British economic demands become unreasonable. Even during the dark days of the 1930's, when Britain did restrict its imports of Argentine meat (but not nearly to the extent that the Australians wanted), the United Kingdom raised no effective tariff or barrier to the free import of Argentine wheat or other cereals. The bottom line was that the British needed Argentina as much as the Argentines needed Britain.

Even more suggestive is the fact that on certain key issues, the British investors urged the Argentine government to adopt a *more* progressive and reformist position than it was willing to take. Because the government's agricultural-research program was so ineffective, the British railways established their own experimental stations in the 1920's, and these stations made notable contributions in plant genetics and seed selection. The British companies also pressed the Argentine government to promote land reform to increase the rural population and thus bolster railway traffic and profits. The railways even indicated their willingness to help finance a pilot land-distribution and colonization project. The weak and ineffective Alvear government failed to follow up on these promising leads, but what is of interest here is the British recognition that Argentine rural society needed reform.

More rewarding than dependency theory as an analytical framework to explain the divergences between the paths of agrarian development in the pampas and the prairies is the concept of the role of the state. And the central importance of government policy is clear when one considers the different land-tenure structures of Argentina and Canada. The most fundamental single difference between the prairie and pampa agricultural sectors was the land-tenure system, and these land-tenure patterns resulted from a series of political events. In the case of Canada, John Macdonald's government passed a liberal homestead act and encouraged immigrant farmers to become citizens because Macdonald realized that any less liberal policy would divert the streams of immigration to the competing states across the border. If the United States had a homestead policy, Canada had to have one too or face the possibility of a barren and depopulated prairie region. In the case of Argentina, the 1853 constitution left public lands in the hands of the provinces and gave the federal government jurisdiction over public land only in the national territories. This meant that Argentina had no national land policy, and that the financially hard-pressed provinces used their lands to raise funds, not to develop a small-farmer class. This Argentine constitutional policy could be interpreted as reflecting the desire of powerful elites to protect their economic positions (and indeed by the 1890's that was true), but in 1853 the land-policy provisions were part of a delicate compromise between the provincial and federal governments aimed at ending decades of civil conflict.

Canada had a national policy of land distribution to the immigrant masses and Argentina did not. From this basic political-historical difference flows many of the other contrasts between these two agrarian societies. One of the reasons why Argentina was plagued by such a primitive grain-marketing system was that no one would build country grain elevators when the rural population was constantly on the move, and when any given part of the pampas might be producing wheat one year and alfalfa for cattle raising the next. Certainly the land-tenure system retarded the growth of agrarian cooperatives in Argentina, and without co-ops the agrarian population had nowhere else to turn but the private grain merchants. One might argue that an equally important cause of the weakness of pampa co-ops was Argentina's lack of basic enabling legislation. This point is suggestive, and it brings up the whole question of the relationship between the land-tenure system and political power in agrarian societies.

In Canada a democratic land system led to progressive and reformist agrarian policy at both the provincial and the federal levels of government. The Liberal government of Laurier (1896–1911) and the Conservative government of Borden (1911–20) both attempted to promote reforms to attract agrarian political support, but the farmers wanted more sweeping change and especially an end to the protective tariff. By 1921 the prairie agrarians had taken control of all three prairie provinces and sent a strong contingent to Ottawa in the form of the new Progressive Party. The Progressives held the balance of power in the House of Commons for several years, and because of this position, they were able to exchange their political support for significant policy concessions from the Mackenzie King government—especially on the crucial issue of railway rates. Prairie historians, resentful of what they see as the exploitation of the West by the Eastern financial-industrial elite, argue that the elite would only tolerate agrarian reforms that benefited the entire national economy. This may well be true in a number of cases, such as the enactment of the Canada Grain Act, but it is also true that prairie political pressure induced the Liberal and Conservative party leadership to move much farther in the direction of reform than they would otherwise have done. Again the railway rates issue provides an important example. Mackenzie King defied the "Government on Wheels"—as the CPR was called—and the Eastern political establishment to restore the Crow's Nest rates after the First World War. And this rate structure helped Canadian wheat compete effectively in world markets.

The land structure of Argentina led to a far less democratic brand of agrarian politics than in Canada. The few thousand large landowners who owned most of the pampas had enormous political power. Well organized in the elite cattlemen's association, the Sociedad Rural, the landowners were a powerful voice in both the Radical and the Conservative Party. Strongly opposed to any plan to redistribute pampa rural property or to alter the main features of the tenant-farming system, the ranching elite subordinated agricultural development to cattle raising. Deprived of the vote by restrictive citizenship laws and, in many cases, not intending to remain permanently in Argentina, the pampa farming population was totally unable to mount a political challenge against the landed elite. The most the farmers were able to do to advance their political interests was to go on strike, but these agrarian strikes achieved little and invited violent government repression. Lacking any political imperative for action, no Argentine government between 1880 and 1930 adopted any serious or consistent policy of agricultural development. The landed elite and the nation's political leadership seemed to feel that government programs to promote agriculture were unnecessary, because nature had endowed the pampas so bountifully. As one writer in the *Anales de la Sociedad Rural Argentina* put it in 1931: "We are optimists, because to be a pessimist in the Argentine Republic is equivalent to opposing nature herself. . . . We are optimists because no other country in the world has such a high proportion of fertile lands, capable of producing with minimum effort the most delicious and varied products, because of the ideal climate, the ample rains, the radiant sun."[1]

The differences between Canadian and Argentine agricultural policies can be aptly summarized by comparing Mackenzie King and Yrigoyen, the dominant political leaders of their countries during the 1920's. Both were consummate politicians (or at least this was true of Yrigoyen until 1930, when he apparently became senile and lost touch). Both identified the cause of their party with moral righteousness and the national good, and both were utterly devoted to building the political strength of the Liberal Party on the one hand and the Radical Party on the other.

But Mackenzie King gave agrarian policy high priority and Yrigoyen did not. It is doubtful that this difference arose, on King's part, from any particularly profound understanding of the problems prairie farmers faced. Rather, he realized that the future of the Liberal Party (which he identified with the future of Canada) required strong political support in the prairie West. Unlike later Liberal leaders, King was unwilling to leave the party in the hands of Ontario financiers and French-Canadian bureau-

TABLE 10.1
"Big Four" Wheat Exports as a Percentage of the World Wheat Trade,
Five-Year Averages, 1922–1982

Period	Argentina	Canada	Australia	United States
1922–27	17.5%	36.9%	11.3%	23.2%
1927–32	20.4	34.6	13.7	17.7
1932–37	24.1	37.4	19.1	9.0
1947–52	8.3	25.9	11.8	44.5
1952–57	10.7	28.7	10.4	31.8
1957–62	5.9	22.5	9.0	37.6
1962–67	7.0	23.7	11.6	37.7
1967–72	3.6	20.2	14.7	34.1
1972–77	4.8	19.5	11.1	46.0
1977–82	4.4	18.7	12.8	45.0

SOURCES: *1922–37.* De Hevesy, pp. 344, 356, 394–95, 665, Appendix 9. *1947–82.* U.S. Dept. of Agriculture, Economic Research Service, *Wheat Situation* (titled *Wheat Outlook and Situation* after 1980), various issues, Oct. 1958–Aug. 1983.

crats. To be successful, the Liberals had to be a true national party, which meant that they had to make political concessions to the prairie agrarians.

Yrigoyen, of course, faced a wholly different set of political circumstances. The key to political power in Argentina was control of the large and populous province of Buenos Aires, which was heavily urbanized because it contained the big suburbs surrounding the Federal District (the city of Buenos Aires).[2] Buenos Aires also contained a large agrarian population, but most of it did not vote. Yrigoyen was well aware that Radical control of the province, along with the capital city and a few large interior provinces, was the path to power, and he formulated his political program carefully to appeal to the native-born working-class and lower-middle-class masses of these provinces. It was a populist program with vague nationalistic appeals and overtones of social reform. Yrigoyen held it all together with the mystical, quasi-divine image he so carefully cultivated. But left out of the Radical leader's political calculations, because they really did not count, were the pampa farmers.

Despite the lack of a national rural-development policy, Argentine agriculture was reasonably successful during the 1920's. As long as masses of impoverished southern Europeans were willing to gamble on pampa tenant farming, the agrarian system functioned effectively. But in 1930 the flow of immigrant labor that had been the cornerstone of Argentine agriculture ceased and never revived. Although Argentina held a large share of the world wheat market in the 1930's (largely because of low U.S. exports during the Dust Bowl era), the collapse of farm prices began to drive the pampa population out of agriculture and toward the cities. After the 1930's Argentine governments became steadily more devoted to industrial development rather than to agriculture, and Argentina lost its once-proud position as a great wheat exporter. No government of this era moved to correct the situation, and all the weaknesses of pampa agriculture that were becoming apparent before 1930 intensified.

Argentine agriculture continued to stagnate after the Second World War. As Table 10.1 shows, Argentina's position as a major wheat exporter virtually collapsed by the 1960's. While in the 1920's and 1930's it exported as much as 20–25 percent of the world's total wheat exports, by the 1960's it accounted only for 4 percent to 7 percent. Canada's share of the world market also declined—from about 35 percent in the late 1920's to about 22 percent in the 1960's, but the difference between the relative Argentine and Canadian performances was that the *quantity* of Canadian wheat exports increased greatly, whereas Argentina exported fewer bushels of wheat in the 1960's and 1970's than it had 40 years earlier (Table 10.2).

What had happened was that total world wheat exports expanded enormously in the postwar era, largely because of the huge U.S. crops that came on the market. Because the trade was growing so rapidly, the quantity of Canadian wheat exports could grow but still decline as a percentage of the world market. But in the case of Argentina, the market share collapsed *and* the quantity of wheat exported fell. It was a dismal performance for a country that had once been a leader in the world wheat trade. And the Argentine export problem involved more than wheat. The nation's overall position as an exporter of rural products declined seriously between the 1920's and the 1960's. Production and exports of all major cereal crops and oilseeds stagnated or fell.[3] This catastrophic drop in exports took place not because of lack of markets—indeed, the world wheat trade was expanding rapidly. Pampa agriculture was in crisis because it had become unprofitable.

TABLE 10.2
Wheat Exports of the "Big Four" Countries, Five-Year Averages, 1922–1982
(millions of bushels)

Period	Argentina	Canada	Australia	United States	World
1922–27	135.9	286.8	87.9	180.9	776.9
1927–32	163.3	277.8	110.1	144.1	808.1
1932–37	138.6	214.5	109.8	1.9	571.9
1947–52	77.2	246.2	111.5	425.6	971.2
1952–57	96.3	273.3	97.7	340.3	1,061.2
1957–62	83.1	320.4	135.2	547.6	1,469.2
1962–67	146.2	482.1	233.0	758.5	2,058.0
1967–72	77.2	337.3	255.0	632.1	1,866.7
1972–77	82.4	471.1	263.8	1,023.8	2,345.2
1977–82	148.5	546.1	380.2	1,242.1	2,913.7

SOURCES: Same as Table 10.1.

While governments in Canada and the United States moved ahead rapidly after the Second World War to promote agricultural research and development, Argentine governments taxed agriculture heavily but did very little to develop the rural sector. The tradition of neglecting agriculture that had already become apparent before 1930 intensified in the postwar years, when Argentina embarked on an ill-advised large-scale industrialization program (largely financed by various taxes on the agrarian sector). While Canada after the war expanded agricultural *and* industrial production, in Argentina industrial output grew while agricultural production and exports stagnated or declined. Since Argentine industrial products could not compete on the world market, the republic still relied on agricultural exports to earn vital foreign exchange. Decades of neglect, however, had left pampa farming in no position to expand production. The Radical government of President Raúl Alfonsín, which came to power in 1983, promised to give agriculture high priority and to provide incentives to greater production. And indeed Argentine cereal exports did begin to rise rapidly after Alfonsín took office. But it remains to be seen whether the political leadership can make the reforms necessary to restore Argentina to its position among the leading world cereal exporters.[4]

Although Canadian agricultural economists and agrarian leaders argue that much needs to be done to improve agricultural productivity in the prairies, it is clear that the solid institutional structure that prairie grain growers and the Canadian state established prior to 1930 has provided an excellent basis for an expansion of the wheat economy.[5] While it is certainly true that prairie farmers deeply resented the Liberal Trudeau government (1968–84) as insensitive to the needs of the wheat economy, it is also certainly true that from a long-term comparative perspective Canadian agricultural policy has been remarkably successful.

Statistical Appendix

TABLE A.1

Wheat Agriculture in Argentina and Canada: Area, Production, Yields, and Exports, 1884–1931

Crop year	Area (millions of acres sown)		Production (millions of bushels)		Yields (bushels sown/acre)		Exports (millions of bushels)	
	Argentina	Canada	Argentina	Canada	Argentina	Canada	Argentina	Canada
1884–1889[a]	2.0	2.6	19.9	38.3	10.0	14.8	4.1	4.2
1889–1894[a]	3.5	2.9	47.3	41.0	13.6	14.2	28.0	7.4
1894–1899[a]	6.2	3.6	59.6	45.4	9.6	12.5	29.4	13.7
1899–1904[a]	8.9	4.1	93.3	74.0	10.5	17.9	54.9	25.7
1904–1909[a]	13.9	5.6	158.1	101.2	11.4	17.9	108.9	43.7
1909–1910	14.4	7.8	130.7	166.7	9.1	21.5	75.1	67.8
1910–1911	15.4	8.9	145.7	132.1	9.4	14.8	90.0	62.4
1911–1912	17.0	11.1	165.8	231.2	9.8	20.8	103.3	97.6
1912–1913	17.1	11.0	187.0	224.2	10.9	20.4	109.6	115.7
1913–1914	16.2	11.0	104.5	231.7	7.9	21.1	39.5	135.6
1914–1915	15.5	10.3	168.8	161.3	10.9	15.7	98.1	86.7
1915–1916	16.4	15.1	168.9	393.5	10.3	26.1	91.7	269.2
1916–1917	16.1	15.3	83.9	262.8	5.2	17.2	40.2	174.6
1917–1918	17.9	14.8	234.3	233.7	13.1	15.8	119.1	169.2
1918–1919	17.0	17.3	179.8	189.1	10.6	10.9	137.7	97.0
1919–1920	17.4	19.1	216.5	193.3	12.4	10.1	193.0	92.5
1920–1921	15.0	18.2	155.8	263.2	10.4	14.5	65.8	167.2
1921–1922	14.2	23.3	190.6	300.8	13.4	12.9	145.3	185.7
1922–1923	16.3	22.4	195.8	399.8	12.0	16.7	139.4	279.4
1923–1924	17.2	21.9	247.8	474.2	14.4	21.6	172.9	346.5
1924–1925	17.8	22.1	191.1	262.1	10.7	11.9	125.3	192.7
1925–1926	19.2	20.8	191.1	395.5	10.0	20.7	97.3	324.6
1926–1927	19.3	23.0	230.1	407.1	11.9	17.7	144.4	292.9
1927–1928	20.7	22.5	282.3	479.7	13.6	21.3	178.1	333.0
1928–1929	22.8	24.2	349.1	566.2	15.3	23.4	222.4	407.6
1929–1930	20.5	25.2	162.6	302.2	7.9	12.0	151.0	186.3
1930–1931	21.3	24.9	232.3	420.7	10.9	16.9	124.7	258.7

SOURCE: De Hevesy, pp. 344, 394–95; Canada [6], p. 24.

[a]Five-year averages.

TABLE A.2
Average Wheat Prices in Canada and Argentina, 1900–1931

(1900 = 100)

Year	Canada (cents per bushel, No. 1 Northern at Ft. William–Pt. Arthur)		Argentina (paper pesos per 100 kilos at Buenos Aires terminal)	
	Price	Index	Price	Index
1900	74.6	100.0	5.89	100.0
1901	75.2	100.8	6.35	107.8
1902	72.9	97.7	6.78	115.1
1903	78.8	105.6	6.39	108.5
1904	91.6	122.8	7.02	119.2
1905	77.2	103.5	6.67	113.2
1906	78.6	105.4	6.65	112.9
1907	105.2	141.0	7.86	133.4
1908	109.0	146.1	8.43	143.1
1909	101.1	135.5	9.52	161.6
1910	94.2	126.3	8.71	147.8
1911	100.8	135.1	8.02	136.2
1912	90.1	120.8	8.46	143.6
1913	89.1	119.4	8.30	141.0
1914	132.4	177.5	8.61	146.2
1915	113.6	152.3	12.0	203.8
1916	205.5	274.5	9.56	162.3
1917	221.1	296.4	14.73	250.1
1918	224.2	300.5	12.90	219.0
1919	217.4	291.4	14.93	253.5
1920	199.4	267.3	20.95	355.7
1921	129.7	173.9	16.31	277.0
1922	110.5	148.1	12.42	210.9
1923	107.0	143.4	11.70	198.6
1924	168.5	225.9	12.46	211.5
1925	151.2	202.7	14.50	246.2
1926	146.2	196.0	12.93	219.5
1927	146.3	196.1	11.46	194.6
1928	124.0	166.2	10.59	179.8
1929	124.2	166.5	9.75	165.5
1930	64.2	86.1	8.92	151.4
1931	59.8	80.2	5.69	96.6

SOURCES: Leacy, Series M228; Universidad Nacional del Litoral, *Los precios y los oscilaciones del area sembrada en la Argentina* (Rosario, 1933), p. 22.

NOTE: Because Canada and Argentina used different crop years in their agricultural accounting, I have used statistical series comparing prices during calendar years.

Notes

Notes

Complete authors' names, titles, and publication data for the works cited in short form in the Notes are given in the Bibliography, pp. 264–84. The following abbreviations are used in the Notes:

GGG   *The Grain Growers' Guide.* Winnipeg.
RRP   *The Review of the River Plate.* Buenos Aires.
SD   *Records of the Department of State. Decimal File, 1910–29: Argentina, Internal Affairs.* File M-514. National Archives, Washington, D.C.

Chapter 1

1. See Watkins's important article "A Staple Theory of Economic Growth." The quote is from p. 149.

2. Fogarty, "Staples," p. 22.

3. When the research for this book was well under way, I became aware of Marc A. Blain's essay "Le Rôle de la dépendance externe et des structures sociales dans l'économie frumentaire du Canada et de l'Argentine (1880–1930)." I consulted with Professor Blain, of the University of Quebec at Montreal, and learned that he was no longer actively pursuing research on this topic. In purpose and scope, his article differs substantially from my work, but I would like to acknowledge the seminal importance of his essay.

4. Ferns, *Argentine Republic*, p. 173. Professor Ferns makes numerous comparisons between the patterns of development of Argentina and his native Canada in this book and his other works.

5. Kirby, p. 279. See also Fogarty, "Comparative Method," pp. 426–29.

6. Denoon, p. 227.

7. North, "Location Theory," p. 248. See also Baldwin, p. 177.

8. Numerous Western Canadian authors have sharply criticized Eastern Canada and the federal government at Ottawa for exploiting the prairie, but none was as influential as the agricultural historian Vernon C. Fowke. See his "Political Economy and the Canadian Wheat Grower" and *Canadian Agricultural Policy*, especially p. 272.

9. For an influential statement of this point of view, see Martínez de Hoz, pp. 46–53. See also Jonathan Kandell, "On Argentina's Fertile Pampas, the Bitter Harvest of Neglect," *New York Times*, Jan. 23, 1975.

10. There is a vast literature on dependency theory. For a useful overview, see

Bath & James. For an examination of the theory's weaknesses and shortcomings, see Platt, "Dependency in Nineteenth-Century Latin America."

Chapter 2

1. Alejandro Bunge, "Paralelo económico argentino-canadiense, 1908–1926."
2. Teichman, L. R. Macdonald, pp. 264–75; P. Phillips, p. 6.
3. The standard historical study of Argentine land policy and of the concentration of ownership is Miguel A. Cárcano, *Evolución histórica del régimen de la tierra pública, 1810–1916*. For a more recent treatment of the land-tenure system, see Scobie, *Revolution*, pp. 26–54.
4. Ross, pp. 5–6. A particularly penetrating analysis of the Argentine elite of this period is McGann, pp. 39–60.
5. Ferns, *Britain and Argentina*, p. 144; Huret, p. 9.
6. Teichman, p. 47.
7. Fowke, "Introduction," p. 80.
8. On the ideological background of Canadian economic nationalism and in particular on the impact of List's ideas, see C. D. W. Goodwin, pp. 47–49; and Den Otter.
9. The entire Autumn 1979 issue of the *Journal of Canadian Studies* was devoted to the National Policy and provides a useful review of its components and political bases. See especially the articles by Craven & Traves; and Paul Phillips.
10. As Dales, p. 109, states, "The Canadian euphemism for protection is the National Policy."
11. Fowke, *National Policy*, p. 58.
12. Paul Phillips, p. 7.
13. Fowke, *National Policy*, p. 67, provides a clear expression of this complaint.
14. Ferrer, pp. 91–152. The Argentine tariff was primarily designed to raise revenue for the government, not to protect industry; Dorfman, pp. 317–34.
15. Cornblit et al.; McGann, pp. 39–105.
16. For an outstanding study of the Argentine political heritage, see Romero.
17. An excellent introduction to the Canadian political heritage is W. Morton, *Canadian Identity*.
18. Readers unfamiliar with Argentine history who wish to explore topics that this overview mentions might consult the following works: Rennie, *Argentine Republic*; Ferns, *Argentina*; Scobie, *Argentina*; and Rock, *Politics in Argentina*.
19. Readers unfamiliar with Canadian history who wish to explore topics that the following overview mentions might consult the following works: McNaught, *Pelican History of Canada*; W. Morton, *Critical Years, 1857–1873*; Waite, *Canada, 1874–1896*; Brown & Cook, *Canada, 1896–1921*; and W. Morton, *Progressive Party*.
20. Canada [1], 2: 452; Argentina [6], 2: 403–17. In 1931 55.1% of the Canadian foreign-born population was naturalized, as was 61.0% of the foreign-born population in the three prairie provinces; Canada [2], 2: 793. As late as 1939 there were only 183,000 naturalized foreigners in Argentina or about 15% of the total foreign-born population; Argentina [4], p. 512.
21. Yuzyk, pp. 177–78; Murchie, pp. 87–88.
22. Otto D. Christensen Autobiography, Glenbow-Alberta Archives D920, p. 148. See also D. Smith, *Prairie Liberalism*, pp. 27–49, 58–65; and Yuzyk, p. 179.
23. Boudreau, p. 2; "Personal Naturalization Only," GGG, Jan. 21, 1920, p. 42.

24. Solberg, *Immigration*, p. 43; Baily, "Patterns," p. 14.

25. The citizenship procedure was thoroughly discussed in the Argentine press for many years. See Solberg, *Immigration*, pp. 124–25. On Canadian rural judges, see "Naturalization Vitally Important," *GGG*, May 18, 1921, p. 21.

26. Solberg, *Immigration*, pp. 121–27.

27. For the Socialist position, see Dickmann, pp. 23–34. On the Liga del Sur, see Liebscher, pp. 200–203, 237.

Chapter 3

1. Maizels, p. 533.

2. De Hevesy, foldout facing p. 684; Surface, *International Competition*, p. 2.

3. Burton; Watkins.

4. De Hevesy, pp. 664–65.

5. Ferns, "Britain's Informal Empire"; Hobsbawm, pp. 121–22; Scalabrini Ortiz, p. 143; Robertson, in London *Times*, July 24, 1930.

6. Ferns, *Britain and Argentina*, pp. 1–2. See also Paish, pp. 177–82.

7. Ford, "British Investments," Phelps, pp. 107–9; Aitken, pp. 58, 68–69; Viner, p. 284; Paterson, pp. 9–10.

8. Rollin E. Smith, p. 353. See also Rickard, pp. 269–70; and Malenbaum, pp. 104–5.

9. Olson, pp. 329–30; North, "Ocean Freight Rates," pp. 551–52; Kuczynski.

10. Canada [6], pp. 122–23; Mackintosh, "Crisis," p. 352. See also Stovel, pp. 326–28; and Strong, "Grain Farming," 2: 980.

11. D. Smith, *Regional Decline*, p. 18.

12. Easterbrook & Aitken, p. 405; MacGibbon, *Canadian Grain Trade, 1931–1951*, p. 7.

13. Britnell, pp. 44–58; W. Morton, *Manitoba*, p. 393.

14. Breen, pp. 23–69; Murchie, pp. 51–60.

15. Breen, pp. 215, 224–31.

16. Stegner, p. 137; S. M. Evans; Breen, pp. 118–30.

17. Breen, pp. 136–236.

18. P. H. Smith, pp. 33–34.

19. Murchie, p. 60; Martínez de Hoz, pp. 10–11.

20. Sociedad Rural Argentina, p. 25; Martínez de Hoz, p. 30; Richardson, p. 175.

21. Gravil, "State Intervention," p. 170.

22. Quoted in Britnell, p. 1.

23. Dorfman, p. 12; Díaz-Alejandro, *Essays*, p. 514; Coote, pp. 1, 66, 76.

24. Ferns, *Argentine Republic*, p. 131; Stovel, p. 124; Dorfman, p. 372.

25. On the rapid growth of Canadian industrial production in the late nineteenth century, see Masters, pp. 146, 174; and McDiarmid, pp. 199–242. On the question of railway equipment, see Zalduendo, p. 89; and Platt, *Latin America*, p. 90.

26. Geller, pp. 782–85.

27. Dieguez emphasizes this point. On foreign capital invested in Canadian manufacturing, see Levitt, p. 61.

28. Wythe, pp. 96–97; Dorfman, p. 301.

29. Masters, pp. 63, 104.

30. Quoted in Fowke, *National Policy*, p. 65.

31. Ferns, *Britain and Argentina*, p. 145. See also Ferns, *Argentina*, pp. 125–27, 146.

32. Solberg, "Tariff," pp. 260–84.
33. Strong, "Conditions," 1: 247.

Chapter 4

1. Christie, p. 32. For similar commentary, see Lumsden, p. 94. See also Rutter, p. 47.
2. Binnie-Clark, p. 50.
3. Jefferson, p. 10; Jackman, "Wheatfields," p. 4. See also Rutter, p. 259.
4. Macleod, p. 102. See also Desmond Morton.
5. Riddell, pp. 276–77.
6. For details of the 1872 act and later modifications, see Martin, pp. 396–407; Dawson & Younge, p. 12; and the thorough discussion in Hedley, pp. 45–56.
7. For the quote, see Rollin Smith, p. 153. See also Norrie, "Rate of Settlement"; and Ankli & Litt.
8. McCutcheon, p. 43.
9. The quote is from Cohnstaedt, p. 27. See also Bickersteth, p. 85; and P. F. Sharp, p. 22.
10. Kerr, p. 18.
11. Binnie-Clark, p. 13.
12. Bennett, p. 104; Murchie, p. 127.
13. For the traveler's quote, see Mitchell, p. 155. See also Britnell, p. 179; Mackintosh, *Economic Problems*, pp. 12, 31, 259–66; and Canada [2], 8: lxii. Murchie, p. 78, presents statistics correcting the 1931 census data on mortgages.
14. Canada [2], 8: xlix. For discussions of the move toward tenancy, see Dawson & Younge, pp. 90–93; and Murchie, pp. 117–25.
15. For the Mill quote, see Murchie, p. 93. See also Saskatchewan, *Report*, p. 31. Some observers, like C. H. Craig, pointed out the advantages of tenancy.
16. In 1900 an estimated $1,000 to $1,500 was the minimum capital required to buy the tools, animals, seeds, and shelter to begin farming in the Canadian prairies; see Ankli & Litt, p. 55. On the background and significance of the cooperative movement, see Fay, *Cooperation at Home and Abroad*, pp. 278–83; and Bennett, pp. 207–9, 238–39, 280.
17. Willmott, "Formal Organizations," p. 28.
18. John Stokes to his father, Oct. 29, 1907, and Feb. 6, 1908 (Wood Bay, Manitoba), Glenbow-Alberta Archives, A. S874, files 5 and 6.
19. For the quote on schools, see Lloyd, p. 18. See also Farnalls, pp. 18–19; and Willmott, *Organizations and Social Life*, p. 8.
20. J. F. C. Wright, pp. 35–53; L. J. Wilson.
21. Dawson & Younge, pp. 154–55.
22. Mackintosh, *Economic Problems*, pp. 41, 149–54; Britnell, p. 179; J. E. Williams.
23. On education, see Dawson & Younge, pp. 159, 176, 190; and Britnell, pp. 113, 194. On libraries, see Lloyd, p. 17; Britnell, p. 122; and Woodward, p. 741.
24. Dawson & Younge, pp. 262–64; Britnell, pp. 19, 138.
25. Cánepa, p. 61.
26. The basic historical study of Argentine land policy is Miguel A. Cárcano, *Evolución histórica del régimen de la tierra pública, 1810–1916*. For a thorough and systematic analysis of Santa Fe's policy, see Gallo, "Agricultural Colonization," especially pp. 22–56, 96–107. For a case study of Córdoba policy, see Arcondo, "Tierra y política."
27. Delich, especially p. 125.

28. Ferns, *Britain and Argentina*, pp. 344–47, 426; Zalduendo, pp. 105, 293–95.

29. Gallo, "Agricultural Colonization," pp. 22–24, 115–18; Sternberg, pp. 113–17, 135–38; Arcondo, *Agricultura en Córdoba*, end foldout; Ruggiero, "Italians in Argentina," pp. 166, 241, 271.

30. Gallo, "Agricultural Colonization," pp. 96–107.

31. For a thorough discussion of the 1887 Buenos Aires law and its results, see Girbal de Blacha, *Centros agrícolas*. See also Morris.

32. Clemenceau, p. 161. For the movement of land and cereal prices, see Cortés Conde, pp. 166, 172.

33. *New York Journal of Commerce* clipping of Sept. 20, 1921, in *SD*, 835.52/19. On the difference between land prices in Canada and Argentina, see Leguizamón, "Inmigración," pp. 97–98; and *RRP*, April 1, 1927, p. 7.

34. Nemirovsky, p. 43; Oddone, pp. 182–85; Delich.

35. *RRP*, June 14, 1935, p. 9.

36. Italian immigrants perceived pampa agricultural conditions as much more favorable than those in Italy; Ruggiero, "Gringo and Creole," p. 170. See also Foerster, pp. 235–36.

37. Gallo, "Agricultural Colonization," p. 249, quotes the *Financial Times*. See also pp. 24, 59, 221–29, 247–48.

38. Campbell Ogilvie, p. 9. On the relationship between alfalfa and cereal agriculture, see S. G. Hanson, pp. 113–16.

39. Coni, "Colonización," 1: 363.

40. Teubal, p. 109; Fogarty, "Comparative Method," p. 417.

41. Sternberg, p. 242.

42. Huergo, *Investigación agrícola*, p. 130. The quote is from Larden, p. 78. See also Scobie, *Revolution*, p. 59.

43. Bialet Massé, 1: 120; Coni, "Colonización," p. 350; Scobie, *Revolution*, pp. 87–88. Yield data from De Hevesy, pp. 344, 394–95. For a more complete statistical analysis of trends in wheat yields, see Table A.1.

44. Between 1930 and 1933 the average cash rent and average wheat prices both dropped 40%. For statistical series on land rents, see Argentina [12], p. 503; and Cánepa, p. 80. Data on wheat prices are given in Argentina [4], p. 207.

45. For the extent of mortgage debt, see Strong, "Farm Mortgage Loans." On the moratorium law, see *RRP*, Sept. 22, 1933, p. 5, and Oct. 13, 1933, p. 22. The moratorium, initially imposed for three years, was extended for two more in 1936; *RRP*, Oct. 2, 1936, p. 7.

46. Gallo, "Agricultural Colonization," pp. 74–75, summarizes these arguments.

47. For a study of cadastral records and land subdivision, see Zamboraín, pp. 39–58. See also Jefferson, p. 142; and Taylor, p. 192.

48. No date is given for this quote, which is from García Serrano, p. 51. See also Gallo, "Agricultural Colonization," pp. 324–34.

49. Ferns, *Argentine Republic*, p. 94.

50. Sternberg, p. 71. The other quotes are drawn from Coni, "Cuestiones agrarias," p. 371; and Gibson, *Land*, p. 171. See also Delich, pp. 132–33.

51. Greca, p. 31.

52. On the Argentine cooperatives, see Bórea, *Tratado*, p. 276; Horne, *Política*, pp. 120, 249–50; and *La Prensa*, July 13, 1930, p. 13. Canadian wheat pool membership is reported in Canada [7], 1: 20.

53. On the question of isolation, see Scobie, *Revolution*, pp. 61–62; and Hirst, p. 110.

54. Ruggiero, "Italians in Argentina," p. 269; Larden, pp. 78–79; Gibson, *Land*, p. 16. See also Miatello, "Hogar agrícola," pp. 565–67.
55. Dreier, pp. 189–90; Miatello, "Hogar agrícola," p. 565; Argentina [11], 1: 757–62.
56. See, for example, Scardin, pp. 232, 408; and Jefferson, p. 172.
57. The quote is from William Goodwin, p. 23.
58. This poem, "María de Alcorta," is found in J. M. García, p. 69.
59. Alejandro Bunge, "Gastos de la educación."
60. R. Cárcano, *800.000 analfabetos*, p. 9. The title of this book refers to the number of illiterates among school-age children in Argentina in 1932. See also Augusto Bunge, pp. 182–84; Mariano Vélez, p. 119; and Argentina [11], 2: 114.
61. Diecidue, p. 117. See also Cánepa, pp. 117–18. According to Nemirovsky, p. 230, the circulation of *La Tierra* was 18,506 in 1926. This compares with the *Guide's* circulation of 120,000 that year.
62. Scardin, pp. 244–45; Slatta, "Gaucho and Rural Life," p. 89.
63. Gudiño Kramer, p. 26.
64. Germani, p. 44; Taylor, p. 334. See also Dreier, pp. 18–19.
65. Campbell Ogilvie, pp. 99–100. For similar comments by other foreign observers, see William Goodwin, p. 26; and Scardin, pp. 270, 273.
66. "Watch Canada!," *RRP*, Feb. 11, 1927, p. 11.
67. *La Prensa*, March 30, 1927, p. 11.
68. Hobbs, pp. 7, 18. The quote is from Haslam.
69. Jackman, "Wheatfields," p. 35. See also Jackman, *Wheat Growing*, pp. 10–12.
70. Bórea, *Mutualidad*, p. 145.
71. Buenos Aires *Herald*, June 5, 1923, in *SD*, 835.00/324.
72. Reported in *SD*, 835.52/34, Nov. 4, 1924.

Chapter 5

1. For the Cartwright quote, see Porter, p. 34. See also pp. 30–31.
2. Willcox, 1: 546.
3. Argentina [10], p. 31. See also Argentina [5], pp. 74–75; and Norman Macdonald, p. 116.
4. Scobie, *Revolution*, p. 123; Jefferson, p. 180.
5. See, for example, "Colonización e inmigración," *Anales de la Sociedad Rural Argentina*, Nov.–Dec. 1910, pp. 142–43.
6. Norman Macdonald, pp. 45–46.
7. McCutcheon, p. 36; Wallace Clegg (Russell, Manitoba) to "Uncle," April 10, 1914, Glenbow-Alberta Archives, A. C624. See also Troper, pp. 79–86.
8. Dafoe, pp. 103–43; Troper, pp. 79–99; Donald Avery, pp. 99–105.
9. On Canada's response to British "Empire Settlement" plans and on King's immigration policies, see Drummond, *Imperial Economic Policy*, pp. 93–108; and Drummond, *British Economic Policy*, pp. 82–84.
10. Ceppi, p. 171; Porter, p. 62.
11. Solberg, *Immigration*, pp. 7–14; Alsina, pp. 205–6.
12. Norman Macdonald, pp. 30–48. On the attitude and policy toward U.S. blacks, see Troper, pp. 121–45; and Colin Thomson.
13. Porter, p. 64.
14. Berger, p. 151; Porter, p. 67.
15. Calderwood.
16. O. D. Skelton, quoted in Stovel, p. 107.

17. W. Morton, *Manitoba*, pp. 177–79, 224–25; Swanson & Armstrong, pp. 14, 16.

18. Silver.

19. For an outstanding general history of the decline of the gaucho, see Slatta, *Gauchos and the Vanishing Frontier.*

20. Larden, p. 67.

21. Lewis H. Thomas, "British Visitors' Perceptions," pp. 183–85. See also Carrothers, p. 246.

22. Kipling, p. 169. See also Sloan, p. 3; and West, p. 18.

23. Reynolds, pp. 26, 34–35, 79; Carrothers, p. 246.

24. Larden, p. 80. For a detailed analysis of nineteenth-century Irish immigration, see Sábato & Korol. See also Platt, "British Agricultural Colonization."

25. Seymour, pp. 56–74; Willcox, 1: 543.

26. John P. Bailey; Jakubs.

27. Vásquez-Presedo, p. 162. On the Welsh colony in Chubut, see G. Williams.

28. Owen, pp. 52–62; Leslie H. Thomas.

29. Owen, pp. 66, 68, 95; Johnson, pp. 92–93; Solberg, *Immigration,* pp. 24–25, 137; Glyn Williams, p. 70.

30. The 1921 Canadian census figures on the U.S.-born are reported in Canada [3], *1933,* p. 126. A classic work on U.S. migration to Canada is Hansen, *The Mingling of the Canadian and American Peoples,* especially Chap. 10. See also P. F. Sharp, pp. 4–16; Troper, pp. 148–54; and Alsina, p. 22.

31. A. S. Morton, pp. 170–71. See also Crerar, p. 31; and Troper, pp. 38–42. On the Scandinavians, see Wonders.

32. "Bennett's Attack on American Settlers," *GGG,* March 12, 1913, p. 26. See also Sloan, pp. 6–7.

33. The classic work on Italian emigration to Argentina appears in Foerster, pp. 223–78. See also Willcox, 1: 543; Ratti, p. 446; and Vásquez-Presedo, p. 103.

34. William Goodwin, p. 12.

35. Rollin Smith, pp. 143, 145.

36. Foerster, pp. 265–66. About 65% of all Italian immigrants in Argentina between 1875 and 1895 were from Piedmont, Liguria, Lombardia, and Venice; Vásquez-Presedo, p. 101. See also Scardin, p. 205.

37. Ruggiero, "Italians in Argentina," p. 272. See also Cortés Conde, p. 252; William Goodwin, p. 12; and Guglieri, p. 13.

38. On re-migration, see Willcox, 1: 543.

39. William Goodwin, pp. 17–18; Jefferson, p. 126.

40. Campbell Ogilvie, p. 13; Larden, p. 72. See also Ruggiero, "Gringo and Creole."

41. Canada [2], 1: 1218–1248; Canada [4], pp. xi–xxxvii; Harney & Troper, pp. 2, 51.

42. The quote is from Gualtieri, p. 59. See also pp. 7–11.

43. Willcox, 1: 543; Rahola, p. 273. On the regional origins of Spanish migration, see "Spanish Emigration," *The Geographical Review,* April 1922, p. 310; and especially, Spain, *Emigración española,* pp. 46–49, 380–91.

44. Willcox, 1: 543.

45. Rahola, pp. 124–27, 346; Jefferson, pp. 140–41; Posada, pp. 258–59; Spain, *Emigración española,* pp. 76, 101, 459.

46. Canada [2], 1: 1218; Spain, *Emigración española,* pp. 95, 122–23.

47. Yuzyk, pp. 176–77. 48. Yuzyk, pp. 28–29, 31, 36.

49. Dafoe, pp. 142–43. 50. England, pp. 25–26.

51. C. W. Peterson papers, Glenbow-Alberta Archives, A. P485, files 3 and 4 (speeches at Calgary, June 4, 1926; Red Deer, 1926; Lethbridge, May 1926; Calgary, March 14, 1929). See also Donald Avery, pp. 19, 23, 98; and "The Agrarian Movement in Canada," *Quarterly Review*, 235 (Jan. 1921), p. 91.

52. Yuzyk, pp. 43–45; Lazarowich; Bychinsky.

53. Willcox, 1: 544.

54. Koch, pp. 224–28; Jefferson, pp. 149–53; Reynal O'Connor, pp. 143–46; Wilkie & Wilkie, pp. 4–10, 28.

55. The best summary treatment of the Jewish colonies is Elkin, pp. 125–43. See also Avni; and Winsberg, Part 1.

56. Both quotes are from Elkin, p. 137.

57. Jefferson, p. 156; Winsberg, 2: 427.

58. Jefferson, p. 184; Sternberg, pp. 194–96; Platt, *Latin America*, p. 127. For an official estimate on the number of golondrinas, see Argentina [14], p. 47. See also Scobie, *Revolution*, p. 61; and Foerster, p. 244.

59. *La Prensa*, Aug. 1, 1911, p. 7; Foerster, p. 276.

60. Nario. For an overview of this group, see Solberg, "Farm Workers."

61. Miatello, *Chacra santafecina*, p. 67. See also Rodríguez Tardati, pp. 386–87; and Denis, p. 266.

62. Horne, *Política*, p. 58; Cánepa, p. 106.

63. *El Capital*, Nov. 18, 1920, p. 4; Nario, p. 8.

64. For an overview, see Marotta, 3: 261–63. For details, see *La Vanguardia*, Dec. 4, 1928, p. 1; *La Protesta*, Nov. 23, 1928, p. 1; and "Agitación agraria," Dec. 20, 1928, pp. 557–58.

65. "Agitación agraria," Dec. 5, 1928, p. 529; *RRP*, Nov. 30, 1928, p. 7.

66. Maradona, pp. 20–23; *RRP*, Dec. 7, 1928, p. 9; "Agitación agraria," Dec. 5, 1928, pp. 529–34.

67. Haythorne, p. 536; "The Harvest Help Problem," *GGG*, Aug. 1, 1911, p. 6; "The Demand for Sikhs," *GGG*, Sept. 18, 1912, p. 8; Thompson, "Bringing in the Sheaves," pp. 470–71; De Gelder, p. 53.

68. Haythorne, pp. 534–44; Thompson, "Bringing in the Sheaves," p. 471; Canadian Pacific Railway, *50,000 Harvesters Wanted: Excursions to Manitoba, Saskatchewan, and Alberta* (Montreal, 1926).

69. Thompson, "Bringing in the Sheaves," pp. 486–87; Donald Avery, pp. 106–7; "The British Harvesters," *Country Guide*, Oct. 1, 1928, 8; Cherwinski.

70. Corcoran, p. 154. For a discussion of wages, see Thompson, "Bringing in the Sheaves," p. 482. See also Robert G. Trussler, "Account of a Trip West in 1925 and Harvesting Experience Around New Norway, Alberta," manuscript in Glenbow-Alberta Archives, A. T873A. On abuses, see Thompson, "Bringing in the Sheaves," pp. 480–81; and *GGG*, Nov. 15, 1926, p. 10.

71. Haythorne, p. 544.

Chapter 6

1. L. A. Wood, p. 242; Porritt, pp. 151–55; Rolph, p. 38.

2. Strong, "Conditions in Argentina," 1: 247.

3. Dales, p. 109; Mackintosh, *Economic Background*, pp. 19–20, 34; Caves & Holton, pp. 235–36.

4. Laurier's speech is quoted in Fowke, *Canadian Agricultural Policy*, p. 262. See also McCutcheon, p. 56; and Porritt, pp. 176–77.

5. P. F. Sharp, pp. 43–46; L. A. Wood, pp. 247–61.

6. Van Horne is quoted in Innis, *Problems*, p. 27. See also P. F. Sharp, p. 43. The most complete analysis of the reciprocity dispute is Ellis, *Reciprocity 1911*.

7. "Manufacturers Throw Down Gauntlet," *GGG*, Sept. 28, 1910, p. 5; "Where Are the Men Who Pay?," *GGG*, Dec. 7, 1910, p. 5; "Think It Over," *GGG*, Dec. 28, 1910, p. 5; "Poisoning the Public Mind," *GGG*, March 15, 1911, p. 5.

8. W. Morton, *Progressive Party*, pp. 21–26; P. F. Sharp, pp. 48–50.

9. Alejandro Bunge, "Política aduanera," p. 208; Mackintosh, *Economic Background*, p. 51.

10. Scobie, *Revolution*, p. 129; Peters, p. 71; Soares, 3: 181–90.

11. Pedro Pagés, president of the Sociedad Rural between 1922 and 1926, argued that Argentina must search for new domestic markets for its rural products. See Pagés, *Crisis ganadera*, pp. 100–101. See also *La Epoca*, Aug. 1, 1923, p. 1.

12. Pintos, *Treinta años*, p. 18; Schleh, *Industria*, pp. 13, 16, 39, 94–105, 266.

13. Scobie, *Revolution*, p. 131.

14. See Solberg, "Tariff and Politics," pp. 272–82, for an analysis of political alignments over the tariff issue.

15. Cornblit, p. 230. Díaz Alejandro, pp. 277–300, makes a strong argument against traditional free-trade interpretations of the Argentine economy during this period and demonstrates the importance of the depreciated peso as a protective factor.

16. Diecidue, p. 181.

17. Sherwood Avery, p. 73; Strong, "Agricultural Implements," pp. 313–14.

18. J. D. Woods, pp. 3–6; Mackintosh, *Economic Background*, pp. 19, 51, 91.

19. The quote is from Naylor, 2: 14. Although U.S. farm machinery accounted for about 80% of Argentine imports by the mid-1930's, Canadian machinery held second place at about 10%; Strong, "Agricultural Implements," p. 315. See also McDiarmid, pp. 196–97.

20. Mallon, p. 46.

21. Strong, "Conditions in Argentina," 1: 247; Conti, p. 6.

22. Strong, "Agricultural Implements," p. 328.

23. Scobie, *Revolution*, p. 84; Strong, "Conditions in Argentina," 2: 309. According to *RRP*, Nov. 18, 1932, the cost of harvesting 100 kg of wheat with a combine was 84 Argentine cents compared with 214 cents (2.14 pesos) using traditional methods.

24. W. G. Phillips, p. 50; MacGibbon, *Canadian Grain Trade* (1932 ed.), p. 470; Strong, "Agricultural Implements," p. 319.

25. Lewis H. Thomas, "Early Combines"; Murchie, pp. 296, 328. Isern, pp. 105–11, discusses the development of windrow harvesting.

26. Hedley, pp. 160–64.

27. For an analysis of comparative per capita agricultural budgets, see Videla, p. 201. See also Fowke, *Canadian Agricultural Policy*, pp. 156–57; and Girbal de Blacha, *Historia*, pp. 15–21.

28. Coni, "Ciencia"; Girbal de Blacha, *Historia,* p. 23.

29. Díaz Alejandro, "Argentine State," p. 236.

30. M. A. Cárcano, "Cuestiones agrarias," p. 11; Garbarini Islas, pp. 1008–1009.

31. On the early growth of Argentine agricultural education, see Scobie, *Revolution*, pp. 148–50; and Tomas Amadeo, "Prólogo," in Huergo, *Enseñanza agrícola*, p. 10.

32. University budgets are not included in these figures. See Rodolfo Allen, p. 30. For the government budgets, see Solberg, "Rural Unrest," pp. 31–32.

33. Rodolfo Allen, pp. 106, 128–29.

34. Argentina [4], p. 498; Rodolfo Allen, p. 106.

35. Fowke, *Canadian Agricultural Policy*, p. 249.

36. A. S. Morton, pp. 147–48; Grant MacEwan, "Struggles, Triumphs, and Heartaches with Western Wheat," in Rasporich & Klassen, pp. 137–40; Fowke, *Canadian Agricultural Policy*, pp. 231–36.

37. McCutcheon, pp. 32–33; A. S. Morton, pp. 101–20; Ankli & Litt, pp. 38–39.

38. Lumsden, p. 142.

39. Swanson & Armstrong, pp. 63–67, 282.

40. The first quote is from Holm, p. 210; Holm was a long-time agricultural columnist for the Buenos Aires *Standard*. The FAA's quote is from its paper, *La Tierra*, April 15, 1921, p. 1.

41. A. S. Morton, pp. 134–39; Turner, pp. 93–94.

42. Lochhead, p. 212. For similar views of another Englishman, see Mitchell, p. 72. See also Turner, pp. 93–94.

43. Garbarini Islas, p. 1004; Argentina [2], sesiones ordinarias, 3, Sept. 5, 1928, p. 741.

44. Fraser, p. 48.

45. De Gelder, p. 52.

46. Glazebrook, pp. 445, 452; Mackintosh, *Economic Problems*, p. 243.

47. W. Morton, *Manitoba*, p. 401; Mackintosh, *Economic Problems*, p. 244.

48. Dawson & Younge, pp. 64, 80–81; Glazebrook, p. 446; C. Dawson, *Settlement of Peace River Country*, p. 44.

49. Argentina [12], p. 86; Garbarini Islas, p. 1003.

50. Brady, p. 13; Alejandro Bunge, *Economía argentina*, 4: 85; Argentina [12], p. 86.

51. Argentina [4], p. 461.

52. Renard, p. 21. See also Holm, p. 91; and Alejandro Bunge, "Caminos de acceso."

53. Argentina [2], sesiones ordinarias, 3, June 13, 1932, p. 93.

54. Ibid., 6, Sept. 26–27, 1932, p. 591.

55. Alejandro Bunge, "Aspecto," pp. 276–78.

56. Cuccorese, pp. 171–81; Argentina [2], sesiones ordinarias, 2, June 13, 1917, p. 35; Alejandro Bunge, "Aspecto," p. 275; Holm, pp. 76–93. For a nationalist attack on the Ley-Mitre, see Scalabrini Ortiz, p. 149.

57. L. B. Smith et al., p. 75.

58. Zalduendo, p. 45.

59. Viner, pp. 115, 303; Paish, pp. 180–82.

60. L. B. Smith et al., pp. 74–75; Lewis, p. 415.

61. Paish, pp. 180, 182; Aitken, p. 69; J. Fred Rippy, *British Investments in Latin America, 1822–1949* (Minneapolis: Univ. of Minnesota Press, 1959), pp. 76, 78; Phelps, p. 108.

62. Innis, *Problems*, pp. 40–59; Easterbrook & Aitken, pp. 437–41; Canada [5], pp. 9–11.

63. Innis, *Problems*, p. 55; Canada [5], p. 17.

64. MacGibbon, *Canadian Grain Trade* (1932 ed.), p. 443; Rutter, p. 144; McClure, pp. 3, 37.

65. W. Morton, *Manitoba*, pp. 217, 327, 401, 407; "Why the Hold Up?," *GGG*, April 13, 1910, p. 5; McClure, pp. 49, 75.

66. Argentina [4], p. 448; W. Wright, "Argentine Railways," p. 91.

67. Randall, p. 171.

68. W. Wright, *British-Owned Railways*, pp. 85–86; P. Goodwin, "Politics," p. 268; Ferns, *Argentine Republic*, pp. 107–8.

69. Randall, p. 177, reviews Ortiz's arguments.

70. Alejandro Bunge, *Ferrocarriles*, pp. 97–102; Scalabrini Ortiz, p. 222.

71. P. Goodwin, "Politics," pp. 262–79; Repetto, "Defensa," p. 731. Winthrop Wright misses the 1919 increase in *British-Owned Railways*, p. 123. See P. Goodwin, *Ferrocarriles*, pp. 175–77.

72. *RRP*, 69, Nov. 14, 1930, p. 15; P. Goodwin, "Politics," pp. 279–81.

73. W. Morton, *Manitoba*, pp. 215–38; "Freight Rates and Protection," *GGG*, May 24, 1911, p. 5.

74. Currie, pp. 40–41; Mackintosh, *Economic Background*, p. 33.

75. R. Dawson, *William Lyon Mackenzie King*, pp. 394–96; Currie, pp. 46–52; W. Morton, *Progressive Party*, pp. 156–57. For an analysis of this issue from the Maritime point of view, see Forbes, pp. 91–93.

76. "A Progressive Victory," *GGG*, July 5, 1922, p. 5; Currie, p. 54; Norrie, "National Policy," p. 29.

77. Regehr, "Western Canada," p. 130; *Maclean's*, Oct. 17, 1983, p. 31.

78. Nemirovsky, p. 117; *RRP*, April 24, 1936, p. 5.

79. De Hevesy, foldout facing p. 814. For another cost-of-production analysis, see Vásquez-Presedo, pp. 164–65.

80. Fowke, "Political Economy," pp. 211–13.

Chapter 7

1. Hedley, p. 76; C. B. Macpherson, p. 224. See also Friedmann, pp. 546–48.

2. Paige, pp. 45, 48; Sinclair, pp. 7–10.

3. Paige, pp. 58–59.

4. P. F. Sharp, p. 33; W. Morton, *Progressive Party*, p. 14; L. A. Wood, pp. 13–155.

5. Britnell, pp. 186–87; Dawson & Younge, pp. 219–20.

6. For these quotes, see Richard Allen, "Social Gospel," p. 176; and Richard Allen, *Social Passion*, p. 201.

7. Richard Allen, "Social Gospel," pp. 177–78; "Dr. Bland at Brandon," *GGG*, Jan. 20, 1915.

8. Patton, *Grain Growers' Cooperation*, pp. 14–16; Mackintosh, *Agricultural Cooperation*, pp. 8–9; Lipset, p. 58.

9. MacGibbon, *Canadian Grain Trade* (1932 ed.), p. 33.

10. For the quotes, see ibid., p. 280; and Rotstein, p. 19. See also Patton, *Grain Growers' Cooperation*, p. 16.

11. West, p. 170; Stead, p. 121.

12. For an excellent analysis of the nineteenth-century prairie farmers' movements, see McCutcheon, especially pp. 252, 275.

13. Fowke, "Royal Commissions," pp. 168–69; Patton, *Grain Growers' Cooperation*, pp. 21–30; Clark, p. 9.

14. Moorhouse, p. 46.

15. McCrorie, "Discussion," p. 327.

16. Embree, pp. 1–2; L. A. Wood, pp. 171–81.

17. "Organizer's Report," *GGG*, March 31, 1915, p. 15; "Utilotopia," *GGG*, Sept. 12, 1912, p. 3; W. Morton, *Manitoba*, p. 411.

18. L. J. Wilson, pp. 28–31.

19. L. A. Wood, pp. 293–94; W. Morton, *Manitoba*, p. 288.

20. Colquette, pp. 141–42; Patton, *Grain Growers' Cooperation*, p. 387.

21. Rolph, p. 38; Patton, *Grain Growers' Cooperation*, pp. 387–88. For a full account of the Siege of Ottawa, see Chipman.

22. Both quotes are from "The Agrarian Movement in Canada," *Quarterly Review*, 235 (Jan. 1921), p. 90. See also P. F. Sharp, p. 36; Colquette, p. 82; Patton, *Grain Growers' Cooperation*, p. 329.

23. Colquette, p. 26.

24. "Friend Partridge Is Unwell," *GGG*, June 7, 1911, p. 18; D. Smith, *Prairie Liberalism*, p. 70; Hedlin; Richard Allen, "Social Gospel," p. 178.

25. Moorhouse, p. 66; Mackintosh, *Agricultural Cooperation*, pp. 18–22; Colquette, p. 27; McCutcheon, p. 370.

26. Mackintosh, *Economic Problems*, pp. 46–47; W. Morton, *Manitoba*, p. 288.

27. Mackintosh, *Agricultural Cooperation*, pp. 18–22; the Partridge quote is on p. 19. See also Fowke, *National Policy*, pp. 136–38; and L. A. Wood, pp. 188–92.

28. Patton, *Grain Growers' Cooperation*, pp. 64–65; Griezic, pp. 108–9.

29. Patton, *Grain Growers' Cooperation*, p. 160; Colquette, p. 67; Hedlin, p. 66.

30. MacGibbon, *Canadian Grain Trade, 1931–1951*, p. 193; Colquette, pp. 117, 196; Mackintosh, *Agricultural Cooperation*, pp. 90–100.

31. Colquette, pp. 85–86; E.A.P., "Another Step in the Work of Creating a Perfect Monopoly in the Grain Trade," *GGG*, Aug. 14, 1909, p. 13.

32. Colquette, p. 74; Mackintosh, *Agricultural Cooperation*, p. 34; Patton, *Grain Growers' Cooperation*, pp. 88–95.

33. Hedlin, p. 65.

34. Fowke, *National Policy*, pp. 144–48; Patton, *Grain Growers' Cooperation*, pp. 98–104; Spafford, "Elevator Issue"; Rolph, pp. 32–33.

35. Patton, *Grain Growers' Cooperation*, p. 98.

36. For the quote, see Fowke, *National Policy*, p. 149. See also Fay, *Agricultural Co-operation*, p. 446; Mackintosh, *Agricultural Cooperation*, p. 55; and Patton, "Canadian Wheat Pool," p. 24.

37. Mackintosh, *Economic Problems*, p. 48.

38. The quote is from West, p. 172. See also Fay, *Agricultural Co-operation*, p. 446; and Clark, pp. 20–21.

39. Patton, *Grain Growers' Cooperation*, pp. 130–38.

40. Ibid., p. 140.

41. MacGibbon, *Canadian Grain Trade* (1932 ed.), pp. 380–82; Patton, *Grain Growers' Cooperation*, pp. 145–48.

42. MacGibbon, *Canadian Grain Trade* (1932 ed.), pp. v, 187, 195–201, 380–81; Boyle, pp. 36–45.

43. MacGibbon, *Canadian Grain Trade* (1932 ed.), p. 381.

44. Morgan, p. 248; Gravil, "State Intervention," p. 151.

45. Rutter, p. 204; Holm, p. 29; Weil, p. 161; Ovidio Víctor Schiopetto, "El comercio de granos," *Revista de Ciencias Económicas*, July 1928, p. 2139.

46. Tulchin, p. 391.

47. Holm, p. 120; Ezcurra, p. 434; Girbal de Blacha, *Historia*, pp. 138–39; Tulchin.

48. Tulchin, p. 392; Scardin, p. 253; Slatta, "Gaucho and Rural Life," p. 219; Holm, p. 130.

49. Coni, "Colonización," 1: 360–61.

50. Nemirovsky, p. 142.

51. Nemirovsky, pp. 145–47; Paz, p. 313.

52. Morgan, pp. 39–40; Paz, pp. 304–12; Pérez Brignoli, pp. 11, 13.

53. Scobie, *Revolution*, p. 95; Strong, "Grain Farming," 2: 983; Jackman, "Wheatfields," p. 34.

54. For the quote, see McCrea et al., p. 473. See also Platt, *Latin America*, p. 196.

55. For descriptions of this system, see Rutter, p. 203; Holm, p. 63; and Jefferson, pp. 173–74.

56. Strong, "Grain Farming," 2: 981; Pérez Brignoli, p. 32; Jackman, "Wheatfields," pp. 4–5.

57. Argentina [13], pp. 19–25; "Grain Elevators," *The British Chamber of Commerce in the Argentine Republic Monthly Journal*, July 31, 1929, pp. 25–26; *La Prensa*, July 13, 1930, p. 13.

58. Bórea, *Tratado*, pp. 284, 286.

59. Repetto, "Cooperativas," p. 159; Horne, *Nuestro problema*, p. 67; Elkin, pp. 160–61.

60. *La Prensa*, July 13, 1930, p. 13; Bórea, *Tratado*, pp. 405–512; Argentina [4], p. 284.

61. Coni, "Colonización," 1: 360. See also Repetto, "Cooperativas," p. 163; Etcheverry, pp. 32–37; and Girola, *Coopératives*, p. 5.

62. Bialet Massé, 1: 120.

63. Grela, *Grito*, pp. 197–98; Larden, pp. 76–78; Scobie, *Revolution*, p. 84.

64. Taylor, p. 171; Jefferson, p. 142. The agronomist Ricardo Huergo noted—and lamented—the farmers' speculative fever as early as 1904; Huergo, *Investigación*, pp. 89, 130. See also Ferraro, pp. 80–81.

65. Alvarez, pp. 217, 220; Grela, *Grito*, pp. 46–47.

66. Arcondo, "Conflicto," pp. 362–72.

67. Ibid., pp. 366, 370; Nemirovsky, pp. 216–18; Lahitte, p. 271.

68. Nemirovsky, pp. 216–17.

69. Arcondo, "Conflicto," is a careful and objective interpretation of the events of 1912. Grela, *El grito*, is a comprehensive, although partisan, source. In a later book on the 1912 strike, *Alcorta*, Grela expands on some of the same themes.

70. On the 1893 strike, see Gallo, *Farmers in Revolt*.

71. *La Prensa*, March 24, 1913, p. 12; Diecidue, pp. 14–24. The priest Pascual Netri, who was accused of links to the Italian Black Hand terrorist organization, spent two months in jail; Grela, *Grito*, pp. 60–62; Grela, *Alcorta*, pp. 144–45.

72. Diecidue, pp. 14–49; Grela, *Grito*, pp. 420–24; Tornatore, pp. 22–24.

73. Nemirovsky, p. 230.

74. Grela, *Grito*, pp. 367–70.

75. Walter, *Socialist Party*, pp. 127–28; *La Vanguardia*, Feb. 9, 1913, p. 1; Grela, *Grito*, pp. 150–59; Diecidue, pp. 58–59, 77.

76. Ferraro. Grela, *Grito*, p. 14, and García Serrano, p. 46, both estimate that over 100,000 farmers struck in 1912. These estimates are probably exaggerated, for the 1914 census counted only 104,000 pampa farmers, and the 1912 strike did not touch all parts of the region—nor did all farmers support it.

77. *La Prensa*, July 11, 1912, p. 20, and July 14, 1912, p. 12; Spangemberg, pp. 523–24; Sienrra, p. 137.

78. On the landowners' response, see *La Vanguardia*, July 6, 1912, p. 1, and July 20, 1912, p. 1.

79. *La Prensa*, June 28, 1912, p. 10, Aug. 4, 1912, p. 11.

80. Campolieti, p. 128; Argentina [2], sesiones ordinarias, 1, July 29, 1912, pp. 815–32; Arcondo, "Conflicto," p. 377.

81. M. Cárcano, "Ensayo histórico," pp. 161–68. See also Liebscher, pp. 225–39.

82. Spangemberg, pp. 526–27; Grela, *Grito*, pp. 99, 124–25. For reports of the settlement of the strike, see *La Prensa*, Aug. 6, 1912, p. 14.

83. Phelps, pp. 22–26; Ford, *Gold Standard*, pp. 170–88.

84. *La Prensa*, March 6, 1913, p. 16; *RRP*, March 14, 1913, p. 655.

85. *La Vanguardia*, Jan. 15, 1913, p. 1; *La Prensa*, Feb. 7, 1913, p. 13, and Feb. 20, 1913, p. 14; W. J. Molins, pp. 354–55.

86. Diecidue, pp. 77–120, provides a detailed description and analysis of this troubled period. See also Tornatore, p. 27. On the political background in Santa Fe at this period, see Liebscher, pp. 121–24, 144–49, 243–63.

87. Diecidue, pp. 127–28. De la Plaza too had used the police to expel striking tenants in 1912; *La Prensa*, Aug. 16, 1912, p. 18.

88. Tornatore, p. 54; Diecidue, pp. 134–39.

89. Diecidue, pp. 189, 256; *La Tierra*, Nov. 30, 1917, p. 1.

90. García Serrano, pp. 33–75.

91. Nemirovsky, p. 227.

92. Paoli, pp. 37–38; García Ledesma, p. 55.

Chapter 8

1. M. W. Sharp, pp. 372–73; Surface, *Grain Trade*, p. 19; Beveridge, pp. 87–88.

2. Quoted in O'Connor, p. 217.

3. C. F. Wilson, pp. 72–73; M. W. Sharp, pp. 376–77.

4. Surface, *Grain Trade*, pp. 29–32; M. W. Sharp, pp. 377–78; Beveridge, pp. 87–89.

5. "The Price of Wheat," *GGG*, Sept. 19, 1917, p. 5; "Fixing Wheat Prices," *GGG*, March 21, 1917, p. 5; M. W. Sharp, pp. 384–88; C. F. Wilson, pp. 81–98.

6. Gravil, "Anglo-Argentine Connection," pp. 65–70; *RRP*, June 16, 1916, p. 1309.

7. P. H. Smith, pp. 69–70; Di Tella & Zymelman, p. 299; L. B. Smith et al., p. 24; Havens, p. 650.

8. Gravil, "Anglo-Argentine Connection," pp. 71–72.

9. For Hoover's remarks, see *SD*, 835.6131/12, Nov. 13, 1917, H. Hoover to Alonzo Taylor (U.S. Ambassador, Paris). See also Surface, *Grain Trade*, pp. 187, 295–97; and Gravil, "Anglo-Argentine Connection," pp. 72–74.

10. Gravil, "Anglo-Argentine Connection," pp. 74–75; Surface, *Grain Trade*, pp. 293–98. The exact terms are in Argentina [2], sesiones ordinarias, 8, Jan. 15, 1918, p. 5. For agrarian opinion, see *La Prensa*, Jan. 23, 1918, p. 9.

11. Gibson, *Memorandum*, p. 7. See also *SD*, 835.6131/68, June 18, 1918, de Billy to Auchincloss, and July 12, 1918, Polk to de Billy; 835.6131/67, July 3, 1918, Morgan (Rio de Janeiro) to Secy. of State; and 835.6131/70, June 14, 1918, U.S. Consul (Rosario) to Secy. of State.

12. Gibson, *Memorandum*, p. 7; Gravil, "Anglo-Argentine Connection," p. 75.

13. For *La Tierra*'s statement, see Argentina [2], sesiones extraordinarias, 6, Nov. 26, 1919, p. 447. See also *RRP*, April 18, 1919, p. 917; London *Times*, Jan. 28, 1919, p. 7; and Argentina [3], sesiones extraordinarias, 2, April 8, 1919, pp. 281–302. *La Prensa*, April 7, 1919, pp. 5–6, opposed the second agreement.

14. London *Times*, June 16, 1920, p. 23, Aug. 4, 1920, p. 10; Great Britain, House of Commons, pp. 29–30; Gibson, *Memorandum*, p. 10.

15. C. F. Wilson, pp. 119–72; Fowke, *National Policy*, pp. 171–73; Patton, *Grain Growers' Cooperation*, p. 199; Rolph, pp. 123–25.

16. Holm, p. 12.

17. Pascale, 1: 218–23.

18. For an analysis of the farm labor situation during the war, see *El Capital* (Rosario), Nov. 18, 1920, p. 4, and Dec. 18, 1921, p. 4.

19. On the bag situation, see Argentina [9], pp. 73, 75. For the government's agreement with the royal commission and a defense of its policies, see "Report on the Supply of Bags by the Royal Commission on Wheat Supplies for the Grain Crop of 1918–1919," *RRP*, April 11, 1919, pp. 855–61.

20. *La Prensa*, Jan. 9, 1919, pp. 11–12, Jan. 25, 1919, p. 9, Jan. 29, 1919, p. 7. See also *RRP*, Oct. 18, 1918, p. 977, and Dec. 20, 1918, p. 1551.

21. Argentina [2], sesiones ordinarias, 3, Aug. 6, 1919, pp. 367–68, 401–17; Argentina [3], sesiones, 1, Sept. 16, 1919, pp. 585, 625; *La Vanguardia*, Dec. 12, 1919, p. 2; *La Prensa*, Sept. 27, 1921, p. 10.

22. On the government's fiscal crisis, see Alejandro Bunge, *Riqueza*, p. 194; Peters, pp. 70–71; and Dell'oro Maini, pp. 289, 360. On Yrigoyen's budgets, see Solberg, "Rural Unrest," p. 31.

23. Argentina [2], sesiones extraordinarias, 7, Jan. 3, 1918, pp. 663–73; 7, Jan. 4, 1918, p. 728.

24. "Los impuestos a la exportación," *Anales de la Sociedad Rural Argentina*, Sept. 1931, p. 528, Feb. 1933, p. 62. See also Pintos, "Se necesitan hombres," p. 238.

25. "Derechos a la exportación," *Anales de la Sociedad Rural Argentina*, Nov. 1932, p. 825; Brown & Mace, pp. 9–10, 22.

26. Solberg, "Rural Unrest," pp. 31–32.

27. *RRP*, Oct. 10, 1919, p. 929; S. G. Hanson, p. 255; *La Prensa*, March 31, 1919, p. 5, April 13, 1919, p. 5.

28. The FAA's petition to Yrigoyen is in *La Tierra*, March 7, 1919, p. 1. See also the issues of May 1–2, 1919, p. 1, and May 9, 1919, p. 1. Esteban Piacenza, the FAA president, emphasized the farmers' plight in an interview with *La Prensa*, April 10, 1919, p. 12.

29. Piacenza emphasized his reform goals in an interview with *La Vanguardia*, Jan. 1, 1920, p. 1. Evidence gathered by the University of La Plata's Faculty of Agronomy revealed that almost without exception farmers aspired to become landowners. See Coni, *¿Arrendamiento o propriedad?*, p. 11. *La Vanguardia*, Sept. 19, 1920, p. 2, summarizes the results of this survey.

30. For a superb summary and analysis of the Semana Trágica, see Rock, *Politics*, pp. 157–79.

31. On Demarchi's tour, see *La Epoca*, April 14, 1919, p. 1, April 25, 1919, p. 4; and *La Prensa*, April 13, 1919, p. 5, April 18, 1919, p. 4, April 21, 1919, p. 5. See also *RRP*, April 4, 1919, p. 789.

32. *RRP*, June 20, 1919, p. 1445, reported that the government deported over 300 foreigners from rural Argentina during this strike. See also *La Vanguardia*, April 10, 1919, p. 2, April 16, 1919, p. 1; and *La Prensa*, May 11, 1919, p. 10, May 22, 1919, p. 12.

33. *La Vanguardia*, Aug. 10, 1919, p. 1.

34. *La Prensa*, April 17, 1919, p. 6; *RRP*, April 11, 1919, p. 855.

35. *SD*, 835.00/185, June 21, 1919, "Daily Diary Series," notes by Sumner Welles. See also *La Tierra*, May 16, 1919, p. 1.

36. M. A. Pueyrredón, "Panaceas," *La Razón*, May 5, 1919; Castex, "Malestar agrícola," pp. 199–200.

37. Palacios, pp. 178–79, 399–402, and Justo, pp. 140–89, both provide overviews of the Socialist Party's agrarian program.

38. W. J. Molins, p. 32; *La Tierra*, June 6, 1919, p. 1; De la Torre, 6: 30. For Escalante's comment, see J. M. García, p. 10.

39. Remmer, pp. 195–96. For press criticism of Yrigoyen's attitude in this affair, see *La Prensa*, March 11, 1919, p. 6, and Aug. 4, 1921, p. 10. For Yrigoyen's explanation of his policy, see Argentina [2], sesiones ordinarias, 2, Aug. 9, 1922, pp. 724–31.

40. Argentina [2], sesiones extraordinarias, 6, April 9, 1919, pp. 652–53; sesiones ordinarias, 2, July 16, 1919, pp. 729–30.

41. De la Torre's 1919 speech is in Monteagudo, pp. 83–88. See also De la Torre, 6: 17–27.

42. *RRP*, May 23, 1919, p. 1184; *SD*, 835.52/16, Robertson (Buenos Aires) to Secy. of State.

43. Argentina [2], sesiones ordinarias, 2, July 2, 1919, pp. 612–15; 2, Sept. 29, 1919, pp. 717–19.

44. The Mortgage Bank Law, Ley 10,676, also provided aid to cattlemen, which probably explains its rapid passage. For an analysis of this legislation, see Coni, *Préstamos de colonización*, pp. 4, 14, 42–43; Alfredo Bonino and Pedro del Carril, *La subdivisión de nuestras tierras y la evolución agraria del país* (Buenos Aires: Estudio de Agronomía y Colonización, 1926), p. 39; and Podestá, pp. 101, 114.

45. *La Protesta*, Nov. 19, 1919, p. 3, Dec. 18, 1919, p. 2.

46. These brigades were associated with the ultra right-wing Liga Patriótica Argentina. See *La Prensa*, Dec. 15, 1919, p. 9; *SD*, 835/504.1, Jan. 10, 1920, Robertson (Buenos Aires) to Secy. of State; "Colaboración eficaz de la Bolsa de Cereales en la obra pacificadora de la Liga Patriótica Argentina," in Buenos Aires, Bolsa de Cereales, pp. 112–13; and Castex, "Propaganda anarquista."

47. *La Protesta*, Feb. 1, 1920, pp. 1–2; *La Prensa*, Dec. 19, 1919, p. 12.

48. *La Tierra*, Dec. 26, 1919, p. 1, May 16, 1919, p. 1, Oct. 5, 1920, p. 1.

49. Argentina, sesiones ordinarias, 5, Sept. 21, 1920, pp. 341–77.

50. The newspapers covered the 1921 farmers' march extensively. See, for example, the front-page stories in *La Tierra*, Aug. 18 and Aug. 30; *La Vanguardia*, Aug. 27 and Aug. 28; and *La Prensa*, Aug. 27 and Aug. 28. See also *La Tierra*, March 25, 1922, p. 1.

51. Palacios, pp. 403–5, reprints the text of Law 11,170.

52. *La Tierra*, Oct. 28, 1921, p. 1, Feb. 21, 1922, p. 1; *La Vanguardia*, Nov. 19, 1921, p. 1.

53. Sienrra, p. 140.

54. *La Tierra*, Oct. 3, 1922, p. 1.

55. The following paragraphs rely heavily on the superb monograph by John Herd Thompson, *The Harvests of War*, especially pp. 46–69.

56. Mackintosh, *Economic Problems*, p. 260.

57. Thompson, *Harvests*, p. 67.

58. Ibid., p. 50.

59. L. G. G. Thomas, *Liberal Party*, p. 155. Leacock is quoted in Thompson, *Harvests*, p. 59.

60. Grove, p. 139. On the growth of land mortgage debts during the war, see Mackintosh, *Economic Problems*, p. 260.

61. L. A. Wood, pp. 287–89; Easterbrook, pp. 96–111; Regehr, "Bankers," pp. 320–26.

62. Fowke, *National Policy*, p. 200; Rolph, p. 23.

63. "Farmers' Condition Intolerable," *GGG*, Nov. 9, 1921, p. 19; P. F. Sharp, p. 130.

64. "Canada Is at War," *GGG*, Aug. 12, 1914, p. 5.

65. Partridge, p. 1.

66. "Farmers Rule a State," *GGG*, Nov. 29, 1916, p. 6; P. F. Sharp, p. 78; Thompson, *Harvests*, p. 124.

67. W. Morton, *Progressive Party*, p. 58.

68. Thompson, *Harvests*, pp. 121–22.

69. R. Dawson, *Government of Canada*, p. 351; Thompson, *Harvests*, p. 126.

70. "The War Election Franchise," *GGG*, Sept. 19, 1917, p. 5; More Ginger, "Doubts Wisdom of Too Great Freedom," *GGG*, Feb. 14, 1917, p. 17.

71. "Reason in Immigration," *GGG*, Nov. 3, 1909, p. 13; "Working Among German Farmers," *GGG*, April 2, 1913, p. 10; Bickersteth, p. 88. See also Yuzyk, pp. 50–51.

72. Thompson, *Harvests*, pp. 130–38.

73. Ibid., p. 135; "Union Government Program," *GGG*, Oct. 24, 1917, p. 5.

74. "The Draft," *GGG*, May 22, 1918, p. 4; C. F. Wilson, pp. 184–87; R. Dawson, *William Lyon Mackenzie King*, pp. 276–79; Griezic, pp. 109–12.

75. Rolph, p. 192. See also pp. 3–16; and Mardiros, pp. 84–85.

76. W. Morton, *Progressive Party*, p. 93. There is an extensive literature on the political emergence of the UFA. See Priestley & Swindlehurst, p. 61; C. B. Macpherson, pp. 28–43; and Rolph, pp. 62–74, among others.

77. Moorhouse.

78. Mardiros, pp. 64, 92–96, 112–14, 120–25; Irvine, p. 196.

79. L. G. Thomas, *Liberal Party*, passim, discusses the period of traditional party rule. On the Alberta debt, see Mackintosh, *Economic Problems*, p. 74. See also Flanagan, pp. 139–41.

80. "Medicine Hat," *GGG*, July 6, 1921, p. 5; Mardiros, pp. 109–11; Rolph, pp. 98–101; L. G. Thomas, *Liberal Party*, p. 202.

81. W. Morton, *Progressive Party*, pp. 32–33, 97–99; W. Morton, *Manitoba*, pp. 336–62.

82. W. Morton, *Manitoba*, p. 374, "The Sympathetic Strike," *GGG*, June 11, 1919, p. 6; "Big Strike Ended," *GGG*, July 2, 1919, p. 6. See also P. F. Sharp, p. 62.

83. W. Morton, *Manitoba*, pp. 371–75; McNaught, *Prophet*, p. 150.

84. Kendle, p. 30; W. Morton, *Manitoba*, pp. 362–74, 378–83.

85. W. Morton, *Manitoba*, pp. 400–405.

86. D. Smith, *Prairie Liberalism*, p. 333.

87. W. Morton, *Progressive Party*, pp. 35, 98; Eager, pp. 6–10; D. Smith, *Prairie Liberalism*, pp. 54–55, 161.

88. W. Morton, *Progressive Party*, passim.

89. Hutchison, pp. 12–65.

90. W. Morton, *Progressive Party*, p. 81.

91. R. Dawson, *William Lyon Mackenzie King*, pp. 317–20; Neatby, p. 411; D. Smith, *Regional Decline*, p. 43.

92. R. Dawson, *William Lyon Mackenzie King*, p. 344.

93. C. F. Wilson, pp. 190–91; W. Morton, *Progressive Party*, pp. 94–106; Griezic, pp. 114–18.

94. C. F. Wilson, pp. 189–90.

95. Quoted in W. Morton, *Progressive Party*, pp. 104–5.

96. "The New Premier," GGG, July 14, 1920, p. 5; "Frenzied Politics," GGG, Oct. 6, 1920, p. 5; GGG, Nov. 16, 1921, p. 5.

97. W. Morton, *Progressive Party*, pp. 122–25.

98. P. F. Sharp, pp. 149–51; W. Morton, *Progressive Party*, p. 127.

99. R. Dawson, *William Lyon Mackenzie King*, p. 380; W. Morton, *Progressive Party*, pp. 130–42.

100. R. Dawson, *William Lyon Mackenzie King*, pp. 391–92, 445.

101. W. Morton, *Progressive Party*, p. 157.

102. "Reasons for a Wheat Board," GGG, Feb. 15, 1922, p. 5; Griezic, p. 119.

103. C. F. Wilson, pp. 175–78; Fowke, *Canadian Agricultural Policy*, p. 247.

104. Kendle, pp. 43–45; C. F. Wilson, pp. 179–81; "The Wheat Board Failure," GGG, Aug. 23, 1922, p. 5.

105. Quoted in C. F. Wilson, GGG, p. 203.

Chapter 9

1. For comparative world wheat production figures, by five-year averages, see Malenbaum, pp. 238–39. On world exports, see *Wheat Studies*, Jan. 1937, p. 220.

2. D. R. McIntyre to Alberta Wheat Pool, May 1939, Glenbow-Alberta Archives, File 267. For an analysis of the impact of European protectionism on the wheat trade, see Canada [6], pp. 122–23; and Malenbaum, pp. 158–66.

3. Malenbaum, pp. 244–47.

4. These figures do not include stocks in India or Russia; *Wheat Studies*, Oct. 1940, p. 99.

5. S. G. Hanson, pp. 246–58.

6. "A Political Crime," GGG, June 6, 1923, p. 5.

7. Neatby, pp. 12, 15–21; McNaught, *Prophet*, p. 208; W. Morton, *Progressive Party*, p. 197.

8. Neatby, pp. 24–30, 45–48. The quote is on p. 43.

9. C. F. Wilson, pp. 206–7; Neatby, p. 129.

10. W. Morton, *Progressive Party*, pp. 237–44.

11. Ibid., pp. 246–48. The quote is on p. 247. See also Neatby, pp. 92, 102–22; "The Premier's Return," GGG, Feb. 24, 1926, p. 5; and "The Budget," GGG, May 1, 1926, p. 7.

12. An analysis of these dramatic political events of 1926 is beyond the scope of this study. For more details, see W. Morton, *Progressive Party*, pp. 251–63.

13. Neatby, pp. 159–72. The quote is on p. 171. See also W. Morton, *Progressive Party*, pp. 261–63, 268–71.

14. Irvine is quoted in Priestley & Swindlehurst, p. 95. See also Easterbrook, pp. 134–35, 158–60; and Mackintosh, *Economic Problems*, pp. 269–71.

15. Neatby, p. 296.

16. Shideler, pp. 99–104. The quotes are from pp. 99–100. See also Knapp, pp. 20–21.

17. Larsen & Erdman, pp. 243–57. The quote is from p. 250.

18. For details on the organization and operation of the pools, see Fowke, *National Policy*, pp. 219–21; and Patton, "Observations."

19. Quoted in Mills, p. 146.

20. Spafford, "Left-Wing," p. 46. See also Spafford, "Origin," pp. 89–93; and Patton, *Grain Growers' Cooperation*, pp. 392–93.

21. Innis, *Diary*, pp. 65–66; Yates, p. 53.

22. *Report of Mass Meeting Addressed by Mr. Aaron Sapiro in Third Avenue Methodist Church, Saskatoon, Saskatchewan, on Tuesday, August 7th, 1923*, (Saskatoon: n.p., 1923). The quotes are from pp. 5, 6, 15, 18. See also Rolph, p. 143; and "Aaron Sapiro's Visit," GGG, Aug. 15, 1923, p. 5.

23. Underhill is quoted in Ian Macpherson, p. 61. See also "Wheat Pool Controversy," GGG, Feb. 27, 1924, p. 5; and "Signing the Contract," GGG, Feb. 27, 1924, p. 5.

24. Rolph, pp. 145–48; Davisson, p. 150; Nesbitt, p. 141; Fay, *Agricultural Co-operation*, pp. 460–61.

25. Quoted in W. Morton, *Progressive Party*, p. 28. See also Schulz, pp. 78–79; and Davisson, p. 150.

26. Innis, *Diary*, pp. 55–56.

27. W. Morton, *Manitoba*, p. 392; Fay, *Agricultural Co-operation*, p. 461.

28. Innis, *Diary*, p. 107; Rolph, p. 151; Patton, *Grain Growers' Cooperation*, pp. viii, 221.

29. Alberta Wheat Pool, Glenbow-Alberta Archives, File 690; "Alberta Wheat Pool," GGG, Sept. 15, 1927, p. 20; "Saskatchewan Pool Annual Meeting," GGG, Oct. 28, 1925, p. 3.

30. For an example of farmer opposition to the pools, see H. F. Lawrence, *Kidnapped*, p. 1.

31. C. Dawson, *Group Settlement*, pp. 78, 326; Yuzyk, pp. 50–51; Alberta Wheat Pool, Glenbow-Alberta Archives, File 690.

32. Alberta Wheat Pool, Glenbow-Alberta Archives, File 689; *Proceedings of the Second Conference*, p. 11; Davisson, pp. 42–43; Canada [6], pp. 66–68.

33. R. H. Mahoney, manager of the Manitoba Wheat Pool, makes these remarks in *Proceedings of the Second Conference*, p. 116. See also Patton, *Grain Growers' Cooperation*, pp. 228–29; Yates, p. 121; and Nesbitt, p. 144.

34. "The Farmers' Union Campaign," GGG, March 4, 1925, p. 9.

35. Innis, *Diary*, pp. 86–93; Patton, *Grain Growers' Cooperation*, pp. 228–29; Yates, pp. 125–27; "Pool Buys Co-op Plant," GGG, April 15, 1926, p. 7.

36. Canada [7], 1: 17.

37. Innis, pp. 92–93.

38. Schulz, pp. 80–81; Spafford, "Left-Wing," p. 48.

39. The quote is from J. F. C. Wright, p. 41. See also Mills, passim; McCrorie, *In Union Is Strength*, p. 113; "News from the Organizations," GGG, Jan. 18, 1927, p. 21.

40. Mills, passim; Spafford, "Left-Wing," p. 48.

41. W. S. Evans, pp. 22–23.

42. Quoted in Knapp, p. 123.

43. Innis, *Diary*, p. 143; Larsen & Erdman, p. 265.

44. "International Pooling," GGG, Feb. 25, 1923, p. 25; "Peace Time Patriotism," GGG, Feb. 27, 1924, p. 8; "Signing the Contract," GGG, Feb. 27, 1924, p. 5.

45. "Wheat Pool Controversy," GGG, Feb. 27, 1924, p. 5.

46. St. Paul *Pioneer Press*, Feb. 15, 1926, p. 1.

47. "The Wheat Pool Conference," GGG, Feb. 24, 1926, p. 5; St. Paul *Pioneer Press*, Feb. 16, 1926, p. 1, Feb. 18, 1926, p. 3; Rolph, pp. 164–66; Alberta Wheat Pool, Glenbow-Alberta Archives, File 689.

48. Dunsdorfs, pp. 156, 207–34. See also Rolph, p. 166; "Australian Wheat Pools," *GGG*, Sept. 27, 1922, p. 3; Alberta Wheat Pool, Glenbow-Alberta Archives, File 689; and *Proceedings of the Second Conference*, pp. 62–71.

49. Shideler, p. 251.

50. Knapp, pp. 9–11; Shideler, pp. 110–11, 251; *Proceedings of the Second Conference*, pp. 39–41, 86–95.

51. "The Land of the Pampas," *GGG*, Sept. 1, 1926, p. 10.

52. Innis, *Diary*, pp. 46, 104, 130; C. W. Jackman to Leonard D. Nesbitt, Aug. 10, 1959, Glenbow-Alberta Archives.

53. Report of W. J. Jackman to Board of Directors of the Canadian Co-operative Wheat Producers, Ltd., Jan. 10, 1927, Glenbow-Alberta Archives, Alberta Wheat Pool File 604.

54. Jackman, *Wheat Growing*; *GGG*, Feb. 1, 1927, pp. 11, 23.

55. Jackman, "Wheatfields."

56. Jackman, *Wheat Growing*, p. 10.

57. Jackman, "Wheatfields," p. 5.

58. Report of W. J. Jackman, cited in note 53, above.

59. *Proceedings of the Second Conference*, p. 145.

60. Ibid., pp. 17, 65.   61. Ibid., p. 12.

62. Ibid., p. 57. See also p. 50.   63. Ibid., p. 72.

64. Ibid., p. 114. See also p. 149; and Knapp, pp. 144–45.

65. *Proceedings of the First Conference*, p. 140; *Proceedings of the Second Conference*, p. 145; C. L. Bailey, p. 71.

66. *Proceedings of the First Conference*, pp. 51–53, 126, 131.

67. Quoted in "Grain Elevators," *RRP*, Feb. 25. 1927, p. 11. Also see *RRP*, Oct. 1, 1926, p. 7.

68. "Watch Canada!" *RRP*, Feb. 11, 1927, p. 21; *La Prensa*, March 30, 1927, p. 12.

69. Alejandro Bunge, "Hechos." The quotes are from pp. 91 and 100. Bunge later restated these ideas in "Paralelo económico." See also Falcoff, p. 70.

70. Leiserson, pp. 110, 118–19; Coni, *Verdades*, p. 30. See also Schiopetto, p. 2132.

71. Ortiz Pereyra, p. 126.

72. Luna, pp. 42–49.

73. For details of this schism, see Solberg, *Oil and Nationalism*, pp. 80–81; and Rock, *Politics*, pp. 223–24, 229–31.

74. *RRP*, Feb. 2, 1923, p. 267; Leon M. Estabrook, "Agricultural Statistics in Argentina," *RRP*, Aug. 8, 1924, p. 359; M. Cárcano, "Nuestro régimen," p. 1115.

75. *RRP*, Feb. 2, 1923, p. 267, Oct. 29. 1926, p. 7, March 30, 1928, p. 13. See also Schleh, *Semillas*, pp. 13–15; and *La Tierra*, Aug. 25, 1925, p. 1.

76. P. Smith, *Politics and Beef*, discusses the political power of the cattlemen and the legislative initiatives congressmen made on their behalf during the 1920's.

77. Nemirovsky, p. 227.

78. Jackman, *Wheat Growing*, pp. 16–17; Rodríguez Tardati, p. 392; Taylor, p. 409.

79. García Ledesma, p. 53; Paoli, pp. 37–38.

80. Jackman, *Wheat Growing*, pp. 16–18.

81. Tornatore, p. 41.

82. For detailed analyses of the methods of evading Law 11, 170, see M. Vélez, pp. 90–96; Repetto, "Defensa," pp. 735–36; and Borras, p. 72.

83. Reforms of Law 11,170 were introduced in the Chamber in 1923, 1925, and 1928, and in the Senate in 1927 and 1928.

84. Argentina [2], sesiones ordinarias, 3, Aug. 21, 1924, pp. 480–81.

85. A copy of Estabrook's letter is found in *SD*, 835.00/351, Nov. 18, 1924.

86. *La Tierra*, Aug. 26, 1924, p. 1; Argentina [2], sesiones ordinarias, 3, Aug. 21, 1925, p. 686.

87. Alejandro Bunge, "Hechos," p. 122; Bórea, *Mutualidad*, p. 145; Francisco Pedro Marotta, *Tierra y patria: los argentinos debemos realizar la segunda expedición desierto* (Buenos Aires: Imprenta Mercatali, 1932), p. 135 and passim.

88. *The Financier*, Oct. 25, 1921, enclosure in *SD*, 835.77/74, Jan. 2, 1922. St. Davids reiterated his position in an interview with the U.S. consul general in 1922; see *SD*, 855.6363/168, Aug. 16, 1922. See also London *Times*, Oct. 27, 1923, p. 17.

89. W. Wright, *British-Owned Railways*, p. 129.

90. Buenos Aires *Herald*, June 5, 1923, enclosure in *SD*, 835.00/324. See also Leguizamón, "Inmigración." Leguizamón was a member of the Board of Directors of the Buenos Aires Western Railway.

91. "Colonización oficial," *Anales de la Sociedad Rural Argentina*, Oct. 15, 1924, pp. 1031–1032.

92. E. Kitchell Farrand to Secy. of State, Sept. 18, 1924, *SD*, 835.52/31.

93. *SD*, 835.52/34, Nov. 4, 1924.

94. Bucich Escobar, pp. 500–501; R. A. Molina, pp. 302–9. Thaw (Buenos Aires) to Secy. of State, Sept. 10, 1925, Nov. 5, 1925, *SD*, 835.00/365, 835.00/370.

95. London *Times*, April 21, 1927, p. 11; Buenos Aires *Herald*, May 10, 1927, enclosure in *SD* 835.52/38, May 12, 1927; P. Z. Cable (Buenos Aires) to Secy. of State, March 29, 1927, *SD*, 835.52/37.

96. Messersmith (Buenos Aires) to Secy. of State, Feb. 23, 1929, *SD*, 835.55/72. See also W. Wright, *British-Owned Railways*, p. 130.

97. Ortiz Pereyra, pp. 108, 110–11.

98. Bórea, *Tratado*, pp. 268, 284; Coni, "Colonización," p. 360; Etcheverry, pp. 37–39.

99. Bórea, *Tratado*, pp. 341–43, 506–12.

100. The debates are reprinted in ibid., pp. 366–402.

101. *La Prensa*, July 13, 1930, p. 13; Bórea, *Tratado*, pp. 405–512; Argentina [4], p. 284.

102. On the National Mortgage Bank during the 1920's, see Coni, *Préstamos*, pp. 3–30, 42–43. See also Bórea, *Colonización*, pp. 50, 54.

103. For an early and strong plea from a respected agronomist, see Girola, *Investigación*, p. 303. See also Pérez Brignoli, pp. 17–21.

104. Argentina [2], sesiones ordinarias, 3, July 30, 1924, pp. 7–8.

105. Duhau, "Libre Concurrencia," p. 1762.

106. Duhau, *Elevadores*, p. 46.

107. Ibid., p. 8. See also Duhau, "Libre concurrencia," p. 1763. For another statement of the Sociedad Rural's views, see H. Lawrence, "Elevadores," p. 333.

108. Argentina [13], pp. 5–26. See also Leguizamón, "Elevadores," pp. 93–94; *RRP*, June 4, 1926, p. 23; Confederación Argentina del Comercio, de la Industria, y de la Producción, *Actas de la Tercera Conferencia Económica Nacional (2–12 de julio 1928)*, (Buenos Aires: Cía Impresora Argentina, 1928), pp. 96–97; *RRP*, Feb. 25, 1927, p. 11; Leiserson, *Elevadores*, p. 8; *La Tierra*, Aug. 17, 1930; and Pérez Brignoli, pp. 23–24.

109. Argentina [2], sesiones ordinarias, 4, Sept. 10, 1928, p. 87; Leiserson, *Elevadores*, pp. 19, 62–63.

110. *La Epoca*, Aug. 17, 1928, p. 1.

111. For reports of opposition to further land-tenure reform, see *La Prensa*, Dec. 23, 1929, p. 10; and G. García, "Leyes," p. 340. See also Argentina [2], sesiones ordinarias, 3, Sept. 18, 1929, p. 618; and Argentina [3], sesiones extraordinarias, 3, Jan. 1930, pp. 311–56.

112. For an analysis of the congressional crisis of 1930, see Robert Smith, "Radicalism in the Province of San Juan."

113. Argentina [2], sesiones ordinarias, 3, Sept. 25, 1929, pp. 909–13, 1070.

114. *RRP*, June 21, 1929, p. 15; *La Prensa*, June 18, 1929, p. 17, April 2, 1920, p. 13.

115. *La Tierra*, Nov. 15, 1928, p. 1, Nov. 24, 1928, p. 2; *RRP*, Nov. 30, 1928, p. 7; *La Prensa*, Nov. 29, 1928, p. 19.

116. Yrigoyen's use of the army in December 1928 is narrated in detail in *La Tierra*, Dec. 8, p. 2; and *La Epoca*, Dec. 2, p. 1, Dec. 3, p. 1, Dec. 12, p. 1. For the farmers' opposition to rural labor legislation, see *La Tierra*, June 16, 1928, p. 1, July 28, 1928, p. 1.

117. *La Tierra*, Sept. 10, 1930, p. 4, Sept. 12, 1930, p. 4.

Chapter 10

1. Jenaro García, "Problemas agrarios."

2. Walter, "Politics."

3. Díaz Alejandro, *Essays*, p. 76; Fienup et al., pp. 382–83.

4. Edward Schumacher, "Argentina Courts Farmers," *New York Times*, May 2, 1984, 27; "Economic Growth Program: Impact on the Debt," *Argentine Report*, Sept. 1984, pp. 1–2.

5. Veeman & Veeman, pp. 102–9.

Bibliographical Material

Bibliographical Note

The sharp contrasts between so many aspects of Argentine and Canadian agrarian life are also found in the historiography of the two nations. While there is a rich historiography on the Canadian West in general and on prairie agriculture in particular, Argentine agricultural history has barely been touched in a serious, systematic way. When studying Canada, the researcher can rely on numerous monographs based on thorough archival investigation. The works of W. L. Morton, Vernon Fowke, and W. A. Mackintosh, to name only a few scholars, provide the basic information and interpretations one needs to launch a study of prairie agricultural history.

How different is the situation in Argentine historiography! Although Argentine scholarship in the areas of political economy in general and agrarian development in particular got off to an early and promising start, the political repression and instability that wracked Argentina after 1930 have drained that nation's scholarly world of much of its vitality. And the general neglect of agriculture that has long characterized Argentine government policy is also noticeable in the world of scholarship. Few Argentine scholars have worked on agricultural history, and there is no solid monographic tradition as in Canada. One of the few existing works on pampa agrarian history was written by James Scobie, a North American. Most of the works that have appeared in Argentina are highly ideological or polemical and must be used with care, although the recent fine studies by Aníbal Arcondo and Ezequiel Gallo hold out hope for the future.

The researcher attempting to do a comparative study of these two agrarian societies thus faces special challenges, for while one may refer to an established scholarly tradition in the case of Canada, one must start virtually from scratch in the case of Argentina. For this reason, my Argentine research draws heavily on primary sources. This includes many contemporary accounts in newspapers and journals. These and other one-time citations are given in full in the Notes.

Bibliography

The following abbreviations are used in the Bibliography:

AHR   Alberta Historical Review
BFDC   U.S. Department of Commerce, Bureau of Foreign and Domestic Commerce
CHR   Canadian Historical Review
CJEPS   Canadian Journal of Economics and Political Science
RCE   Revista de Ciencias Económicas
REA   Revista de Economía Argentina
SH   Saskatchewan History

"La agitación agraria en Santa Fe: Informe de un Inspector del Departamento Nacional del Trabajo," *Boletín de Servicios de la Asociación del Trabajo*, Dec. 5, 1928; Dec. 20, 1928.

Aitken, Hugh G. *American Capital and Canadian Resources.* Cambridge, Mass.: Harvard Univ. Press, 1961.

Akenson, Donald H., ed. *Canadian Papers in Rural History.* 2 vols. Gananoque, Ontario: Langdale Press, 1978–80.

Allen, Richard. "The Social Gospel as the Religion of the Agrarian Revolt," in Berger and Cook, eds., *The West and the Nation*, pp. 174–86.

————. *The Social Passion: Religion and Social Reform in Canada, 1914–28.* Toronto: Univ. of Toronto Press, 1971.

Allen, Rodolfo. *Enseñanza agrícola: Documentos orgánicos.* Buenos Aires: Ministerio de Agricultura de la Nación, 1929.

Alsina, Juan A. *La inmigración en el primer siglo de la independencia.* Buenos Aires: F. S. Alsina, 1910.

Alvarez, Juan. *Temas de historia económica argentina.* Buenos Aires: W. M. Jackson, 1929.

Ankli, Robert E., and Robert M. Litt. "The Growth of Prairie Agriculture: Economic Considerations," in Akenson, ed., *Canadian Papers in Rural History*, Vol. 1, pp. 33–64.

Arcondo, Aníbal B. *La agricultura en Córdoba, 1870–1880.* Córdoba: Univ. Nacional de Córdoba, Facultad de Filosofía y Humanidades, 1965.

————. "El conflicto agrario argentino de 1912: Ensayo de interpretación," *Desarrollo Económico*, Oct.–Dec. 1980 (20), pp. 351–81.

————. "Tierra y política de tierras en Córdoba," *Revista de Economía y Estadística* (Córdoba), 13, Nos. 3–4 (1969), pp. 13–44.

Argentina, Official and Semiofficial Publications. (All works published in Buenos Aires.)

[1] Academia Nacional de la Historia. *Historia argentina contemporánea, 1862–1930.* 4 vols. "El Ateneo," 1965–67.

[2] Cámara de Diputados de la Nación. *Diario de sesiones de la Cámara de Diputados.* 1912–32.

[3] Cámara de Senadores de la Nación. *Diario de sesiones de la Cámara de Senadores.* 1912–30.

[4] Comité Nacional de Geografía. *Anuario geográfico argentino.* 1941.

[5] Department of Agriculture. *Sketch of the Argentine Republic as a Country for Immigration (Latest Data).* 2d ed. 1904.

[6] Dirección General de Estadística. *Tercer censo nacional, levantado el 1⁰ de junio de 1914.* 10 vols. 1916–17.

[7] Dirección General de Estadística de la Nación. *El comercio exterior argentino y estadísticas económicas retrospectivas.* No. 219, 1937; No. 223, 1939; No. 225, 1940.

[8] Dirección General de Estadística y Censos de la Nación. *Anuario del comercio exterior de la República Argentina correspondiente a 1943.* 1944.

[9] Ministerio de Agricultura. *Memoria del Ministerio de Agricultura: Aplicación de la Ley No. 10.777.* 1921.

[10] ———. *¿Qué es la Argentina? Como país agrícola. Como país de inmigración.* 1911.

[11] ———, Comisión Nacional del Censo Agropecuario. *Censo nacional agropecuario, año 1937.* 4 vols. 1939–40.

[12] ———, Dirección de Economía Rural y Estadística. *Anuario agropecuario: Año 1935.* 1935.

[13] ———, Régimen de Elevadores de Granos, Comisión Especial. *Informe presentado a S.E. el Sr. Ministro de Agricultura Don Emilio Mihura sobre la implantación de un sistema general de elevadores de granos.* 1928.

[14] Ministerio del Interior. *La desocupación de los obreros en la República Argentina.* 1915.

Avery, Donald. *"Dangerous Foreigners": European Immigrant Workers and Labour Radicalism in Canada, 1896–1932.* Toronto: McClelland & Stewart, 1979.

Avery, Sherwood H. *Markets for Agricultural Implements and Farm Machinery in Argentina and Uruguay.* Washington, D.C.: BFDC, 1925.

Avni, Haim. "La agricultura judía en la Argentina: Exito o fracaso?" *Desarrollo Económico,* Jan.–March 1983 (22), pp. 535–48.

Bailey, Clara L. "International Wheat Agreements." M.A. thesis, George Washington University, 1936.

Bailey, John P. "Inmigración y relaciones étnicas: Los ingleses en la Argentina," *Desarrollo Económico,* Jan.–Mar. 1979 (18), pp. 539–58.

Baily, Samuel L. "Patterns of Assimilation of Italians in Buenos Aires: 1880–1940." Paper presented at the annual meeting of the American Historical Association, Dallas, 1979.

Baldwin, Robert E. "Patterns of Development in Newly Settled Regions," *The Manchester School of Economic and Social Studies,* May 1956 (24), pp. 161–79.

Bath, C. Richard, and Dilmus D. James. "Dependency Theory and Latin America," *Latin American Research Review,* 11 (1976), No. 3, pp. 3–54.

Bennett, John W. *Northern Plainsmen: Adaptive Strategy and Agrarian Life.* Chicago: Aldine, 1969.

Bercuson, David Jay, ed. *Canada and the Burden of Unity.* Toronto: Macmillan, 1977.

Berger, Carl. *The Sense of Power: Studies in the Ideas of Canadian Imperialism, 1867–1914.* Toronto: Univ. of Toronto Press, 1970.

Berger, Carl, and Ramsay Cook, eds. *The West and the Nation: Essays in Honour of W. L. Morton.* Toronto: McClelland & Stewart, 1976.

Beveridge, William H. *British Food Control.* London: Oxford Univ. Press, 1928.

Bialet Massé, Juan. *Informe sobre el estado de las clases obreras en el interior de la República.* 3 vols. Buenos Aires: Adolfo Grau, 1904.

Bickersteth, J. Burgon. *The Land of Open Doors: Being Letters from Western Canada.* London: Wells Gardner, Darton, 1914.

Binnie-Clark, Georgina. *Wheat and Woman* (1914), ed. Susan Jackal. Toronto: Univ. of Toronto Press, 1979.

Blain, Marc-A. "Le Rôle de la dépendance externe et des structures sociales dans l'économie frumentaire du Canada et de l'Argentine (1880–1930)," *Revue d'Histoire de l'Amérique Française,* Sept. 1972 (26), pp. 239–70.

Booth, J. F. "Cooperative Marketing of Grain in Western Canada," U.S. Dept. of Agriculture *Technical Bulletin* 63 (Jan. 1928), pp. 1–116.

Bórea, Domingo. *La colonización oficial y particular en la República Argentina.* Buenos Aires: "Gadola," 1923.

———. *La mutualidad y el cooperativismo en la República Argentina.* Buenos Aires: L. J. Rosso y Cía., 1917.

———. *Tratado de cooperación.* Buenos Aires: "Gadola," 1927.

Borras, Antonio. *Nuestra cuestión agraria: En defensa de la producción y del productor.* Buenos Aires: "La Vanguardia," 1932.

Boudreau, Joseph A. "Western Canada's 'Enemy Aliens' in World War One," *AHR,* Winter 1964 (12), pp. 1–9.

Boyle, James E. *Marketing Canada's Wheat.* Winnipeg: Winnipeg Grain Exchange, 1929.

Brady, George S. *Argentine Market for Motor Vehicles.* Washington, D.C.: BFDC, 1923.

Breen, David H. *The Canadian Prairie West and the Ranching Frontier, 1874–1924.* Toronto: Univ. of Toronto Press, 1983.

Britnell, G. E. *The Wheat Economy.* Toronto: Univ. of Toronto Press, 1939.

Brown, Harold R., and Brian M. Mace, Jr. *Costs of Transportation and Handling of Argentine Wheat.* Washington, D.C.: BFDC, 1926.

Brown, Robert C., and Ramsay Cook. *Canada, 1896–1921: A Nation Transformed.* Toronto: McClelland & Stewart, 1974.

Bucich Escobar, Ismael. *Historia de los presidentes argentinos.* Buenos Aires: Roldán, 1934.

Buenos Aires, Bolsa de Cereales. *Memoria e informe de la Comisión Directiva presentada en la asamblea del 28 de abril de 1921. Ejercicio 1920–1921.* Buenos Aires: C. Girard, 1921.

Bunge, Alejandro E. "Un aspecto del problema de las carreteras en la Argentina," *REA,* April 1929 (22), pp. 275–80.

———. "Los caminos de acceso a las estaciones," *La Industria Azucarera,* June 1927 (33), p. 549.

———. *La economía argentina.* 4 vols. Buenos Aires: Agencia Gen. de Lib. y Publ., 1928–30.

———. *Ferrocarriles argentinos: Contribución al estudio del patrimonio nacional.* Buenos Aires: Mercatali, 1918.

———. "Los gastos de la educación en la Argentina," *REA,* Jan. 1937 (36), p. 269.

————. "Los hechos económicos y financieros del año 1926. Su significado en la vida económica de la república," *REA*, Feb. 1927 (18), pp. 91–139.

————. *Una nueva Argentina*. Buenos Aires: Guillermo Kraft, 1940.

————. "Paralelo económico argentino–canadiense, 1908–1926," *REA*, Feb. 1929 (22), pp. 113–20.

————. "Política aduanera," *REA*, March 1924 (12), pp. 207–10.

————. *Riqueza y renta de la Argentina: Su distribución y su capacidad contributiva*. Buenos Aires: Agencia Gen. de Lib. y Publ., 1919.

————. "La tragedia de la ganadería argentina," *REA*, Nov. 1933, pp. 349–50.

Bunge, Augusto. *Una Argentina sin analfabetos*. Buenos Aires: Agencia Gen. de Lib. y Publ., 1917.

Burton, F. W. "Wheat in Canadian History," *CJEPS*, May 1937 (3), pp. 210–17.

Bychinsky, Anna. "Ukrainians' Pioneering," *Grain Growers' Guide*, Sept. 1, 1920, pp. 34–35.

Calderwood, William. "Religious Reactions to the Ku Klux Klan in Saskatchewan," *SH*, Autumn 1973 (26), pp. 103–14.

Campbell Olgivie, P. *Argentina from a British Point of View and Notes on Argentine Life*. London: Werthheimer, Lea, 1910.

Campolieti, Roberto. *Política agraria internacional*. Buenos Aires: Tor, 1936.

Canada, Official and Semiofficial Publications. (The place of publication, where given, is Ottawa.)

[1] Census and Statistics Office. *Fifth Census of Canada, 1911*. 6 vols. 1912.

[2] Dominion Bureau of Statistics. *Seventh Census of Canada, 1931*. 11 vols. 1933–36.

[3] ————, General Statistics Branch. *The Canada Year Book*. 1914–36.

[4] Royal Commission Appointed to Inquire into the Immigration of Italian Labourers to Montreal and the Alleged Fraudulent Practices of Employment Agencies. *Report of Commissioner and Evidence*. 1905.

[5] Royal Commission to Inquire into Railways and Transportation in Canada, 1931–2. *Report*. 1932.

[6] Royal Grain Inquiry Commission. *Report*. 1938

[7] ————. *Submission by the Pool Organizations of Alberta, Saskatchewan, and Manitoba*. 2 vols. n.p., [1938?].

Cánepa, Luis Rodolfo. *Economía agraria argentina*. Buenos Aires: "El Ateneo," 1942.

Cárcano, Miguel Angel. "Cuestiones agrarias," *RCE*, Aug. 1921 (17), pp. 5–17.

————. "Ensayo histórico sobre la presidencia de Roque Sáenz Peña," in Argentina [1], Vol. 1, Part 2, pp. 135–91.

————. *Evolución histórica del régimen de la tierra pública, 1810–1916*. Buenos Aires: Lib. Mendesky, 1917.

————. "Nuestro régimen agrario," *RCE*, Oct. 1927 (29), pp. 1112–1118.

Cárcano, Ramón J. *800.000 analfabetos. Aldeas escolares*. Buenos Aires: Roldán, 1933.

Carrothers, W. A. *Emigration from the British Isles with Special Reference to the Development of the Overseas Dominions* (1929). New York: Augustus M. Kelley, 1966.

Castex, Alberto E. "El malestar agrícola," *Anales de la Sociedad Rural Argentina*, April 1919 (53), pp. 197–201.

————. "Propaganda anarquista en la campaña," *Boletín de la Sociedad Rural Argentina*, Dec. 1919 (53), p. 1017.

Caves, Richard E., and Richard H. Holton. *The Canadian Economy: Prospect and Retrospect*. Cambridge, Mass.: Harvard Univ. Press, 1959.

Ceppi, José. (pseud. Aníbal Latino). *Los factores del progreso de la República Argentina*. Buenos Aires: Lib. Nacional, 1910.

Cherwinski, W. J.C. "'Misfits, Malingerers, and Miscontents': The British Harvester Movement of 1928," in Foster, ed., *The Developing West*, pp. 271–302.

Chipman, George. *The Siege of Ottawa*. Winnipeg: Grain Growers' Guide, 1911.

Christie, F. W. *Six Thousand Miles in Canada as "Dundee Courier" Commissioner*. Dundee: W. & D. C. Thomson, 1904.

Clark, W. C. *The Country Elevator in the Canadian West*. Kingston: Queen's Univ., 1916.

Clemenceau, Georges. *South America To-Day*. New York: Putnam's Sons, 1911.

Cohnstaedt, Wilhelm. *Western Canada 1909: Travel Letters by Wilhelm Cohnstaedt*, tr. Herta Holle-Scherer. Regina: Univ. of Regina, Canadian Plains Research Centre, 1976.

Colquette, R. D. *The First Fifty Years: A History of United Grain Growers Limited*. Winnipeg: Public Press, 1957.

Coni, Emilio A. *¿Arrendamiento o propiedad?* La Plata: Univ. Nacional de la Plata, 1920.

———. "La ciencia y la técnica en la agricultura argentina," *REA*, Sept. 1933 (31), p. 207.

———. "La colonización," *RCE*, 2 parts, Vol. 11 (1923). Part 1, June, pp. 348–68; Part 2, Aug.–Sept., pp. 90–100.

———. "Cuestiones agrarias," *RCE*, June 1926 (26), 366–81.

———. *Los préstamos de colonización del Banco Hipotecario Nacional*. Buenos Aires: "El Ateneo," 1931.

———. *Verdades y falacias cooperativistas*. Buenos Aires: Confederación Argentina del Comercio, de la Industria y de la Producción, 1928.

Conti, Marcelo. *Lo que deben conocer nuestros agricultures sobre cosecha de trigo*. Buenos Aires: Univ. de Buenos Aires, 1929.

Coote, J. A. *A Graphical Survey of the Canadian Textile Industries*. Montreal: McGill Univ. Press, 1936.

Corcoran, Edward. "My Experiences as a Farm Hand in Canada," *United Empire*, March 1929 (20), pp. 149–58.

Cornblit, Oscar E. "European Immigrants in Argentine Industry and Politics," in Veliz, ed., *The Politics of Conformity*, pp. 221–48.

Cornblit, Oscar E., Ezequiel Gallo (h.), and Alfredo A. O'Connell. "La generación del 80 y su proyecto: Antecedentes y consecuencias," in Di Tella et al., *Argentina, sociedad de masas*, pp. 18–58.

Cortés Conde, Roberto. *El progreso argentino, 1880–1914*. Buenos Aires: Sudamericana, 1979.

Craig, G. H. "Land Settlement and Tenancy in the Lomond and Vulcan Districts, Alberta, " *Economic Annalist*, April 1937 (7), pp. 22–23.

Craven, Paul, and Tom Traves. "The Class Politics of the National Policy, 1872–1933," *Journal of Canadian Studies*, Fall 1979 (14), pp. 14–38.

Crerar, T. A. "Agriculture After the War," *Grain Growers' Guide*, Jan. 24, 1917, p. 31.

Cuccorese, Horacio Juan. *Historia de los ferrocarriles en la Argentina*. Buenos Aires: Ediciones Macchi, 1969.

Currie, A. W. "Freight Rates on Grain in Western Canada," *CHR*, March 1940 (21), pp. 40–55.

Dafoe, John W. *Clifford Sifton in Relation to His Times.* Toronto: Macmillan, 1931.

Dales, J. H. *The Protective Tariff in Canada's Development.* Toronto: Univ. of Toronto Press, 1966.

Davisson, Walter P. *Pooling Wheat in Canada.* Ottawa: Graphic Publishers, 1927.

Dawson, C. A. *Group Settlement: Ethnic Communities in Western Canada.* Toronto: Macmillan, 1936.

———. *The Settlement of the Peace River Country: A Study of a Pioneer Area.* Toronto: Macmillan, 1934.

Dawson, C. A., and Eva R. Younge. *Pioneering in the Prairie Provinces: The Social Side of the Settlement Process.* Toronto: Macmillan, 1940.

Dawson, R. MacGregor. *The Government of Canada.* 4th ed. Toronto: Univ. of Toronto Press, 1963.

———. *William Lyon Mackenzie King: A Political Biography, 1874–1923.* Toronto: Univ. of Toronto Press, 1958.

De Gelder, Willem. *A Dutch Homesteader on the Prairies,* tr. Herman Ganzevoort. Toronto: Univ. of Toronto Press, 1973.

De Hevesy, Paul. *World Wheat Planning and Economic Planning in General.* London: Oxford Univ. Press, 1940.

De la Torre, Lisandro. *Obras de Lisandro de la Torre.* 6 vols. Buenos Aires: Hemisferio, 1954.

Del Carril, Pedro. *La subdivisión de nuestras tierras y la evolución agraria del país.* Buenos Aires: Estudio de Agronomía y Colonización, 1926.

Delich, Francisco J. "Empresas agrícolas en gran escala," *Revista Paraguaya de Sociología,* May–Aug. 1973 (10), pp. 119–76.

Dell'oro Maini, Atilio. *El impuesto sobre la renta y su aplicación en la República Argentina.* Buenos Aires: A. de Martiro, 1920.

Denis, Pierre. *The Argentine Republic: Its Development and Progress,* tr. Joseph McCabe. London: Unwin, 1922.

Denoon, Donald. *Settler Capitalism: The Dynamics of Dependent Development in the Southern Hemisphere.* Oxford: Clarendon Press, 1983.

Den Otter, A. A. "Alexander Galt, the 1859 Tariff, and Canadian Economic Nationalism," *CHR,* June 1982 (63), pp. 151–78.

Díaz Alejandro, Carlos F. "The Argentine State and Economic Growth: A Historical Overview," in Ranis, ed., *Government and Economic Development,* pp. 216–50.

———. *Essays on the Economic History of the Argentine Republic.* New Haven, Conn.: Yale Univ. Press, 1970.

Dickmann, Adolfo. *Los argentinos naturalizados en la política.* Buenos Aires: Juan Perrotti, 1915.

Diecidue, Antonio. *Netri: Líder y mártir de una gran causa.* Rosario: Federación Agraria Argentina, 1969.

Diéguez, Héctor L. "Argentina y Canadá: Un comentario," *Desarrollo Económico,* July–Sept. 1981 (21), pp. 271–76.

Di Tella, Guido, and Manuel Zymelman. *Las etapas del desarrollo económico argentino.* Buenos Aires: Univ. de Buenos Aires, 1967.

Di Tella, Torcuato S., et al. *Argentina, sociedad de masas.* Buenos Aires: Univ. de Buenos Aires, 1965.

Dorfman, Adolfo. *Evolución industrial argentina.* Buenos Aires: Losada, 1942.

Dreier, Katherine S. *Five Months in the Argentine from a Woman's Point of View.* New York: F. F. Sherman, 1920.

Drummond, Ian M. *British Economic Policy and the Empire, 1919–1939.* London: Allen & Unwin, 1972.

———. *Imperial Economic Policy, 1917–1939: Expansion and Protection.* London: Allen & Unwin, 1974.

Duhau, Luis. *Los elevadores de granos en el Canadá.* Buenos Aires: Sociedad Rural Argentina, 1928.

———. "La libre concurrencia y el mercado de cereales: El Canadá y la Argentina," *RCE,* March–April 1928 (30), pp. 1749–1763.

Dunsdorfs, Edgars. *The Australian Wheat-Growing Industry.* Melbourne: Melbourne Univ. Press, 1956.

Eager, Evelyn. "The Conservatism of the Saskatchewan Electorate," in Ward and Spafford, eds., *Politics in Saskatchewan,* pp. 1–19.

Easterbrook, W. T. *Farm Credit in Canada.* Toronto: Univ. of Toronto Press, 1938.

Easterbrook, W. T., and Hugh G. J. Aitken. *Canadian Economic History.* Toronto: Macmillan, 1956.

Elkin, Judith Laikin. *Jews in the Latin American Republics.* Chapel Hill: Univ. of North Carolina Press, 1979.

Ellis, L. Ethan. *Reciprocity 1911: A Study in Canadian-American Relations.* New Haven, Conn.: Yale Univ. Press, 1939.

Embree, David Grant. "Rise of the United Farmers of Alberta," *AHR,* Autumn 1957 (5), pp. 1–5.

England, Robert. *The Central European Immigrant in Canada.* Toronto: Macmillan, 1929.

Estabrook, Leon M. "Agricultural Statistics in Argentina," *The Review of the River Plate,* Aug. 8, 1924, p. 359.

Etcheverry, Victor D. *Las cooperativas agrícolas en Entre Ríos.* Concepción del Uruguay: Univ. Nacional de La Plata, Facultad de Agronomía y Veterinaria, 1914.

Evans, Simon M. "Canadian Beef for Victorian Britain," *Agricultural History,* Oct. 1979 (53), pp. 748–62.

Evans, W. Sanford. *The Canadian Wheat Pool.* [Winnipeg?]: Dawson Richardson Publications, [1926?].

Ezcurra, Mariano de. *Cuestión social, cuestión rural.* Buenos Aires: Antonio Prudent y Cía., 1923.

Falcoff, Mark. "Economic Dependency in a Conservative Mirror: Alejandro Bunge and the Argentine Frustration, 1919–1943," *Inter-American Economic Affairs,* Spring 1982 (35), pp. 57–75.

Farnalls, Paul L. *Memories of Life in Alberta.* [Calgary?]: Commercial Printers, [1960?].

Fay, C. R. *Agricultural Co-operation in the Canadian West.* London: P. S. King & Son, 1925.

———. *Co-operation at Home and Abroad: A Description and Analysis.* London: P. S. King & Son, 1908.

Ferns, H. S. *Argentina.* New York: Praeger, 1969.

———. *The Argentine Republic, 1516–1971.* New York: Barnes & Noble, 1973.

———. *Britain and Argentina in the Nineteenth Century.* Oxford: Clarendon Press, 1960.

———. "Britain's Informal Empire in Argentina, 1806–1914," *Past and Present,* 1953, No. 4 (Nov.), pp. 60–75.

Ferraro, Roberto A. "El grito de Alcorta en Córdoba," *Todo es Historia,* July 1974 (8), pp. 78–93.

Ferrer, Aldo. *La economía argentina: Las etapas de su desarrollo y los problemas actuales.* 2d ed. México: Fondo de Cultura Económica, 1963.

Fienup, Darrell F., Russell H. Brannon, and Frank A. Fender. *The Agricultural Development of Argentina: A Policy and Development Perspective.* New York: Praeger, 1969.

Flanagan, Thomas. "Political Geography and the United Farmers of Alberta," in Trofimenkoff, ed., *The Twenties in Western Canada,* pp. 138–69.

Foerster, Robert F. *The Italian Emigration of Our Times.* Cambridge, Mass.: Harvard Univ. Press, 1919.

Fogarty, John P. "The Comparative Method and the Nineteenth Century Regions of Recent Settlement," *Historical Studies,* April 1981 (19), pp. 412–29.

———. "Staples, Super-Staples and the Limits of Staple Theory: The Experiences of Argentina, Australia and Canada Compared," in D. C. M. Platt and Guido di Tella, eds., *Argentina, Australia, and Canada: Studies in Comparative Development,* pp. 19–36. London: Macmillan, 1985.

Fogarty, John P., et al. *Argentina y Australia.* Buenos Aires: Instituto Torcuato di Tella, 1979.

Forbes, Ernest R. *The Maritime Rights Movement, 1919–1927: A Study in Canadian Regionalism.* Montreal: McGill-Queen's Univ. Press, 1979.

Ford, A. G. "British Investments and Argentine Economic Development, 1880–1914," in Rock, ed., *Argentina in the Twentieth Century,* pp. 12–40.

———. *The Gold Standard, 1880–1914: Britain and Argentina.* Oxford: Clarendon Press, 1962.

Foster, John E., ed. *The Developing West: Essays on Canadian History in Honor of Lewis H. Thomas.* Edmonton: Univ. of Alberta Press, 1983.

Fowke, Vernon C. *Canadian Agricultural Policy: The Historical Pattern.* Toronto: Univ. of Toronto Press, 1947.

———. "An Introduction to Canadian Agricultural History," *Agricultural History,* April 1942 (16), pp. 79–90.

———. *The National Policy and the Wheat Economy.* Toronto: Univ. of Toronto Press, 1957.

———. "Royal Commissions and Canadian Agricultural Policy," *CJEPS,* May 1948 (14), pp. 163–75.

Fowke, Vernon C., and Donald Fowke. "Political Economy and the Canadian Wheat Grower," in Ward and Spafford, eds., *Politics in Saskatchewan,* pp. 207–20.

Fraser, John Foster. *The Amazing Argentine: A New Land of Enterprise.* London: Cassell, 1914.

Friedmann, Harriet. "World Market, State, and Family Farm: Social Bases of Household Production in the Era of Wage Labor," *Comparative Studies in Society and History,* Oct. 1978 (20), pp. 545–86.

Friesen, Gerald. *The Canadian Prairies, a History.* Lincoln: Univ. of Nebraska Press, 1984.

Gallo, Ezequiel. "Agricultural Colonization and Society in Argentina: The Province of Santa Fe, 1870–1895," Ph.D. dissertation, Oxford University, 1970.

———. *Farmers in Revolt: The Revolutions of 1893 in the Province of Santa Fe, Argentina.* London: Athlone, 1976.

Garbarini Islas, Guillermo. "La mala situación del país y de las industrias rurales en particular requieren algunas medidas urgentes," *REA,* Oct. 1930, pp. 243–62.

García, Genaro. "Leyes agrarias peligrosas," *REA,* May 1930 (24), pp. 335–41.

———. "Problemas agrarios," *Anales de la Sociedad Rural Argentina*, April 1931 (65), p. 215.

García, José María. *El campo argentino a 60 años del grito de Alcorta*. Buenos Aires: "Centro de Estudios," 1972.

García Ledesma, H. *Lisandro de la Torre y la pampa gringa*. Buenos Aires: Indoamérica, 1954.

García Serrano, Tomás. *Esteban Piacenza: Apuntes biográficos*. 2d ed. Rosario: Lib. y Ed. Ruiz, 1967.

Geller, Lucio. "El crecimiento industrial argentino hasta 1914 y la teoría del bien primario exportable," *El Trimestre Económico*, Oct.–Dec. 1970 (37), pp. 763–811.

Germani, Gino. *Estructura social de la Argentina*. Buenos Aires: Raigal, 1955.

Gibson, Herbert. *The Land We Live On*. Buenos Aires: R. Grant & Co., 1914.

———. *Memorandum by Sir Herbert Gibson, K.B.E., A Commissioner of the Royal Committee on Wheat Supplies on the Purchase and Shipment of Cereals in the Argentine and Uruguayan Republics on Behalf of His Majesty's Government and Those of Allied Countries*. Buenos Aires: Biblioteca Tornquist, 1921.

Girbal de Blacha, Noemí M. *Los centros agrícolas en la Provincia de Buenos Aires*. Buenos Aires: Consejo Nacional de Investigaciones Científicas y Técnicas, 1980.

———. *Historia de la agricultura argentina a fines del siglo XIX (1890–1900)*. Buenos Aires: Fundación para la Educación, la Ciencia y la Cultura, 1982.

Girola, Carlos D. *Les coopératives agricoles en Argentine*. Buenos Aires: P. Gadola, 1913.

———. *Investigación agrícola en la República Argentina*. Buenos Aires: Cía. Sud-Americana de Billetes de Banco, 1904.

Glazebrook, G. P. de T. *A History of Transportation in Canada*. Toronto: Ryerson Press, 1938.

Goodwin, Crawford D. W. *Canadian Economic Thought: The Political Economy of a Developing Nation*. Durham, N.C.: Duke Univ. Press, 1961.

Goodwin, Paul B. *Los ferrocarriles británicos y la Unión Cívica Radical (1916–1930)*. Buenos Aires: La Bastilla, 1974.

———. "The Politics of Rate-Making: The British-Owned Railways and the Unión Cívica Radical, 1921–1928," *Journal of Latin American Studies*, Nov. 1974 (6), pp. 257–87.

Goodwin, William. *Wheat Growing in the Argentine Republic*. Liverpool: Northern Publishing Co., 1895.

Gras, Mario Cesar. *Los gauchos colonos: Novela agraria argentina*. Buenos Aires: L. J. Rosso, 1928.

Gravil, Roger. "The Anglo-Argentine Connection and the War of 1914–1918," *Journal of Latin American Studies*, 9, No. 1 (May 1977), pp. 59–89.

———. "State Intervention in Argentina's Export Trade Between the Wars," *Journal of Latin American Studies*, 2, No. 2 (Nov. 1970), pp. 147–73.

Great Britain, House of Commons. *Sessional Papers*. "First Report of the Royal Commission on Wheat Supplies with Appendices. Presented to Parliament by Command of His Majesty." 1921 (Cmd. 1544).

Greca, Alcides. *La pampa gringa: Novela del sur santafecino*. Santiago de Chile: Ediciones Ercilla, 1936.

Grela, Placido. *Alcorta: Origen y desarrollo del pueblo y de la rebelión agraria de 1912*. Rosario: Litoral Ediciones, 1975.

———. *El grito de Alcorta: Historia de la rebelión campesina de 1912*. Rosario: Tierra Nuestra, 1958.

Griezic, Foster J. K. "The Honourable Thomas Alexander Crerar: The Political Career of a Western Liberal Progressive in the 1920s," in Trofimenkoff, ed., *The Twenties in Western Canada*, pp. 107–37.

Grove, Frederick P. *Fruits of the Earth* (1933). Toronto: McClelland & Stewart, 1965.

Gualtieri, Francesco M. *We Italians: A Study in Italian Immigration in Canada*. Toronto: Italian World War Veterans' Assn., 1929.

Gudiño Kramer, Luis. *Tierra ajena*. Buenos Aires: Lautaro, 1943.

Guglieri, Pablo. *Las memorias de un hombre de campo*. Buenos Aires: n.p., 1913.

Hansen, Marcus Lee. *The Mingling of the Canadian and American Peoples*. New Haven, Conn.: Yale Univ. Press, 1940.

Hanson, Simon G. *Argentine Meat and the British Market*. Stanford, Calif.: Stanford Univ. Press, 1938.

Harney, Robert N., and Harold Troper. *Immigrants: A Portrait of the Urban Experience, 1890–1930*. Toronto: Van Nostrand Reinhold, 1975.

Haslam, J. H. "The Future of Wheat Growing," *Grain Growers' Guide*, May 7, 1919, p. 38.

Havens, V. L. "Fuel in Argentina," *Power*, Oct. 26, 1920 (52), pp. 649–50.

Haythorne, George V. "Harvest Labor in Western Canada: An Episode in Economic Planning," *The Quarterly Journal of Economics*, May 1933 (47), pp. 533–44.

Hedley, Max J. "The Social Conditions of Production and the Dynamics of Tradition: Independent Commodity Production in Canadian Agriculture," Ph.D. dissertation, University of Alberta, 1976.

Hedlin, Ralph. "Edmund A. Partridge," *Transactions of the Historical and Scientific Society of Manitoba*, ser. 3, No. 15 (1960), pp. 59–68.

Hines, Norman E., ed. *Economics, Sociology and the Modern World*. Cambridge, Mass.: Harvard Univ. Press, 1935.

Hirst, John. "Respuesta de John Hirst a los comentarios," in Fogarty et al., *Argentina y Australia*, pp. 109–12.

Hobbs, Allan B. "Agriculture in Argentina," *Grain Growers' Guide*, Dec. 3, 1913, pp. 7–18.

Hobsbawm, E. J. *Industry and Empire*. New York: Pantheon Books, 1968.

Holm, Gert T. *The Argentine Grain Grower's Grievances*. Buenos Aires: Rugeroni Hnos., 1919.

Horne, Bernardino C. *Nuestro problema agrario*. Buenos Aires: "La Facultad," 1937.

———. *Política agraria y regulación económica*. Buenos Aires: Losada, 1942.

Huergo, Ricardo H. *La enseñanza agrícola*. Buenos Aires: Ministerio de Agricultura, 1909.

———. *Investigación agrícola en la región septentrional de la Provincia de Buenos Aires*. Bueno Aires: Cía. Sud-Americana de Billetes de Banco, 1904.

Huret, Jules. *La Argentina del Plata a la Cordillera de los Andes*, tr. E. Gómez Carrillo. Paris: Louis-Michaud, 1910.

Hutchison, Bruce. *The Incredible Canadian*. Toronto: Longmans, Green, 1952.

Innis, Harold A. *Problems of Staple Production in Canada*. Toronto: Ryerson Press, 1933.

———, ed. *The Diary of Alexander James McPhail*. Toronto: Univ. of Toronto Press, 1940.

Irvine, William. *The Farmers in Politics*. Toronto: McClelland & Stewart, 1920.

Isern, Thomas D. "Adoption of the Combine on the Northern Plains," *South Dakota History*, Spring 1960 (10), pp. 101–18.

Jackman, W. J. *Wheat Growing and Rural Economic Conditions in the Argentine Republic.* Winnipeg: Canadian Co-operative Wheat Producers, 1927.

———. "Wheatfields Under the Southern Cross," *Country Guide,* May 15, 1928 (21), pp. 4–5, 34–35.

Jakubs, Deborah Lynn. "A Community of Interests: A Social History of the British in Buenos Aires, 1860–1914," Ph.D. dissertation, Stanford University, Dec. 1985.

Jefferson, Mark. *Peopling the Argentine Pampa.* New York: American Geographical Society, 1926.

Johnson, Gilbert. "The Patagonia-Welsh," *SH,* Autumn 1963 (16), pp. 90–94.

Justo, Juan B. *Discursos y escritos políticos.* Buenos Aires: "El Ateneo," 1933.

Kendle, John. *John Bracken: A Political Biography.* Toronto: Univ. of Toronto Press, 1979.

Kerr, D. C. "Saskatoon 1910–1913: Ideology of the Boomtime," *SH,* Winter 1979 (32), pp. 16–27.

Kipling, Rudyard. *Letters of Travel, 1892–1913.* Garden City, N.Y.: Doubleday, Page, 1920.

Kirby, John. "On the Viability of Small Countries: Uruguay and New Zealand Compared," *Journal of Inter-American Studies and World Affairs,* Aug. 1975 (17), pp. 259–80.

Knapp, Joseph G. *The Hard Winter Wheat Pools: An Experiment in Agricultural Marketing Integration.* Chicago: Univ. of Chicago Press, 1933.

Koch, Fred C. *The Volga Germans: In Russia and the Americas, from 1763 to the Present.* University Park: Pennsylvania State Univ. Press, 1977.

Kubat, Daniel, and David Thornton. *A Statistical Portrait of Canadian Society.* Toronto: McGraw Hill–Ryerson, 1974.

Kuczynski, Robert René. "Freight Rates on Argentine and North American Wheat," *Journal of Political Economy,* June 1902 (10), pp. 333–60.

Lahitte, Emilio. "El conflicto agrario," *Revista de Derecho, Historia y Letras,* 43 (Oct. 1912), pp. 263–71.

Larden, Walter. *Estancia Life: Agricultural, Economic, and Cultural Aspects of Argentine Farming.* London: Unwin, 1911.

Larsen, Grace H., and Henry E. Erdman. "Aaron Sapiro: Genius of Farm Cooperative Promotion," *Mississippi Valley Historical Review,* 49 (Sept. 1962), pp. 242–68.

Lawrence, H. F. *Kidnapped.* Pine Lake, Alberta: n.p., 1923.

Lawrence, H. R. "Elevadores de granos," *Anales de la Sociedad Rural Argentina,* April 1, 1927 (61), pp. 331–33.

Lazarowich, P. J. "Ukrainian Pioneers in Western Canada," *AHR,* Autumn 1957 (15), pp. 17–27.

Leacy, F. H., ed. *Historical Statistics of Canada.* 2d ed. Ottawa: Statistics Canada, 1983.

Leguizamón, Guillermo E. "Los elevadores de granos en el Canadá," *REA,* Feb. 1928 (20), pp. 93–95.

———. "La inmigración colonizadora en el Canadá," *REA,* Feb. 1928 (20), pp. 97–100.

Leiserson, Samson. *Elevadores de granos y cooperación.* Buenos Aires: "La Vanguardia," 1927.

Levitt, Kari. *Silent Surrender: The Multinational Corporation in Canada.* Toronto: Macmillan, 1970.

Lewis, Colin. "British Railway Companies and the Argentine Government," in Platt, ed., *Business Imperialism*, pp. 395–427.

Liebscher, Arthur Francis. "Commercial Expansion and Political Change: Santa Fe Province, 1897–1916," Ph.D. dissertation, Indiana University, 1975.

Lipset, Seymour M. *Agrarian Socialism: The Cooperative Commonwealth Federation in Saskatchewan.* Rev. ed. Berkeley: Univ. of California Press, 1971.

Lloyd, Lewis L. *Memories of a Co-operative Statesman.* [Calgary? 1967?]

Lochhead, W. "Agricultural Colleges and Agricultural Development in Canada," *United Empire*, April 1924 (15), pp. 209–17.

Lumsden, James. *Through Canada in Harvest Time: A Study of Life and Labour in the Golden West.* London: Unwin, 1903.

Luna, Felix. *Alvear.* Buenos Aires: Libros Argentinos, 1958.

McClure, Russell Schee. "The Hudson Bay Wheat Route," M.A. thesis, University of Washington, 1939.

McCrea, Roswell C., Thurman W. Van Metre, and George Jackson Eder, "International Competition in the Trade of Argentina," *International Conciliation*, No. 271 (June 1931), pp. 321–487.

McCrorie, James M. "Discussion of 'The Farmer as a Social Class in the Prairie Region,' by Earl J. Tyler," in Tremblay and Anderson, eds., *Rural Canada in Transition*, pp. 322–37.

McCrorie, James Napier. *In Union is Strength.* Saskatoon: Univ. of Saskatchewan, 1964.

McCutcheon, Brian Robert. "The Economic and Social Structure of Political Agrarianism in Manitoba, 1870–1900," Ph.D. dissertation, University of British Columbia, 1974.

McDiarmid, Orville J. *Commercial Policy in the Canadian Economy.* Cambridge, Mass.: Harvard Univ. Press, 1946.

Macdonald, L. R. "Merchants Against Industry: An Idea and Its Origins," *CHR*, Sept. 1975 (56), pp. 263–81.

Macdonald, Norman. *Canada: Immigration and Colonization, 1841–1903.* Toronto: Macmillan, 1966.

McGann, Thomas F. *Argentina, Estados Unidos y el sistema interamericano, 1880–1914*, tr. Germán O. Tjarks. Buenos Aires: Univ. de Buenos Aires, 1960.

MacGibbon, Duncan Alexander. *The Canadian Grain Trade.* Toronto: Macmillan, 1932.

———. *The Canadian Grain Trade, 1931–1951.* Toronto: Univ. of Toronto Press, 1952.

Mackintosh, W. A. *Agricultural Cooperation in Western Canada.* Kingston: Queen's University, 1924.

———. "The Crisis in Wheat," *Queen's Quarterly*, Autumn 1939 (46), pp. 348–59.

———. *The Economic Background of Dominion-Provincial Relations.* Ottawa: King's Printer, 1939.

———. *Economic Problems of the Prairie Provinces.* Toronto: Macmillan, 1935.

Macleod, R.C. "Canadianizing the West: The North-West Mounted Police as Agents of the National Policy, 1873–1905," in Lewis H. Thomas, ed., *Essays*, pp. 101–10.

McNaught, Kenneth. *The Pelican History of Canada.* Rev. ed. New York: Penguin Books, 1982.

———. *A Prophet in Politics: A Biography of J. S. Woodsworth.* Toronto: Univ. of Toronto Press, 1959.

Macpherson, C. B. *Democracy in Alberta: Social Credit and the Party System*. 2d ed. Toronto: Univ. of Toronto Press, 1962.

Macpherson, Ian. "The Co-operative Union of Canada and the Prairies, 1919–1929," in Trofimenkoff, ed., *The Twenties in Western Canada*, pp. 50–74.

Maizels, Alfred. *Industrial Growth and World Trade*. Cambridge: Cambridge Univ. Press, 1963.

Malenbaum, Wilfred. *The World Wheat Economy, 1885–1939*. Cambridge, Mass.: Harvard Univ. Press, 1953.

Mallon, Richard D. *Economic Policymaking in a Conflict Society: The Argentine Case*. Cambridge, Mass.: Harvard Univ. Press, 1975.

Maradona, Waldino B. *En defensa de los trabajadores del campo*. Buenos Aires: La Vanguardia, 1946.

Mardiros, Anthony. *William Irvine: The Life of a Prairie Radical*. Toronto: James Lorimer, 1979.

Marotta, Sebastián. *El movimiento sindical argentino: Su genesis y desarrollo*. 3 vols. Buenos Aires: "Lacio," 1960–73.

Martin, Chester. *"Dominion Lands" Policy*. Toronto: Macmillan, 1938.

Martínez de Hoz, Jose Alfredo (h.), *La agricultura y la ganadería argentina en el período 1930–1960*. Buenos Aires: Sudamericana, 1967.

Masters, D. C. *The Rise of Toronto, 1850–1890*. Toronto: Univ. of Toronto Press, 1947.

Miatello, Hugo. *La chacra santafecina en 1905*. Buenos Aires: Cía. Sud-Americana de Billetes de Banco, 1905.

———. "El hogar agrícola," *Boletín del Museo Social Argentino*, 3, Nos. 35–36 (1914), pp. 541–81.

Mills, Ivor J. *Stout Hearts Stand Tall: Biographical Sketch of a Militant Saskatchewan Farmer, the Late Hopkin Evan Mills*. Vancouver: Evergreen Press, 1971.

Mitchell, E. B. *In Western Canada Before the War: A Study of Communities*. London: John Murray, 1915.

Molina, Raúl A. "Presidencia de Marcelo T. De Alvear," in Argentina [1], Vol. 1, Part 2, pp. 271–345.

Molins, W. Jaime. *La Pampa*. Buenos Aires: "Oceana," 1918.

Monteagudo, Pío Isaac. *Migraciones internas en la Argentina y las utopias revolucionarias de L. de la Torre: Problemas agrarios*. Buenos Aires: Comisión Nacional de Homenaje a Lisandro de la Torre, 1956.

Moorhouse, Hopkins. *Deep Furrows*. Toronto: George J. McLeod, 1918.

Morgan, Dan. *Merchants of Grain*. New York: Viking, 1979.

Morris, A. S. "The Failure of Small Farmer Settlement in Buenos Aires Province, Argentina," *Revista Geográfica*, June 1977 (85), pp. 63–78.

Morton, Arthur S. *History of Prairie Settlement*. Toronto: Macmillan, 1938.

Morton, Desmond. "Cavalry or Police: Keeping the Peace on Two Adjacent Frontiers, 1870–1900," *Journal of Canadian Studies*, Spring 1977 (12), pp. 27–37.

Morton, W. L. *The Canadian Identity*. 2d ed. Toronto: Univ. of Toronto Press, 1972.

———. *The Critical Years, 1857–1873*. Toronto: McClelland & Stewart, 1964.

———. *Manitoba: A History*. 2d ed. Toronto: Univ. of Toronto Press, 1967.

———. *The Progressive Party in Canada*. Toronto: Univ. of Toronto Press, 1950.

Murchie, R. W. *Agricultural Progress on the Prairie Frontier*. Toronto: Macmillan, 1936.

Nario, Hugo. "Los crotos," *Todos es Historia*, July 1980 (12), pp. 6–19.

Nayor, R. T. *The History of Canadian Business, 1867–1914.* 2 vols. Toronto: James Lorimer, 1975.

Neatby, H. Blair. *William Lyon Mackenzie King, 1924–1932: The Lonely Heights.* Toronto: Univ. of Toronto Press, 1963.

Nemirovsky, Lázaro. *Estructura económica y orientación política de la agricultura en la República Argentina.* Buenos Aires: Jesús Menéndez, 1933.

Nesbitt, Leonard D. *Tides in the West.* Saskatoon: Modern Press, n.d.

Norrie, Kenneth H. "The National Policy and Prairie Economic Discrimination, 1870–1930," in Akenson, ed., *Canadian Papers in Rural History*, Vol. 1, pp. 13–32.

———. "The Rate of Settlement of the Canadian Prairies, 1870–1911," *Journal of Economic History*, June 1975 (35), pp. 410–27.

North, Douglass C. "Location Theory and Regional Economic Growth," *Journal of Political Economy*, June 1955 (63), pp. 243–58.

———. "Ocean Freight Rates and Economic Development, 1750–1913," *Journal of Economic History*, Dec. 1958 (18), pp. 537–55.

O'Connor, Richard. *The Oil Barons: Men of Greed and Grandeur.* Boston: Little, Brown, 1971.

Oddone, Jacinto. *La burguesía terrateniente argentina.* 3d ed. Buenos Aires: Ediciones Populares Argentinos, 1956.

Olson, Mancur. "The United Kingdom and the World Market in Wheat and Other Primary Products, 1885–1914," *Explorations in Economic History*, Summer 1974 (11), pp. 325–55.

Ortiz Pereyra, Manuel. *La tercera emancipación: Actualidad económica y social de la República Argentina.* Buenos Aires: J. Lajouane & Cía., 1926.

Owen, Geraint Dyfnallt. *Crisis in Chubut: A Chapter in the History of the Welsh Colony in Patagonia.* Swansea: Christopher Davies, 1977.

Pagés, Pedro T. *Crisis ganadera argentina.* Buenos Aires: Comité Nacional de Defensa de la Producción, 1922.

Paige, Jeffrey M. *Agrarian Revolution: Social Movement and Export Agriculture in the Underdeveloped World.* New York: Free Press, 1975.

Paish, George. "Great Britain's Capital Investments in Individual Colonial and Foreign Countries," *Journal of the Royal Statistical Society*, Jan. 1911 (74), pp. 167–84.

Palacios, Alfredo L. *La justicia social.* Buenos Aires: Claridad, 1954.

Paoli, Pedro de. *Defrauden! (La quiebra escandalosa de la Federación Agraria Argentina).* Buenos Aires: n.p., 1935.

Partridge, E. A. *Manifesto of the No-Party League of Western Canada.* Winnipeg: De Monfort Press, 1913.

Pascale, Silvio. "El factor 'arrendamiento' en el problema agrícola," *RCE*, 2 parts, Vol. 19. Part 1, March 1931, pp. 201–27; Part 2, April 1931, pp. 291–312.

Paterson, D. G. *British Direct Investment in Canada, 1890–1914.* Toronto: Univ. of Toronto Press, 1976.

Patton, Harald S. "The Canadian Wheat Pool in Prosperity and Depression," in Norman E. Hines, ed., *Economics, Sociology and the Modern World.* Cambridge, Mass.: Harvard Univ. Press, 1935, pp. 23–44.

———. *Grain Growers' Cooperation in Western Canada.* Cambridge, Mass.: Harvard Univ. Press, 1928.

———. "Observations on Canadian Wheat Policy Since the World War," *CJEPS*, May 1937 (3), pp. 218–33.

Paz, Roque. "El grupo Bunge y Born en la economía nacional," *Argumentos: Revista Mensual de Estudios Sociales*, Feb. 1939 (1), pp. 301–16.

Pérez Brignoli, Héctor. "Los intereses comerciales en la agricultura argentina de exportación (1880–1955)." Paper presented at the Fifth Symposium on the Economic History of Latin America, Lima, 1978.

Peters, H. E. *The Foreign Debt of the Argentine Republic.* Baltimore: Johns Hopkins Univ. Press, 1934.

Phelps, Vernon L. *The International Economic Position of Argentina.* Philadelphia: Univ. of Pennsylvania Press, 1938.

Phillips, Paul. "The National Policy Revisited," *Journal of Canadian Studies*, Fall 1979 (14), pp. 3–13.

Phillips, W. G. *The Agricultural Implement Industry in Canada.* Toronto: Univ. of Toronto Press, 1956.

Pintos, Guillermo. "Se necesitan hombres de gobierno en hacienda," *Revista de Derecho, Historia y Letras*, Oct. 1918 (61), pp. 237–49.

———. *Treinta años de proteccionismo excesivo.* Buenos Aires: José Tragant, 1917.

Platt, D. C. M. "British Agricultural Colonization in Latin America," *Inter-American Economic Affairs*, 2 parts. Part 1, Winter 1964 (18), pp. 3–38; Part 2, Summer 1965 (19), pp. 23–42.

———. "Dependency in Nineteenth Century Latin America: An Historian Objects," *Latin American Research Review*, 15, No. 1 (1980), pp. 113–30.

———. *Latin America and British Trade, 1806–1914.* London: Adam & Charles Black, 1972.

———, ed. *Business Imperialism, 1840–1930: An Inquiry Based on British Experience in Latin America.* Oxford: Clarendon Press, 1977.

Platt, D. C. M., and Guido di Tella, eds., *Argentina, Australia, and Canada: Studies in Comparative Development.* London: Macmillan, 1985.

Podestá, José P. "La pequeña propiedad rural en la República Argentina," in Univ. Nacional de Buenos Aries, Facultad de Ciencias Económicas, *Investigaciones de Seminario*, 3 (1923), pp. 1–144.

Porritt, Edward. *The Revolt in Canada Against the New Feudalism.* London: Cobden Club, 1911.

Porter, John. *The Vertical Mosaic: An Analysis of Social Class and Power in Canada.* Toronto: Univ. of Toronto Press, 1965.

Posada, Adolfo. *Pueblos y campos argentinos: Sensaciones y recuerdos.* Madrid: Caro Raggio, 1926.

Priestley, Norman F., and Edward B. Swindlehurst. *Furrows, Faith and Fellowship.* Edmonton: Co-op Press, 1967.

Proceedings of the First International Pool Conference which Includes the Third International Wheat Pool Conference. Held at Regina, Saskatchewan, June 5th, 6th and 7th, 1928. [Winnipeg, 1928?]

Proceedings of the Second International Co-operative Wheat Pool Conference Held at Kansas City, Missouri, May 5th, 6th and 7th, 1927. [Winnipeg, 1927?]

Rahola, Federico. *Sangre nueva: Impresiones de un viaje a la América del Sud.* Barcelona: "La Académica," 1905.

Randall, Laura. *An Economic History of Argentina in the Twentieth Century.* New York: Columbia Univ. Press, 1978.

Ranis, Gustav, ed., *Government and Economic Development.* New Haven, Conn.: Yale Univ. Press, 1971.

Rasporich, Anthony W., ed. *Western Canada: Past and Present.* Calgary: McClelland & Stewart, 1975.

Rasporich, Anthony W., and Henry C. Klassen, eds. *Prairie Perspectives 2.* Toronto: Holt, Rinehart & Winston, 1973.

Ratti, Anna Maria. "Italian Migration Movements, 1876 to 1926," in Willcox, ed., *International Migrations,* Vol. 2, pp. 440–90.

Regehr, T. D. "Bankers and Farmers in Western Canada, 1900–39," in Foster, ed., *The Developing West,* pp. 303–36.

———. "Western Canada and the Burden of National Transportation Policies," in Bercuson, ed., *Canada and the Burden of Unity,* pp. 115–41.

Remmer, Karen L. *Party Competition in Argentina and Chile: Political Recruitment and Public Policy, 1890–1930.* Lincoln: Univ. of Nebraska Press, 1984.

Renard, A. H. "Argentina's First Road Congress," *Comments on Argentine Trade,* June 1922 (2), p. 21.

Rennie, Ysabel F. *The Argentine Republic.* New York: Macmillan, 1945.

Repetto, Nicolás. "Cooperativas agrícolas," *RCE,* Feb.–April 1921 (9), pp. 122–65.

———. "La defensa de la producción agrícola nacional," *RCE,* Sept. 1929 (17) pp. 729–45.

Reynal O'Connor, Arturo. *Por las colonias.* Buenos Aires: L. J. Ross y Cía., 1921.

Reynolds, Lloyd G. *The British Immigrant: His Social and Economic Adjustment in Canada.* Toronto: Oxford Univ. Press, 1935.

Richardson, J. Henry. *British Economic Foreign Policy.* London: Allen & Unwin, 1936.

Rickard, Bruce. "The North Atlantic Triangle and Changes in the Wheat Trade Before the Great War," *Dalhousie Review,* 55 (Summer 1975), pp. 263–71.

Riddell, R. G. "A Cycle in the Development of the Canadian West," *CHR,* Sept. 1940 (21), pp. 268–84.

Rock, David. *Politics in Argentina, 1890–1930: The Rise and Fall of Radicalism.* London: Cambridge Univ. Press, 1975.

———, ed. *Argentina in the Twentieth Century.* Pittsburgh: Univ. of Pittsburgh Press, 1975.

Rodríguez Tardati, José. "Los trabajadores del campo," *RCE,* April 1926 (14), pp. 382–97.

Rolph, William Kirby. *Henry Wise Wood of Alberta.* Toronto: Univ. of Toronto Press, 1950.

Romero, José Luis. *A History of Argentine Political Thought,* tr. Thomas F. McGann. 3d ed. Stanford, Calif.: Stanford Univ. Press, 1963.

Ross, Gordon. *Argentina and Uruguay.* New York: Macmillan, 1916.

Rotstein, Abraham. "Innis: The Alchemy of Fur and Wheat," *Journal of Canadian Studies,* Winter 1977 (12), pp. 6–31.

Ruggiero, Kristin. "Gringo and Creole: Foreign and Native Values in a Rural Argentine Community," *Journal of Inter-American Studies and World Affairs,* May 1982 (24), pp. 163–82.

———. "Italians in Argentina: The Waldenses at Colonia San Gustavo, 1850–1910," Ph.D. dissertation, Indiana University, 1978.

Rutter, W. P. *Wheat-Growing in Canada, the United States and the Argentine.* London: Adam & Charles Black, 1911.

Sábato, Hilda, and Juan Carlos Korol. *Cómo fue la inmigración irlandesa en Argentina.* Buenos Aires: Plus Ultra, 1981.

Saskatchewan, Royal Commission on Immigration and Settlement. *Report.* Regina: Ministry of Natural Resources, 1930.

Scalabrini Ortiz, Raúl. *Política británica en el Río de la Plata*. Buenos Aires: Reconquista, 1940.

Scardin, Francisco. *La Argentina y el trabajo: Impresiones y notas*. Buenos Aires: Jacobo Peuser, 1908.

Schiopetto, Ovidio Víctor. "El comercio de granas," *RCE*, July 1928 (16), pp. 2108–42.

Schleh, Emilio J. *La industria azucarera en su primer centenario, 1821–1921*. Buenos Aires: Ferrari Hnos., 1921.

———. *Semillas para el agricultor*. Buenos Aires: Ferrari Hnos., 1924.

Schulz, Jacob. *Rise and Fall of Canadian Farm Organizations*. Winnipeg: Privately published, 1955.

Scobie, James R. *Argentina: A City and a Nation*. New York: Oxford Univ. Press, 1964.

———. *Revolution on the Pampas: A Social History of Argentine Wheat, 1860–1910*. Austin: Univ. of Texas Press, 1964.

Seymour, Richard Arthur. *Pioneering in the Pampas or the First Four Years of a Settler's Experience in the La Plata Camps*. London: Longmans, Green, 1869.

Sharp, Mitchell W. "Allied Wheat Buying in Relationship to Canadian Marketing Policy, 1914–18," *CJEPS*, Aug. 1940 (6), pp. 372–89.

Sharp, Paul F. *The Agrarian Revolt in Western Canada*. Minneapolis: Univ. of Minnesota Press, 1948.

Shideler, James H. *Farm Crisis, 1919–1923*. Berkeley: Univ. of California Press, 1957.

Sienrra, Celestino. *Campo y ciudad: El problema agrario argentino*. 2d ed. Rosario: Privately published, 1972.

Silver, A. I. "French Canada and the Prairie Frontier, 1870–1890," *CHR*, March 1969 (50), pp. 11–36.

Sinclair, Peter R. "Class Structure and Populist Protest: The Case of Western Canada," *The Canadian Journal of Sociology*, Spring 1975, pp. 1–17.

Slatta, Richard Wayne. "The Gaucho and Rural Life in Nineteenth-Century Buenos Aires Province, Argentina," Ph.D. dissertation, University of Texas–Austin, 1980.

———. *Gauchos and the Vanishing Frontier*. Lincoln: Univ. of Nebraska Press, 1983.

Sloan, Robert J. "The Canadian West: Americanization or Canadianization?," *AHR*, Winter 1968 (16), pp. 1–7.

Smith, David E. *Prairie Liberalism: The Liberal Party in Saskatchewan, 1905–1971*. Toronto: Univ. of Toronto Press, 1975.

———. *The Regional Decline of a National Party: Liberals on the Prairies*. Toronto: Univ. of Toronto Press, 1981.

Smith, L. Brewster, Harry T. Collings, and Elizabeth Murphey. *The Economic Position of Argentina During the War*. Washington, D.C.: BFDC, 1920.

Smith, Peter H. *Politics and Beef in Argentina: Patterns of Conflict and Change*. New York: Columbia Univ. Press, 1969.

Smith, Robert. "Radicalism in the Province of San Juan: The Saga of Federico Cantoni (1916–34)," Ph.D. dissertation, University of California–Los Angeles, 1970.

Smith, Rollin E. *Wheat Fields and Markets of the World*. St. Louis, Mo.: Modern Miller Company, 1908.

Soares, Carlos F. *Economía y finanzas de la nación argentina*. 3 vols. Buenos Aires: Rodríguez Giles, 1916–32.

Sociedad Rural Argentina. *Anuario de la Sociedad Rural Argentina: Estadísticas Económicas y Agrarias.* Buenos Aires, 1928.

Solberg, Carl E. "Farm Workers and the Myth of Export-Led Development in Argentina," *The Americas,* Oct. 1974 (31), pp. 121–38.

———. *Immigration and Nationalism: Argentina and Chile, 1890–1914.* Austin: Univ. of Texas Press, 1970.

———. *Oil and Nationalism in Argentina: A History.* Stanford, Calif.: Stanford Univ. Press, 1979.

———. "Rural Unrest and Agrarian Policy in Argentina, 1912–1930," *Journal of Inter-American Studies and World Affairs,* Jan. 1971 (13), pp. 18–52.

———. "The Tariff and Politics in Argentina, 1916–1930," *Hispanic American Historical Review,* May 1973 (53), pp. 252–74.

Spafford, D. S., "The Elevator Issue, The Organized Farmer and the Government, 1908–11," *SH,* Autumn 1962 (15), pp. 81–92.

———. "The 'Left-Wing,' 1921–1931," in Ward and Spafford, eds., *Politics in Saskatchewan,* pp. 44–58.

———. "The Origin of the Farmers' Union of Canada," *SH,* Autumn 1965 (18), pp. 89–98.

Spain, Consejo Superior de Emigración. *La emigración española transoceánica.* Madrid: Hijos de T. Minuesa de los Ríos, 1916.

Spangemberg, Silvio. "El conflicto agrario del sud de Santa Fe," *Boletín Mensual del Museo Social Argentina,* Sept. 1912, pp. 522–31.

Stead, Robert J. C. *Grain* (1926). Toronto: McClelland & Stewart, 1963.

Stegner, Wallace. *Wolf Willow: A History, A Story and a Memory of the Last Prairie Frontier* (1955). Toronto: Macmillan, 1977.

Sternberg, Rolf. "Farms and Farmers in an Estanciero World," Ph.D. dissertation, Syracuse University, 1971.

Stovel, John A. *Canada in the World Economy.* Cambridge, Mass.: Harvard Univ. Press, 1959.

Strong, J. A. "Agricultural Implements: Trends in Consumption Costs to the Farmer and Methods of Distribution in Argentina," *Commercial Intelligence Journal* (Ottawa), Aug. 15, 1936 (55), pp. 313–28.

———. "Conditions in Argentina in 1938," *Commercial Intelligence Journal,* 2 parts, Vol. 60 (1939). Part 1, Feb. 25, pp. 241–48; Part 2, March 4, pp. 304–12.

———. "Farm Mortgage Loans in Argentina," *Commercial Intelligence Journal,* June 25, 1938 (58), pp. 1044–1046.

———. "Grain Farming in Argentina," *Commercial Intelligence Journal,* 2 parts, Vol. 54 (1936). Part 1, May 16, pp. 897–903; Part 2, May 23, pp. 978–87.

Surface, Frank M. *The Grain Trade During the World War.* New York: Macmillan, 1928.

———. *International Competition in the Production of Wheat for Export.* Washington, D.C.: BFDC, 1924.

Swanson, W. W., and P. C. Armstrong. *Wheat.* Toronto: Macmillan, 1930.

Sweet, Dana Royden. "A History of United States–Argentine Commercial Relations, 1918–1933: A Study of Competitive Farm Economies," Ph.D. dissertation, Syracuse University, 1971.

Taylor, Carl C. *Rural Life in Argentina.* Baton Rouge: Louisiana State Univ. Press, 1948.

Teichman, Judith. "Businessmen and Politics in the Process of Economic Development: Argentina and Canada," *Canadian Journal of Political Science,* March 1982 (15), pp. 47–66.

Teubal, Miguel. "Policy and Performance of Agriculture in Economic Development: The Case of Argentina," Ph.D. dissertation, University of California–Berkeley, 1975.

Thomas, L. G. *The Liberal Party in Alberta: A History of Politics in the Province of Alberta, 1905–21*. Toronto: Univ. of Toronto Press, 1959.

Thomas, Leslie H. "From the Pampas to the Prairies: The Welsh Migration of 1902," *SH*, Winter 1971 (24), pp. 1–12.

Thomas, Lewis H. "British Visitors' Perceptions of the West, 1885–1914," in Rasporich and Klassen, eds., *Prairie Perspectives*, pp. 181–96.

————. "Early Combines in Saskatchewan," *SH*, Winter 1955 (8), pp. 1–5.

————, ed. *Essays on Western History*. Edmonton: Univ. of Alberta Press, 1976.

Thompson, John Herd. "Bringing in the Sheaves: The Harvest Excursionists, 1890–1929," *CHR*, Dec. 1978 (59), pp. 467–89.

————. *The Harvests of War: The Prairie West, 1914–1918*. Toronto: McClelland & Stewart, 1978.

Thomson, Colin A. "Dark Spots in Alberta," *Alberta History*, Autumn 1977 (25), pp. 31–36.

Tornatore, Alejandro. *Historia de la evolución y revolución agraria en la Argentina y de la creación de la Federación Agraria Argentina, según uno de sus fundadores*. Salto (Buenos Aires province): Privately published, 1967.

Tornquist, Ernesto, & Cía. *El desarrollo económico de la República Argentina en los últimos cincuenta años*. Buenos Aires, 1920.

Tremblay, Marc-Adélard, and Walton J. Anderson, eds. *Rural Canada in Transition*. Ottawa: Agricultural Economics Research Council of Canada, 1970.

Trofimenkoff, S. M., ed. *The Twenties in Western Canada*. Ottawa: National Museums of Canada, 1972.

Troper, Harold Martin. *Only Farmers Need Apply: Official Canadian Government Encouragement of Immigration from the United States, 1896–1911*. Toronto: Griffin House, 1972.

Tulchin, Joseph S. "El crédito agrario en la Argentina, 1910–1926," *Desarrollo Económico*, Oct.–Dec. 1978 (18), pp. 381–408.

Turner, Allan R. "W. R. Motherwell and Agricultural Education, 1905–1918," *SH*, Autumn 1959 (12), pp. 81–96.

United States, Dept. of Commerce, Bureau of the Census. *Historical Statistics of the United States: Colonial Times to 1970*. 2 vols. Washington, D.C., 1975.

Urquhart, M. C., and K. A. H. Buckley. *Historical Statistics of Canada*. Toronto: Macmillan, 1965.

Vásquez-Presedo, Vicente. *El caso argentino: Migración de factores, comercio exterior y desarrollo, 1875–1914*. Buenos Aires: Univ. de Buenos Aires, 1971.

Veeman, Terry, and Michele Veeman. *The Future of Grain: Canada's Prospects for Grains, Oilseeds and Related Industries*. Toronto: Canadian Institute for Economic Policy, 1984.

Vélez, Mariano. *La situacion agrícola de La Pampa*. Buenos Aires: "La Vanguardia," 1934.

Veliz, Claudio, ed. *The Politics of Conformity in Latin America*. New York: Oxford Univ. Press, 1967.

Videla, Ricardo. "La cuestión agraria argentina," *REA*, March 1922 (8), pp. 197–204.

Viner, Jacob. *Canada's Balance of International Indebtedness, 1900–1913*. Cambridge, Mass.: Harvard Univ. Press, 1924.

Waite, Peter B. *Canada, 1874–1896: Arduous Destiny*. Toronto: McClelland & Stewart, 1971.

Walter, Richard J. "Politics, Parties, and Elections in Argentina's Province of Buenos Aires, 1912–42," *Hispanic American Historical Review*, Nov. 1984 (64), pp. 707–36.

———. *The Socialist Party of Argentina, 1890–1930*. Austin: Univ. of Texas Press, 1977.

Ward, Norman, and Duff Spafford, eds. *Politics in Saskatchewan*. Don Mills, Ont.: Longmans, 1968.

Watkins, Melville H. "A Staple Theory of Economic Growth," *CJEPS*, May 1963 (29), pp. 141–58.

Weil, Felix J. *Argentine Riddle*. New York: John Day, 1944.

West, Edward. *Homesteading: Two Prairie Seasons*. London: Unwin, 1918.

Wilkie, Richard W., and Jane Riblett Wilkie. *Migration and an Argentine Rural Community in Transition*. Amherst: Univ. of Massachusetts, International Area Studies Programs, 1980.

Willcox, Walter F., ed. *International Migrations*. 2 vols. New York: National Bureau of Economic Research, 1929.

Williams, Glyn. *The Desert and the Dream: A Study of Welsh Colonization in Chubut, 1865–1915*. Cardiff: Univ. of Wales Press, 1975.

Williams, J. Earl. "Origin and Development of Public Telephones in Alberta," *AHR*, Spring 1963 (11), pp. 8–12.

Willmott, Donald E. "The Formal Organizations of Saskatchewan Farmers, 1900–65," in Rasporich, ed., *Western Canada*, pp. 28–41.

———. *Organizations and Social Life of Farm Families in a Prairie Municipality*. Saskatoon: Univ. of Saskatchewan, Centre for Community Studies, 1964.

Wilson, C. F. *A Century of Canadian Grain: Government Policy to 1951*. Saskatoon: Western Producer Prairie Books, 1978.

Wilson, L. J. "Educational Role of the United Farm Women of Alberta," *Alberta History*, Spring 1977 (25), pp. 28–36.

Winsberg, M. D. "Jewish Agricultural Colonization in Entre Ríos, Argentina," *The American Journal of Economics and Sociology*, 2 parts, Vol. 27 (1968). Part 1, July, pp. 285–95; Part 2, Oct., pp. 423–28.

Woltman, Harry Raymond. "The Decline of Argentina's Agricultural Trade: Problems and Policies, 1929–54," Ph.D. dissertation, Stanford University, 1959.

Wonders, William C. "Scandinavian Homesteaders," *Alberta History*, Summer 1976 (24), pp. 1–4.

Wood, Louis Aubrey. *A History of Farmers' Movements in Canada: The Origins and Development of Agrarian Protest, 1872–1924*. Toronto: Univ. of Toronto Press, 1975.

Woods, J. D. *The Canadian Agricultural Machinery Industry*. Ottawa: Royal Commission on Canada's Economic Prospects, 1956.

Woodward, J. S. "Wheat and Politics on the Prairies," *Queen's Quarterly*, Autumn 1931 (38), pp. 733–44.

Wright, J. F. C. *The Louise Lucas Story: This Time Tomorrow*. Montreal: Harvest House, 1965.

Wright, Winthrop Robins. "Argentine Railways and the Growth of Nationalism," Ph.D. dissertation, University of Pennsylvania, 1964.

———. *British-Owned Railways in Argentina: Their Effects on Economic Nationalism, 1854–1948*. Austin: Univ. of Texas Press, 1974.

Wythe, George. *Industry in Latin America*. New York: Columbia Univ. Press, 1945.

Yates, S. W. *The Saskatchewan Wheat Pool: Its Origin, Organization and Progress, 1924–1935*. Saskatoon: United Farmers of Canada, 1947.

Yuzyk, Paul. *The Ukrainians in Manitoba: A Social History*. Toronto: Univ. of Toronto Press, 1953.

Zalduendo, Eduardo A. *Libras y rieles: Las inversiones británicas para el desarrollo de los ferrocarriles en Argentina, Brasil, Canadá, y India durante el siglo XIX*. Buenos Aires: El Coloquio, 1975.

Zamboraín, Saturnino M. *La verdad sobre la propiedad de la tierra en la Argentina*. Buenos Aires: Sociedad Rural Argentina, n.d.

Index

Index

Library of Congress Cataloging-in-Publication Data

Solberg, Carl E.
  The prairies and the pampas.

  (Comparative studies in history, institutions, and public policy)
  Bibliography: p.
  Includes index.
  1. Wheat trade—Government policy—Canada—History.
  2. Wheat trade—Government policy—Argentina—History.
  I. Title.   II. Series.
  HD9049.W5C297   1987   338.1'8   86-27854
  ISBN 0-8047-1346-4 (alk. paper)